OKANAGAN UNIV/COLLEGE LIBRARY
D0575571

Environmental Remote Sensing from Regional to Global Scales

Environmental Remote Sensing From Regional to Global Scales

edited by

Giles M. Foody and Paul J. Curran

JOHN WILEY & SONS
Chichester • New York • Brisbane • Toronto • Singapore

Published 1994 by John Wiley & Sons Ltd,
Baffins Lane, Chichester,
West Sussex PO19 1UD, England

Telephone (+44) 243 779777

Other Wiley Editorial Offices

John Wiley & Sons, Inc., 605 Third Avenue,
New York, NY 10158-0012, USA

Jacaranda Wiley Ltd, 33 Park Road, Milton,
Queensland 4064, Australia

John Wiley & Sons (Canada) Ltd, 22 Worcester Road,
Rexdale, Ontario M9W 1L1, Canada

John Wiley & Sons (SEA) Pte Ltd, 37 Jalan Pemimpin #05-04,
Block B, Union Industrial Building, Singapore 2057

Library of Congress Cataloging-in-Publication Data

Environmental remote sensing from regional to global scales / edited by
Giles Foody and Paul Curran.
p. cm.
Includes bibliographical references and index.
ISBN 0-471-94434-3
1. Environmental monitoring – Remote sensing.
I. Foody, Giles M., 1962– . II. Curran, Paul J., 1955–
GE70.E583 1993
628–dc20 93-5434
CIP

British Library Cataloguing in Publication Data

A catalogue record for this book is available from the British Library

ISBN 0-471-94434-3

Typeset in 10/11pt Sabon by Mayhew Typesetting, Rhayader, Powys

Printed in Great Britain by
Antony Rowe Ltd, Chippenham, Wiltshire

Contents

List of Contributors vii

Preface xi

Acknowledgements xii

1 Environmental Issues at Regional to Global Scales 1
Paul J. Curran and Giles M. Foody

2 Remote Sensing and Terrestrial Global Environment Research – the TIGER Programme 8
S.A. Briggs

3 Explaining and Monitoring Land Cover Dynamics in Drylands Using Multi-temporal Analysis of NOAA AVHRR Imagery 16
A.C. Millington, J. Wellens, J.J. Settle and R.J. Saull

4 The Use of Remote Sensing to Characterise the Regenerative States of Tropical Forests 44
Paul J. Curran and Giles M. Foody

5 Global Land Cover: Comparison of Ground-based Data Sets to Classifications with AVHRR Data 84
R.S. DeFries and J.R.G. Townshend

6 Snow Monitoring in the United Kingdom Using NOAA AVHRR Imagery 111
Richard M. Lucas and Andrew R. Harrison

7 A Near-real-time Heat Source Monitoring System Using NOAA Polar Orbiting Meteorological Satellites 131
Geoffrey M. Smith and Robin A. Vaughan

8 PER-ZONE CLASSIFICATION OF URBAN LAND COVER FOR URBAN POPULATION ESTIMATION 142
Christiane Weber

9 ATTEMPTS TO DRIVE ECOSYSTEM SIMULATION MODELS AT LOCAL TO REGIONAL SCALES 149
Paul J. Curran

10 REMOTE SENSING AND THE DETERMINATION OF GEOPHYSICAL PARAMETERS FOR INPUT TO GLOBAL MODELS 167
Arthur P. Cracknell

11 ENVIRONMENTAL MONITORING USING MULTIPLE-VIEW-ANGLE (MVA) REMOTELY-SENSED DATA 181
M.J. Barnsley

12 EARTH OBSERVATION DATA – OR INFORMATION? 202
Paul M. Mather

13 SPATIAL DATA: DATA TYPES, DATA APPLICATIONS AND REASONS FOR PARTIAL ADOPTION AND NON-INTEGRATION 214
J.A. Allan

14 SCALE AND ENVIRONMENTAL REMOTE SENSING 223
Giles M. Foody and Paul J. Curran

INDEX 233

List of Contributors

Professor Tony Allan is interested mainly in the monitoring and evaluation of renewable natural resources in arid and semi-arid regions, and in the Earth observation techniques which contribute to the economical survey and their evaluation. He is based at the School of Oriental and African Studies in the University of London where he is coordinator of remote sensing and GIS in the University. He is a past Chairman and President of the Remote Sensing Society and is Chairman of the United Kingdom Committee for Photogrammetry and Remote Sensing. He acts as consultant to a number of agencies including the World Bank, the European Community, Food and Agriculture Organisation and governments in the Middle East and Africa on water resources and Earth observation issues.

Dr. Mike Barnsley holds B.A. and Ph.D. degrees from the University of Reading. He is currently lecturer in remote sensing and GIS at the University College of London. His research interests include the derivation of land surface parameters from an analysis of the bidirectional reflectance properties of Earth surface materials, issues on scaling and generalisation in the production of small scale land cover maps and the development of kernel-based and object-based spatial re-classification procedures for urban land cover identification.

Dr. Stephen Briggs studied for a B.Sc. degree in Astronomy at the University College of London before carrying out Ph.D. and postdoctoral research into the dynamics and physics of active galaxies. After a period as lecturer in astrophysics at Queen Mary College, Dr Briggs joined the NERC in 1983, working in the analysis of Earth observation imagery. He was appointed Head of the NERC Remote Sensing Applications Development Unit in 1986. His present interests lie in the use of remotely sensed data in the study of terrestrial surface processes and their implementation in operational applications. He is a member of numerous scientific expert groups, Principal Investigator on a range of programmes and adviser on Earth observation to the European Space Agency, Commission of the European Community and UK government departments.

Professor Arthur Cracknell is currently Carnegie Professor of Physics and Head of the Department of Applied Physics and Electronic and Manufacturing Engineering at Dundee University. He graduated with a degree in Natural Sciences from the University of Cambridge before completing a D.Phil. in Metallurgy at the University of Oxford. He held posts at the Universities of Singapore and Essex before moving to Dundee. His interests in remote sensing developed in the late 1970s, and focus mainly on marine systems and the extraction of geophysical parameters from remotely sensed data. He established the biennial Dundee summer schools on remote sensing, the M.Sc. course on Remote Sensing, Image Processing and Applications, and is editor of the *International Journal of Remote Sensing.*

Professor Paul Curran received the B.Sc. degree in Geography from the University of Sheffield in 1976 and the Ph.D. and D.Sc. degrees from the University of Bristol in 1979 and 1991 respectively. His primary research interest is in the use of airborne imaging spectrometry and broad-band satellite sensors to drive ecological models. Professor Curran has been Principal Investigator on numerous grants and contracts and sits on many committees concerned with remote sensing and ecology. He has published widely and is editor of *Remote Sensing Letters*. He is currently Professor of Physical Geography at the University of Southampton, having previously held a similar post at the University College of Swansea from 1990 to 1993.

Dr. Ruth DeFries is currently an associate research scholar in the Department of Geography, University of Maryland, College Park, in the USA. She holds a Ph.D. in Physical Geography from The Johns Hopkins University in Baltimore. Previous positions were held at the National Academy of Sciences and the Indian Institute of Technology in Bombay. She has edited and authored several books on the subject of global environmental change.

Dr. Giles Foody graduated with a B.Sc. degree in Geography from the University of Sheffield in 1983. He stayed in Sheffield to complete a Ph.D. on radar remote sensing in 1986. His main research interests are in environmental remote sensing, particularly in relation to the characterisation and mapping of terrestrial land cover. He is currently a lecturer in the Department of Geography at the University College of Swansea.

Dr. Andrew Harrison received the B.A. degree in Geography from Portsmouth Polytechnic in 1979 and the Ph.D. degree from the University of Reading in 1984. He carried out research on the development of large environmental databases at the NERC Thematic Information Services during 1982–4 before taking-up his present post of lecturer in the Department of Geography at Bristol University. He has published widely in the fields of remote sensing and GIS and leads a research group working on environmental applications of remote sensing.

Dr. Richard Lucas studied at the University of Bristol, from where he graduated with a B.Sc. degree in Biology and Geography in 1986 and a Ph.D. degree in 1989. In 1990 and 1991 he continued his research in Australia and is now working in the Department of Geography at the University College of Swansea.

Professor Paul Mather is a graduate of the Universities of Cambridge and Nottingham. He is currently Professor of Geography at the University of Nottingham, where he established a Remote Sensing Unit in 1981. He has published widely on computational methods of analysis of spatial data and played a major role in the Remote Sensing Society, of which he is currently vice-president.

Professor Andrew Millington is currently with the Geography Department at Leicester University and was previously based in the Geography Department at Reading University. He has degrees from the Universities of Hull, Colorado and Sussex. His research interests include monitoring and mapping vegetation in the wet and dry tropics and land degradation issues. He has extensive experience of work in Pakistan, the Middle East, Africa and Latin and Central America.

Dr. Richard Saull is co-founder and co-director of IS Ltd, an image processing consultancy company based in Bristol. He graduated with the degrees of B.Sc. and Ph.D. from the Universities of Sheffield and Reading respectively. His main research interests are in image analysis techniques for environmental applications.

Dr. Jeff Settle has been with the NERC Unit for Thematic Information Systems at the University of Reading since 1985, where he is currently a Senior Research Fellow. He has degrees in Mathematics and Astrophysics from the Universities of Cambridge and London. His main research interests are in the remote sensing of the land surface in the optical and thermal regions of the electromagnetic spectrum.

Geoffrey Smith graduated in Environmental Science from Lancaster University. He has since pursued graduate studies in Applied Geophysics at Birmingham University and Remote Sensing at Dundee University. He is currently involved in research for a Ph.D. concerned with the remote sensing of foliar chemistry and based in the Department of Geography at the University College of Swansea.

Professor John Townshend is currently Chairman of the Department of Geography at the University of Maryland, College Park, USA. Previously he was Director of the NERC Unit for Thematic Information Systems at the University of Reading. He has held visiting positions at NASA's Goddard Space Flight Centre and the Graduate School of Geography, Clark University. He has held a number of Principal Investigatorships associated with the use of remote sensing for the monitoring of vegetation at small scales, and with the application of Landsat TM data for land-cover mapping and the monitoring of sediment transfer in semi-arid areas. Currently he is Chair of the International Geosphere Biosphere's Committee on Land Cover Change and the NASA/NOAA's AVHRR Land Pathfinder Science Working Group. He also serves on the National Academy of Science's Committee on Geophysical and Environmental Data. He has published widely and been a member of the editorial board of the *International Journal of Remote Sensing* since its inception.

Dr. Robin Vaughan is a senior lecturer in Physics at the University of Dundee. Since 1984 he has been course leader of the M.Sc. degree in Remote Sensing, Image Processing and Applications. He has been a member of the editorial board of the *International Journal of Remote Sensing* since 1985. Dr Vaughan has been active within the Remote Sensing Society, holding posts of News Editor from 1985 to 1989 and Honorary General Secretary from 1991. He is also treasurer of the European Association of Remote Sensing Laboratories.

Dr. Christiane Weber received a Ph.D. degree for her research on a comparative ecological factorial analysis of cities in the Rhine Valley in 1982. Since 1985 she has been a full-time researcher at the Centre National de la Recherche Scientifique (CNRS) and works at the Geography Department at the Louis Pasteur University, Strasbourg, France. She has wide experience in remote sensing and GIS applications, particularly in relation to urban studies.

Dr. Jane Wellens graduated from the School of Development Studies at the University of East Anglia. She then moved to the Geography Department at Reading University where she undertook research for a Ph.D. on the use of multi-temporal satellite sensor imagery for monitoring vegetation dynamics in Tunisia. She is currently a research officer at Reading University where she is investigating the use of remotely-sensed data for the identification of forest types in Latin America.

Preface

Concern over our changing environment has led increasingly to a need to observe and characterise a range of environmental phenomena at regional to global scales. At these scales, remote sensing is often the only practical approach to data acquisition. Furthermore, remote sensing typically offers digital data in a 'map-like' format over time periods that are relevant to the detection and monitoring of environmental change. This book arose from a session on remote sensing at regional to global scales at the Institute of British Geographers (IBG) annual conference at the University College of Swansea. It took place at the onset of International Space Year, 1992, and brought together people with a broad range of interests. The overt aim of the session was to illustrate the rôle remote sensing can play in the provision of data and also information on the environment at regional to global scales. Perhaps the more important aim was to initiate a dialogue between those who are facing up to regional to global scale issues and those who are using remotely sensed data at regional to global scales. As the chapters reveal, this dialogue concentrated on the specifics of particular applications but against a backdrop of one powerful and pervasive theme:

> The technology that carries us away from earth provides us with a unique vantage point from which to look back at it. This has profoundly changed our perception of our planet, forcing us to confront new realities: first that earth is indeed a planet – and a rather small one at that. Second, that it is very much alone, adrift in a black, and empty space with no friendly neighbors nearby. And finally, that it is all we have. (Dotto, L., 1991, *Blue Planet*, Smithsonian Institute, Washington D.C., p. 5.)

Acknowledgements

The production of a book such as this invariably involves a large number of people. In addition to the contributors and those they have acknowledged in their own chapters, there are many who have helped produce the book before you. It would be impractical to name everyone but we do wish to mention a few key helpers. First, there were those who assisted with the conference itself. Here we are most grateful to the Institute of British Geographers (IBG) conference organisation committee, the Remote Sensing Society and Quantitative Methods Study Group of the IBG for promotion and support. The conference session itself ran smoothly thanks to Laurence Moore and Colin Thomas who allowed us to do essentially what we wanted and Doreen Boyd for mastering the many audio-visual aids. Ultimately the success of the session was dependent on the speakers and we are grateful to them for their presentations. This includes Ian Dowman and Kathie Bowden, who were unfortunately unable to prepare chapters but whose lectures were a valuable part of the session. Converting the conference presentations into a book required a lot of goodwill and help, and the final production of the book would have floundered without the assistance of the Geography Department of the University College of Swansea. We must, however, stress our gratitude to the chapter authors for preparing their work and responding positively to editorial comments, and acknowledge that we benefited from the word-processing skills of David Reynolds, David Robinson and Eileen Baker, the cartographic expertise of Guy Lewis and Nicola Jones, and some computer wizardry from John Duigenan. Last, we wish to thank the publishers for their patience in the final stages of production.

1. Environmental Issues at Regional to Global Scales

Paul J. Curran and Giles M. Foody

'The human brain now holds the key to our future.
We have to recall the image of the planet from
outer space: a single entity in which
air, water and continents are
interconnected. That is our home.'
(Suzuki, 1991, p. 19.)

Environmental Change

The geological record bears witness to dramatic environmental changes that have affected all regions of Earth. The rather pleasant seaside city in which this text was written has, in the geological past, switched between environments ranging from deep ocean to sandy desert and dense hot forest to post-glacial wasteland. These environmental changes have been in response to phenomena such as the migration of continents, the building and erosion of mountains, the reorganisation of oceans, variations in solar output and even the catastrophic effects of large meteorites. These phenomena have led to changes on local scales (few km^2), to regional scales (few 10^4 km^2), to global scales (5×10^8 km^2), with speeds that have been as slow as millennia (e.g., plate tectonics, soil development) to as fast as minutes (e.g., fires, storms) (Johnson, 1989; Kemp, 1990). These environmental changes are beyond human control and provide a rather humbling backdrop to the very recent environmental changes that result from human activity (Miller and Jacobson, 1992; Chambers, 1993) (Table 1.1). These recent changes have tended to be local to regional in scale and include modification of: the atmosphere by industry; the pattern of erosion and deposition by construction; biota by chemical pollution and the hydrological cycle by agriculture (Mungall and McLaren, 1990; Turner *et al.*, 1990; Mannion and Bowlby, 1992). As the World's population grows and technology develops, so human activity will become an increasingly potent agent for environmental change and these changes are likely to be ever more rapid at the scale of the region to the globe (NRC, 1988; NERC, 1989).

Table 1.1 Points in Earth history.

Global event	Time before present: Years	Time before present: Artificially compressed time scale with 30 years = 1 second
Age of Earth	4,600,000,000	5.0 years
Life	600,000,000	33.1 weeks
Dinosaurs become extinct	60,000,000	3.3 weeks
Early homonids	4,000,000	1.5 days
Permanent agriculture	10,000	5.5 hours
Metal smelting	7,000	3.8 hours
Pyramids, Stonehenge	4,500	2.5 hours
Rapid industrialisation	150	5.0 sec
Satellite remote sensing	30	1.0 sec

Sources: NRC (1988), NASA (1988), Burnett (1990), Mannion (1992). The artificially compressed time scale gives perspective to the literal time scale. If, on this artificially compressed time scale, the Earth were formed five years ago, then dinosaurs would have walked the planet a month ago, the main impacts of humans would have occurred during the last five seconds and satellite remote sensing would be only a second old.

Evidence accumulated primarily since the late 1960s has indicated not just that there are human-induced environmental changes (Houghton *et al.*, 1990) but that these changes are the result of complex interactions between human and natural systems (NASA, 1988; IGBP, 1990; Williamson, 1992). At the local scale our understanding of these interactions is sound and is based on centuries of observation and speculation. At the regional to global scale our understanding of these interactions is non-existent to poor, as it is based on a very limited amount of research and is lacking both robust models and *relevant data* (NERC, 1989).

Environmental research at regional to global scales is being undertaken within most science subjects and topic areas within those subjects. The seven US topic areas of climate and hydrologic systems, biochemical dynamics, ecological systems and dynamics, Earth system history, human interactions, solid Earth processes and solar influences listed in Figure 1.1 illustrates the breadth of this research. Figure 1.1 also emphasises just how much of this environmental research 'leans on spaceborne observations for its global perspective' (NRC, 1986, p. iv).

.... REMOTE SENSING AND ENVIRONMENTAL CHANGE

Arguments for the use of remotely sensed data in the study of environmental change at regional to global scales are compelling (Perry, 1986; Graetz, 1989; Curran, 1989; Mooney and Hobbs, 1989). Perhaps the most powerful and certainly the politically most appealing argument is that, 'Global problems ... require global solutions and these in turn require information on a global

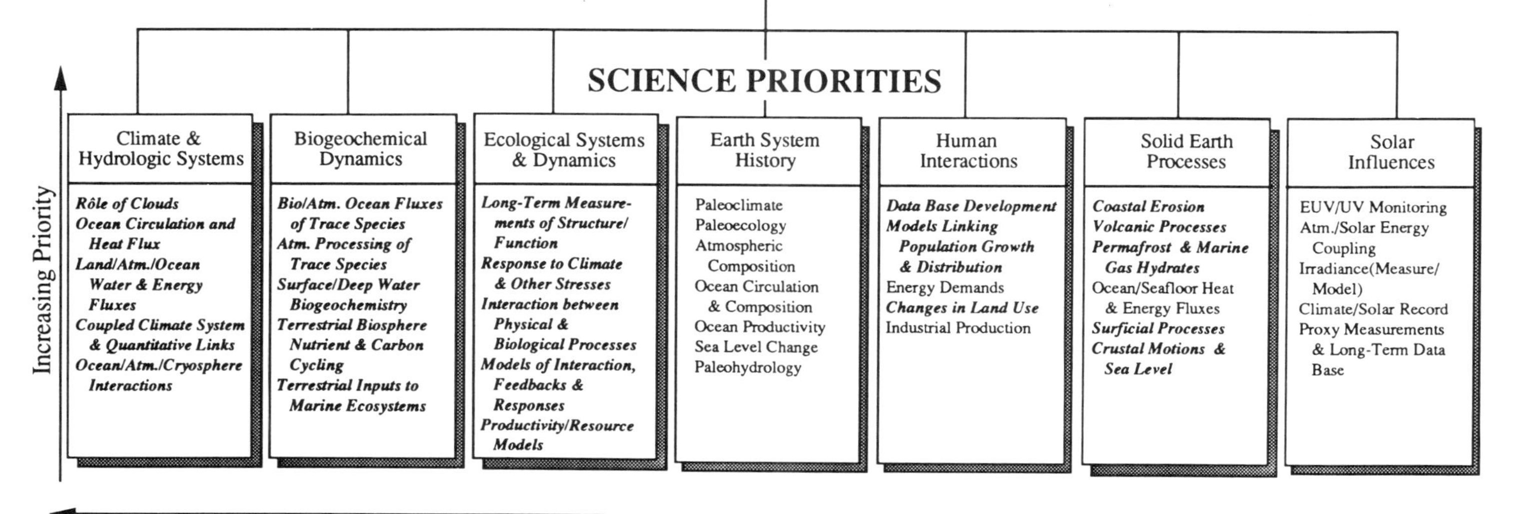

Figure 1.1 The proposed US global change programme (Johnson, 1989). The research areas that draw upon remotely sensed data or information derived from those data are shown in bold italics.

Table 1.2 The four methods for using remotely sensed data in environmental research at regional to global scales.

Research method	Example
Direct estimation of a physical variable independently of other data	Remotely sensed estimates of sea surface temperature
Inference from the measurement of a related variable	Estimates of methane flux using maps of wetland derived from remotely sensed data
Extrapolation of local scale measurements to estimates for large areas	Point measurements of evapotranspiration extrapolated to a continent via the correlation between evapo-transpiration and remotely sensed estimates of biomass
Modelling environmental processes	Use of remotely sensed data to parameterise and drive environmental models

Source: Modified from Briggs (1991).

scale. Earth observation from space offers unique opportunities to obtain that information' (Heseltine, 1992).

This new potential rôle for remote sensing has changed the scale at which those involved with remote sensing are undertaking some of their research. Early work in remote sensing concentrated on the local scale and provided information that could also be collected by traditional means, with the emphasis being on convenience, accuracy and cost. Today remote sensing activities are moving from the field and forest to the landscape and the continent where it can provide direct estimates of certain phenomena. More importantly, remotely sensed data can be used to infer the unmeasurable, extrapolate point data to areas and parameterise and drive environmental models (Table 1.2). In other words remotely sensed data at the local scale are *useful* but at the regional to global scale they offer a unique perspective and are *vital*: a distinction that underpins much of the discussion within this book.

As a result of good luck and good management there are currently twelve major satellites that are carrying sensors of relevance to the research outlined in Figure 1.1 (Table 1.3). These satellites will be joined by around forty further satellites carrying environmental sensors during the next ten years (CEOS, 1992; NASA, 1992). The most important of these satellites will be the Polar Orbiting Environmental Mission (Envisat) and the Earth Observing Mission with a morning orbit (EOS AM-1) in 1998 and the Earth Observing Mission with an afternoon orbit (EOS PM-1) in 2000 (Moore and Dozier, 1992).

As we move further into the 1990s the environmental questions at regional to global scales are being redefined and refined in the light of developments in environmental understanding and technological capability. Increasingly, the specifications for the data and data products from satellite sensors are derived from the environmental programmes of the World Meteorological Organisation (WMO), the International Council of Scientific Unions (ICSU), the Intergovernmental Oceanographic Commission (IOC),

Table 1.3 Current satellites carrying sensors that are suitable for environmental research at regional to global scales.

Application area	Satellite (acronym)
Land	Landsat
	System Probatoire d'Observation de la Terre (SPOT)
	Indian Remote Sensing Satellite (IRS)
	European Remote Sensing Satellite (ERS)
	Fuyo (previously named the Japanese Earth Resources Satellite, JERS)
Ocean/Atmosphere	National Oceanic and Atmospheric Administration (NOAA)
	Geostationary Operational Environmental Satellite (GEOS)
	Geostationary Meteorological Satellite (GMS)
	Insat
	Meteosat
	Marine Observational Satellite (MOS)
	Upper Atmosphere Research Satellite (UARS)

the United Nations Environment Programme (UNEP), the Food and Agriculture Organisation (FAO) and the Commission of the European Communities (CEC). They also draw from major international, interagency programmes such as the Global Climate Observing System (GCOS), the Global Ocean Observing System (GOOS), the World Climate Research Programme (WCRP) and the International Geosphere-Biosphere Programme (IGBP) (CEOS, 1992). The twelve satellites in orbit (Table 1.3) carry sensors that fulfil the regional to global scale research requirements of part of these programmes. For example, the chemical properties of the stratosphere is being estimated by sensors on UARS; land cover is being monitored by NOAA AVHRR, SPOT HRV, Landsat MSS and Landsat TM; land cover in cloudy regions and the extent of sea ice is being monitored by ERS-1 SAR and Fuyo-1 (JERS-1) SAR, and land and sea surface temperature and albedo are being estimated by meteorological satellite sensors and through a number of sensors on ERS-1 (ESA, 1991; CEOS, 1992; Townshend, 1992; Mather 1992). There are, however, areas where remotely sensed data do not fulfil the regional to global scale research requirements of these programmes. For example, remotely sensed data are unable to provide accurate estimates of precipitation, tropospheric chemistry, foliar chemistry, aerosols, soil water concentration and vertical profiles of temperature and water (Johnson, 1989). It is hoped that the satellite sensors of the late 1990s will overcome some of these deficiencies.

.... CONCLUDING COMMENT

This book discusses the use of remotely sensed data for the study of the environment at regional to global scales. The research while urgent is immature; Landsat has just celebrated its twentieth birthday, global data sets have only recently been made routinely available to researchers and many of the Earth observation satellites designed for the needs of regional to global scale research are still on the drawing board. This book is written at a time of transition, not only in technology but also in the minds of politicians and those who are involved with remote sensing. It is the change in the perceptions of these two groups that will ensure that remote sensing develops from a fledgling technique into an essential and unifying tool for the study of our environment at regional to global scales.

.... REFERENCES

Briggs, S.A., 1991, *Note on the relevance of future Earth observation system options to terrestrial science*, Natural Environment Research Council/British National Space Centre, Remote Sensing Applications Development Unit, Abbots Ripton, 19pp.

Burnett, J., 1990, 'The evolution of the biosphere and the ascent of industrial man', in Polunin, N., Burnett, J.H. (editors), *Maintenance of the biosphere*, Edinburgh University Press, Edinburgh: 2–11

CEOS, 1992, *The relevance of satellite missions to the study of the global environment*, Committee on Earth Observation Satellites, British National Space Centre, London

Chambers, F.M. (editor), 1993, *Climate change and human impact on the landscape*, Chapman and Hall, London

Curran, P.J., 1989, 'Editorial: global change and remote sensing', *International Journal of Remote Sensing*, **10**: 1459–60

ESA, 1991, *Report of the Earth observation user consultation meeting*, European Space Agency SP-1143, Noordwijk

Graetz, R.D., 1989, 'Remote sensing of terrestrial ecosystem structure: An ecologists pragmatic view', in Hobbs, R.J., Mooney, H.A. (editors), *Remote sensing of biosphere functioning*, Springer Verlag, New York: 5–30

Heseltine, M., 1992, 'Forward', in CEOS (editors), *The relevance of satellite missions to the study of the global environment*, Committee on Earth Observation Satellites, British National Space Centre, London

IGBP, 1990, *The International Geosphere Biosphere Programme: A study of global change*, International Geosphere-Biosphere Programme, Stockholm

Houghton, J.T., Jenkins, G.J., Ephraums, J.J., 1990, *Climate change*, Cambridge University Press, Cambridge

Johnson, R.G. (editor), 1989, *The changing planet: The FY 1990 research plan*, Committee on Earth Sciences, Washington, D.C.

Kemp, D.D., 1990, *Global environmental issues*, Routledge, London

Mannion, A.M., 1992, 'Environmental change: lessons from the past', in Mannion, A.M., Bowlby, S.R. (editors), *Environmental issues in the 1990s*, Wiley, Chichester: 39–59

Mannion, A.M., Bowlby, S.R. (editors), 1992, *Environmental issues in the 1990s*, Wiley, Chichester

Mather, P.M. (editor), 1992, *TERRA-1 understanding the terrestrial environment*, Taylor and Francis, London

Miller, R.B., Jacobson, H.K., 1992, 'Research on the human components of global change', *Global Environmental Change*, 2: 170–82

Mooney, H.A., Hobbs, R.J., 1989, 'Introduction', in Hobbs, R.J., Mooney, H.A. (editors), *Remote sensing of biosphere functioning*, Springer Verlag, New York: 1–4

Moore III, B., Dozier, J., 1992, 'Adapting the Earth Observing System to the projected $8 billion budget: Recommendations from the EOS investigators', *The Earth Observer*, **4**: 3–10

Mungall, C., McLaren, D.J. (editors), 1990, *Planet under stress*, Oxford University Press, Oxford

NASA, 1988, *Earth System Science: A closer view*, National Aeronautics and Space Administration, Washington, D.C.

NASA, 1992, *EOS data and information systems (EOS DIS)*, National Aeronautics and Space Administration, Washington, D.C.

NERC, 1989, *Our future World – global environmental research*, Natural Environment Research Council, Swindon

NRC, 1986, *Global change in the geosphere-biosphere: Initial priorities for an IGBP*, National Research Council, National Academy Press, Washington, D.C.

NRC, 1988, *Toward an understanding of global change*, National Academy of Sciences, National Academy Press, Washington, D.C.

Perry, M.J., 1986, 'Assessing marine primary production from space', *BioScience*, **36**: 461–7

Suzuki, D.T., 1991, 'Awakening', in Porritt, J. (editor), *Save the Earth*, Dorling Kindersley, London: 19

Townshend, J.R.G., 1992, *Improved global data for land applications*, The International Geosphere-Biosphere Programme, Stockholm

Turner, II, B.L., Clark, W.C., Kates, R.W., Richards, J.F., Matthews, J.T., Meyer, W.B. (editors), 1990, *The Earth as transformed by human action*, Cambridge University Press, Cambridge

Williamson, P., 1992, *Global change: reducing uncertainties*, International Geosphere–Biosphere Programme, Stockholm

2. Remote Sensing and Terrestrial Global Environment Research – the TIGER Programme

S.A. Briggs

.... INTRODUCTION

There has been an increasing recognition in recent years of the importance of global research to address the processes which affect the behaviour of the Earth as a system. This has resulted in the development of a number of large scale umbrella programmes such as the World Climate Research Programme (WCRP) (WMO, 1984) and the International Geosphere-Biosphere Programme (IGBP) (IGBP, 1986), within which core projects address more specific science questions. Some of these are well advanced, for example the World Ocean Circulation Experiment (WOCE), while others are still under consideration. Many of these programmes will make extensive use of data acquired from Earth observation systems. The proposed Global Climate Observing System will co-ordinate the collection of information on the state of the Earth's climate, with satellite systems again playing a vital rôle in acquisition and relaying of climate data. These programmes provide a framework for the co-ordination of national research activities.

Until recently, there has been greater emphasis on the oceanographic and atmospheric aspects of global environmental research but the terrestrial element, through such projects as Global Change and Terrestrial Ecosystems (GCTE), is becoming increasingly important. Within the UK environmental research community, the main contribution to such work is through the Terrestrial Initiative in Global Environmental Research (TIGER) programme of the Natural Environment Research Council. Elements of the TIGER programme contribute significantly to the UK effort in support of IGBP core projects, together with the WCRP Global Energy and Water Cycle Experiment (GEWEX).

The TIGER programme is defined in terms of the priority areas for terrestrial aspects of climate change without any specific reference to the rôle of Earth observation. It is, however, quite clear that Earth observation will be a key technology to address many of the problem areas included in the TIGER programme. An outline of the programme is given below, with a fuller description of those areas where remote sensing techniques are particularly relevant.

.... THE TIGER PROGRAMME

The objective of the TIGER programme is to study the function of the terrestrial biosphere in global change processes. The specific foci which have been chosen in the programme reflect the particular research interests and experience of the UK terrestrial science community. The TIGER programme has four main themes; the carbon cycle on land, trace gas exchange with the biosphere, global energy and water balance, and impacts of climate change on terrestrial ecosystems.

Each of these areas is related strategically to continuing research interests in the UK. The tropical forest biome is of particular importance, due both to its physical rôle in primary productivity and evapotranspiration processes and to its immense biodiversity. The former of these two aspects is a key study area for the carbon balance element of the TIGER programme. Results from this work will contribute to the GCTE and Global Analysis, Interpretation and Modelling (GAIM) IGBP core projects. The exchange of trace greenhouse gases with the terrestrial biosphere is the second TIGER focus, concentrating on the identification and quantification of the sources of methane and nitrous oxide in a range of environments. The programme will also include the development of new technologies for direct airborne measurements of relevant chemical species. Many of the research activities contained within this focus contribute to the International Global Atmospheric Chemistry project (IGAC).

The third focus of the TIGER programme concerns the exchange of water and energy between the land surface and the boundary layer. The objective is to study energy and water balance for selected temperate, arid and tropical areas and to derive aggregated values for inclusion in regional and global models. This element of the programme relates strongly to activities in the Biospheric Aspects of the Hydrological Cycle (BAHC) core project and to GEWEX, with particular reference to the Mississippi Basin continental scale modelling programme.

The fourth and final area in which the TIGER programme research is concentrated lies in the impacts of global change on terrestrial ecosystems. The aim is to develop a capacity to predict the biological, ecological and physical effects of elevated carbon dioxide concentration and climate change on terrestrial ecosystems, including soils and freshwater. Work in this focus is particularly relevant to the IGBP GCTE core project.

.... TIGER I – THE CARBON CYCLE ON LAND

The first focus of the TIGER programme looks at a number of aspects of the terrestrial carbon cycle, both independently and in conjunction with related work in water/energy Soil-Vegetation-Atmosphere-Transfer (SVAT) modelling.

One element of the programme will develop models of carbon cycling in vegetation which can couple to Global Circulation Models (GCMs). This is closely related to the hydrological modelling of TIGER III and will be linked to the GCMs of the UK Hadley Centre for Climate Research. An input to the model will be the global distribution of major relevant biomes, segmented both through taxonomic criteria and through functional type. Data from Earth observation systems will be assessed to determine their ability to assist in the definition and identification of major cover types according to both sets of criteria (e.g., Holdridge, 1964). However, the major use of remotely sensed data in the study of the terrestrial carbon cycle will be in the characterisation of tropical forest regenerative status through both passive visible/near infrared and synthetic aperture radar microwave imagery, as described below.

The carbon balance of tropical forests is a key factor in estimating the global terrestrial carbon balance. Tropical forests represent about one-third of the total terrestrial store of carbon but are also undergoing considerable change as a result of land practices in the tropics. Field measurements will be made of CO_2 fluxes at sites in Brazil and at a long term experimental site in Cameroon, West Africa. Detailed measurements of the forest status will also be made at the field sites, including measures of standing biomass, together with the carbon content of soil and root systems. A number of vegetational successional stages of importance in characterising the carbon content of the forest will be derived from imagery obtained from a number of satellite sensors. The detailed field measurements will then be related to the synoptic distributions of vegetation status derived from the remotely sensed data to provide regional scale and above estimates of carbon balance. The use of remotely sensed data to provide a basis for scaling detailed point measures will be a crucial element in this work.

Radiance data at visible and infrared wavelengths will be acquired from the NOAA Advanced Very High Resolution Radiometer (AVHRR), while synthetic aperture radar (SAR) data will be acquired from the European Remote Sensing satellite (ERS-1) and through principal investigatorships in the Japanese Earth Resources Satellite Fuyo-1 (formerly JERS-1) and SAREX programmes. Models of backscatter as a function of above-ground biomass will be developed as a part of the programme, with further investigation of the ability of multiple frequency synthesised images to estimate tropical forest biomass. These data will be particularly important due to the all-weather operation of SAR systems in a region of near continuous cloud cover. The research is related to ongoing collaboration within the EC International Forest Inventory Team (IFIT) programme.

Preliminary investigations using the Jet Propulsion Laboratory's multi-polarisation, multi-frequency airborne AIRSAR have indicated that these data can be used to estimate above-ground biomass in forest stands (Durden *et al.*, 1993). An example of the relationship between backscattered power and stand age is shown in Figure 2.1, derived from a flight with the AIRSAR over a well-characterised plantation in Thetford Forest, UK (Baker and Mitchell, 1992).

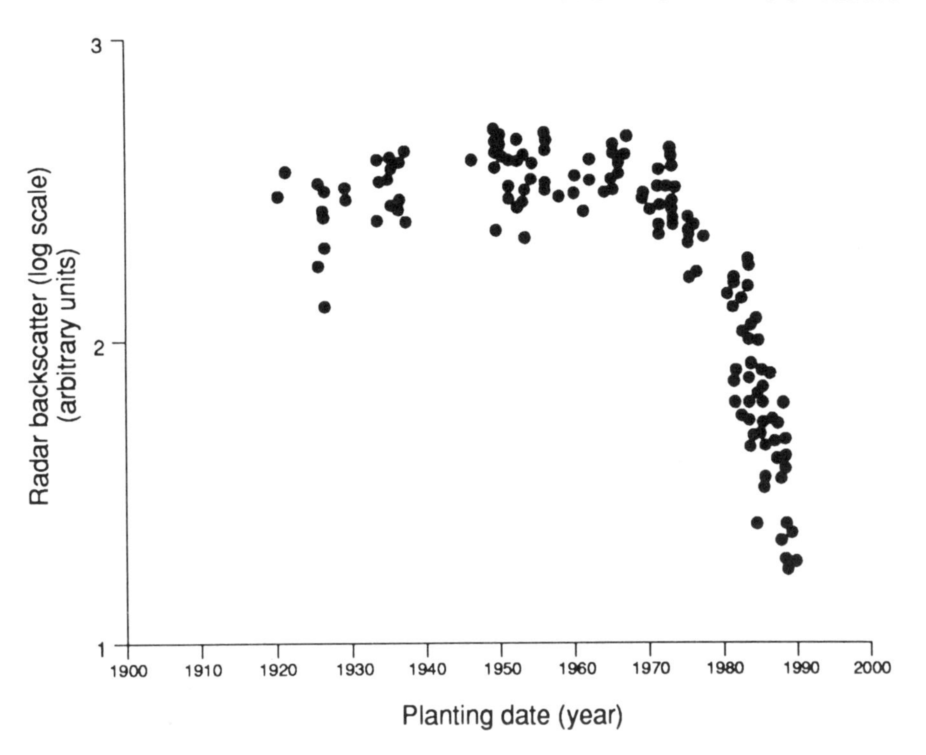

Figure 2.1 Relationship between backscattered L-band power and planting date for a forest stand.

.... TIGER II – TRACE GREENHOUSE GASES

The extent to which Earth observation is able to contribute to our understanding of tropospheric chemistry is at present very limited. Future systems which will begin to study relevant chemical concentrations in the lower troposphere include the Global Ozone Monitoring Experiment (GOME) on ERS-2, and the Tropospheric Emission Spectrometer (TES) and Measurements of Pollution In The Troposphere (MOPITT) instruments planned for the NASA Earth Observing System. These will have a key rôle to play in our understanding of tropospheric chemistry.

The current TIGER sub-programme is focused on *in situ* measures of atmospheric concentrations of tropospheric greenhouse gases and the processes by which they are created and destroyed. These include studies of photochemistry in the troposphere, investigation of rates of methane release from wetlands and the development of global models of the behaviour of methane and related gases, related to the UK Universities Global Atmosphere Modelling Programme (UGAMP).

.... TIGER III – WATER AND ENERGY BALANCE

Earth observation is a key input into a number of components of the TIGER

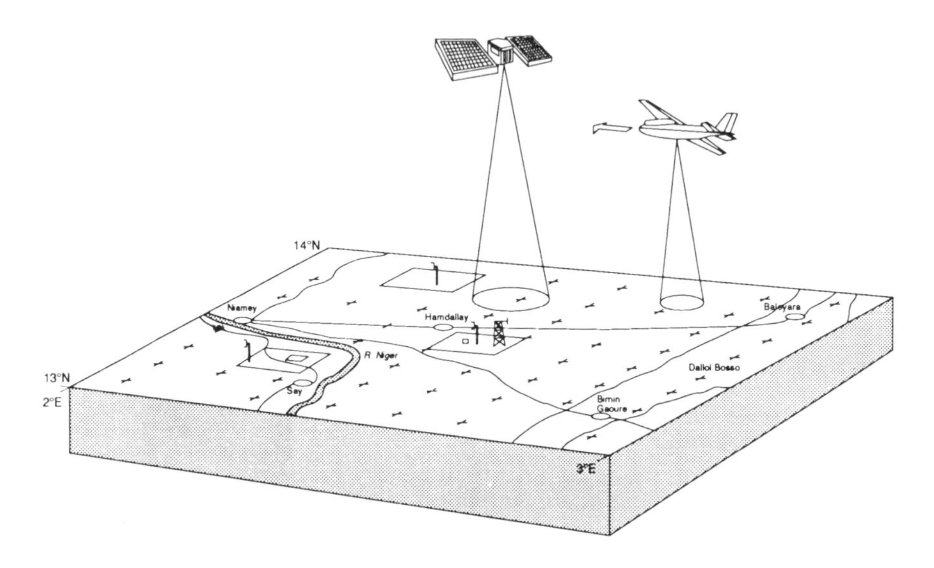

Figure 2.2 Schematic illustration of airborne, satellite-borne and ground-based measurements in the HAPEX-Sahel experiment, Niger.

III sub-programme. A major contribution to the SVAT modelling activities of TIGER III will be based on the HAPEX-Sahel experiment of 1992 (Hoepffner *et al.*, 1990). Detailed measurements of bidirectional reflectance distributions will be made at the HAPEX field site over a number of different cover types. The same instrumentation will also be used to characterise the radiative properties of the local atmosphere, allowing a more accurate estimate of the total surface radiation loading to be derived. This work will be carried out in conjunction with simultaneous micrometeorological measurements to develop a complete SVAT model for the field site (Figure 2.2).

In addition to the field measurements, models of both incident and reflected radiation fields will be established. For single cover types, simple reflectance models such as the Minnaert model (Woodham and Gray, 1987) will be used, while for vegetation only the Camillo model (Camillo, 1987), based on an exact solution to the radiative transfer equation applied to plant canopies, will be used. Atmospheric aerosol distributions will also be estimated to allow accurate modelling of the incident radiation field. Further field measures of the reflectance properties of the components (leaves, stems, etc.) of individual plants will be made to build a model of radiative transfer in the canopy. The HAPEX-Sahel experiment includes the acquisition of a number of airborne datasets; of particular relevance is the deployment of an airborne system which will acquire multiple view angle imagery of the field site. Flight plans for this instrument are being developed to link with the TIGER field experiment. This will allow the bi-directional reflectance distribution functions (BRDFs) built up from the field measures to be related to the reflectances from the aircraft, which in turn are related to satellite sensor measurements of radiative loading. The rôle of remote sensing in scaling-up local measures of radiative loading and evapotranspiration will be developed through such experiments.

The TIGER III sub-programme also includes a component which will be proposed as the UK contribution to the GEWEX Continental-scale International Programme (GCIP) (WCRP, 1990). The objective of the TIGER work is to derive a model capable of predicting hydrological consequences of global change at the large regional scale. Such models will also provide the land–atmosphere interactions for future GCMs, so that water and energy pathways at the land surface are modelled more realistically. Remotely sensed data will be used to develop distributions of land cover type and topography as inputs to the hydrological model, allowing its general application at any terrestrial location. Model structures will be developed in terms of parameters which can be derived from current and proposed remote sensing systems.

Other elements of the TIGER III sub-programme will address energy and water balance in tropical forests, in tandem with the work on carbon budgets described above, and the development of more sophisticated SVAT models.

.... TIGER IV – IMPACTS ON ECOSYSTEMS

The goal of this component of the TIGER programme is to develop a capacity to predict the biological, ecological and physical effects of elevated carbon dioxide concentration and climate change on terrestrial ecosystems. The general approach will be to identify and quantify sensitivities of biota and ecosystem processes.

The framework of the TIGER IV sub-programme is hierarchical with a series of projects which move from consideration of single organisms at a physiological level, through population dynamics, to the structure (species composition) and function (processes) of ecosystems. The spatial disposition and sensitivities of biota and soils will provide a basis for predicting impacts of changes in atmospheric carbon dioxide, climate and land use at various scales, from local patch scale responses to the distribution of global biomes. It is clear that the contribution to be made by Earth observation will be most important in the latter cases, and the projects which are outlined here describe work undertaken at the regional and biome scale. In the following, the terminology of patch, landscape, biome etc. is that of IGBP/GCTE, with which this programme is closely linked (Walker, 1992).

In terms of regional scale impacts, remotely sensed data form an input to models of landscape change primarily through information on land cover and, where relevant, inferred land use. The proposed programme will compile a database of climate, soil, hydrology, agriculture, land use and species distributions at a patch, national and continental scale. The database will be used to model the response surfaces of these environmental variables in relation to present climate and, by applying climate change scenarios, to predict new equilibrium landscapes and species distributions. Dispersal and invasive processes will then be included to propagate the likely changes in equilibrium distributions to their conclusion.

A second component proposed within the TIGER IV sub-programme where

Earth observation is a key methodology will study the potential change in biome distributions as a function of climate change. In the same way as dispersal and invasive mechanisms are important at the regional scale, wholesale changes in the distribution of biomes will depend on these mechanisms and their likely success, in addition to the controlling influence of climate.

Existing mechanistic models predict biome distributions in relation to climatic scenarios. The models are static, in that they incorporate no simulations of the factors which determine the sensitivities of biomes to climatic change. It takes no account of the ability of species to respond dynamically to changes in climate, nor of the competing influences upon them. Mechanisms of dispersal and invasion are not considered. It is, however, likely that the distribution of biomes will not respond as readily to climate change without adequate dispersal/invasive mechanisms. A key element in allowing the transition of biomes will be the rate of disturbance within the biome, as a measure of susceptibility to change. The rate of these disturbances, and their distribution in time and space over which they occur, will hence be a key factor in determining the rate of response of biome distributions to climate change (Hall *et al.*, 1991).

The proposed work will take visible/near infrared data over a range of spatial resolutions from SPOT HRV panchromatic data (10m) to AVHRR Local Area Coverage data (c.1.1km) and investigate the distributions of disturbance within selected key biomes, as a function of spatial scale. By incorporating these data as a measure of susceptibility to change, mechanistic static models of biome-climate interaction can be improved to take account of invasion and dispersal mechanisms.

.... CONCLUSION

The programme which has been described here will make a significant contribution to the international research effort in understanding the interaction of climate change and the terrestrial environment. It is focused on key areas where there is particular UK interest and experience. Earth observation will play an important rôle in three of the four programme areas identified. This chapter has outlined the main elements of the programme, and described in greater detail the particular contribution which it is intended Earth observation will make to the scientific achievement of the TIGER programme.

.... ACKNOWLEDGEMENTS

This chapter summarises proposed work in the TIGER programme to be carried out by a wide range of scientists in the field of remote sensing. Foremost among these are Prof. Paul Curran (University of Southampton), Dr Giles Foody (University College of Swansea), Dr John Baker (British National Space Centre, BNSC), Dr Geoff Groom (Institute of Terrestrial Ecology, ITE), Prof. Robert Gurney and Dr Jeff Settle (NERC Unit for Thematic Information

Systems), Andrew Wilson (BNSC), Prof. Peter Muller and Dr Mike Barnsley (University College London), and Dr Barry Wyatt (ITE), in the order of the occurrence of relevant work described in this chapter.

.... REFERENCES

Baker, J.R., Mitchell, P.L., 1992, 'The UK element of the MAESTRO-1 SAR campaign', *International Journal of Remote Sensing*, **13**: 1593–1608

Camillo, P.J., 1987, 'A canopy reflectance model based on an analytical solution to the multiple scattering equation', *Remote Sensing of Environment*, **23**: 453–77

Durden, S.L., Freeman, A., Klein, J.D., Oren, R., Vane, G., Zebker, H.A., Zimmerman, R., 1993, 'Multi-frequency polarimetric radar measurements of a tropical rainforest', *IEEE Transactions on Geoscience and Remote Sensing*, in press

Hall, F.G., Botkin, D.B., Strebel, D.E., Woods, K.D., Goetz, S.J., 1991, 'Large scale patterns of forest succession as determined by remote sensing', *Ecology*, **72**: 628–40

Hoepffner, M., Goutorbe, J-P., Sellers, P., Tinga, A., 1990, *HAPEX-Sahel experiment plan*, ORSTROM internal report

Holdridge, L.R., 1964, *Life zone ecology*, Tropical Science Centre, San Jose, Costa Rica

IGBP, 1986, *The International Geosphere-Biosphere Programme. A study of global change – a plan for action.* No. 1, International Geosphere-Biosphere Programme, Stockholm

Walker, B.H., 1992, 'Global models of ecosystem change: a status report', *European International Space Year Conference Proceedings*, European Space Agency, ESTEC Noordwijk, **1**: 3–8

WCRP, 1990, *Scientific Plan for GEWEX*, World Climate Research Programme publication, Geneva

WMO, 1984, *Scientific Plan for the World Climate Research Program*, World Meteorological Organisation, World Climate Research Programme Series No. 2, WMO/TD No. 6, Geneva

Woodham, R.J., Gray, M.H., 1987, 'An analytical method for radiometric correction of satellite multispectral data', *IEEE Transactions on Geoscience and Remote Sensing*, **25**: 258–71

3. Explaining and Monitoring Land Cover Dynamics in Drylands Using Multi-temporal Analysis of NOAA AVHRR Imagery

A.C. Millington, J. Wellens, J.J. Settle and R.J. Saull

.... INTRODUCTION

Dryland ecosystems are characterised by dynamic land cover behaviour, caused by inter-annual variability of climate and land use conversion (e.g., urban expansion and desert reclamation). This is mainly a result of the considerable spatial and temporal variations in moisture availability. Furthermore, overexploitation and mismanagement of dryland vegetation and soils leads to significant changes in land use, and therefore in land cover, because of the associated environmental degradation. The increased awareness of environmental issues and, in particular, the linkages between drought, food production and famine, suggests that monitoring dryland degradation is an important issue at the present time. Specifically, those areas at greatest risk from land degradation due to changes in vegetated land cover need to be identified. This chapter examines the potential of imagery acquired by the AVHRR sensor carried on board the NOAA series of satellites in this context.

Since the early 1980s, data from the AVHRR have been used extensively in studies of dryland vegetation and land cover (e.g., Diallo *et al.*, 1991; Wylie *et al.*, 1991; Prince, 1991; Franklin and Hiernaux, 1991; Prince and Tucker, 1986; Eidenshink and Haas, 1992). In particular, data from AVHRR channels 1 and 2 have been used to generate the Normalised Difference Vegetation Index [NDVI = (channel 2 – channel 1)/(channel 2 + channel 1)] (Holben, 1986), a ratio which is sensitive to vegetation growth. The use of multi-date AVHRR-NDVI to discriminate between different vegetation types based upon phenological variations in the AVHRR-NDVI has enabled land cover to be mapped at regional and continental scales (Tucker *et al.*, 1985a; Townshend *et al.*, 1987; Millington *et al.*, 1989; 1993). Values of the AVHRR-NDVI integrated over the growing season have been found to be correlated with seasonal primary productivity (Tucker *et al.*, 1985b; Tucker and Sellers, 1986; Prince and Tucker, 1986; Prince, 1991) and end-of-growing season AVHRR-NDVI has been correlated with above-ground biomass (Kennedy, 1989).

Many of the studies which employ the AVHRR-NDVI rely on data from only one or two growing seasons (e.g., Justice and Hiernaux, 1986; Tucker *et al.*, 1986; Eidenshink and Haas, 1992). However, as the number of AVHRR receiving stations expands, and regular archiving of the relatively fine spatial resolution Local Area Coverage (LAC) data and High Resolution Picture Transmission (HRPT) data has taken place, the use of datasets which span several growing seasons has increased (Prince 1991; Wylie *et al.*, 1991; Tucker *et al.*, 1984; Saull *et al.*, 1991; Schneider *et al.*, 1985; Millington *et al.*, in prep.). These studies highlight the large inter- and intra-annual variations that occur in dryland vegetation cover.

Consequently, an understanding of the factors which enable such variations to be monitored by remote sensing, particularly the relationship between AVHRR-NDVI and moisture availability, is fundamental to any remote sensing programme which aims to monitor environmental degradation. The AVHRR-NDVI moisture availability relationship has been examined using data from Tunisia and Pakistan. In addition, a new approach to monitoring, mapping and quantifying changes in land cover using multi-year AVHRR-NDVI data is presented.

Most studies of land cover mapping using AVHRR-NDVI data have focused on the spatial patterns. However, an analysis of trends in AVHRR-NDVI data from specific locations is fundamental to such mapping (Robinson, 1991). This is because analysis of point data enables the factors which influence the AVHRR-NDVI phenologies used to map land cover to be fully understood before mapping commences. Moreover, such an analysis is important to understanding the mechanisms behind medium- and long-term trends in AVHRR-NDVI and land cover.

.... AVHRR-NDVI AND MOISTURE AVAILABILITY RELATIONSHIPS

RELATIONSHIPS BETWEEN AVHRR-NDVI, RAINFALL AND IRRIGATION RECORDS

The importance of analysing AVHRR-NDVI point data can be illustrated by examining AVHRR-NDVI and moisture availability relationships. Examples of such data from locations in Tunisia and Pakistan (Figure 3.1) show clearly the seasonal and inter-annual variations in AVHRR-NDVI (Figures 3.2–3.5). In addition, they reveal significant variations in the annual cycle of AVHRR-NDVI between different climatic zones in Tunisia and in Pakistan, between rain-fed and irrigated cropland.

AVHRR-NDVI curves for two locations in Tunisia, Bizerte (which lies in the sub-humid zone) and Thala (which lies in the semi-arid zone), are illustrated in Figures 3.2 and 3.3. In both figures the temporal profiles are of mean LAC-derived AVHRR-NDVI for a 3 × 3 pixel area between November 1984 and August 1988. The Bizerte site is dominated by garrigue vegetation consisting of short (30–50 cm tall) sclerophyllous shrubs such as *Quercus*

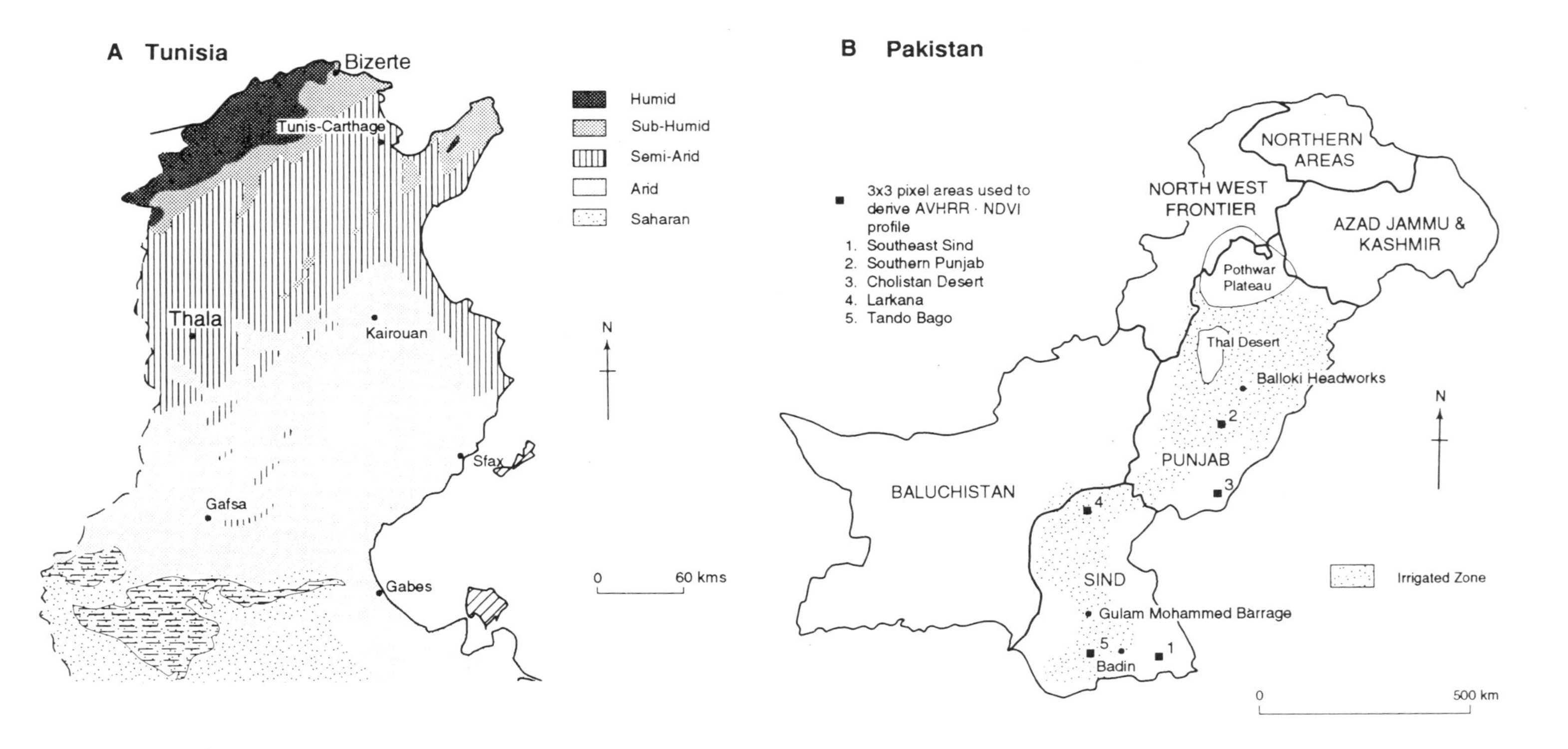

Figure 3.1 Location maps: (a) sites in Tunisia, (b) sites in Pakistan.

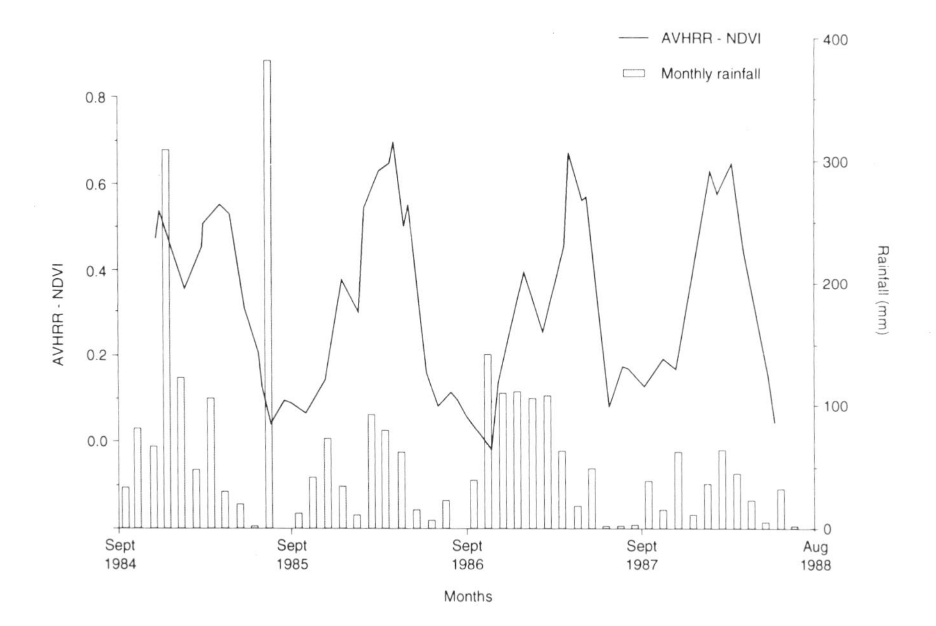

Figure 3.2 Bizerte (North-East Tunisia): mean AVHRR-NDVI profile for agricultural years 1984/5–1987/8 (starting in November 1984), and mean monthly rainfall totals, September 1984–August 1988.

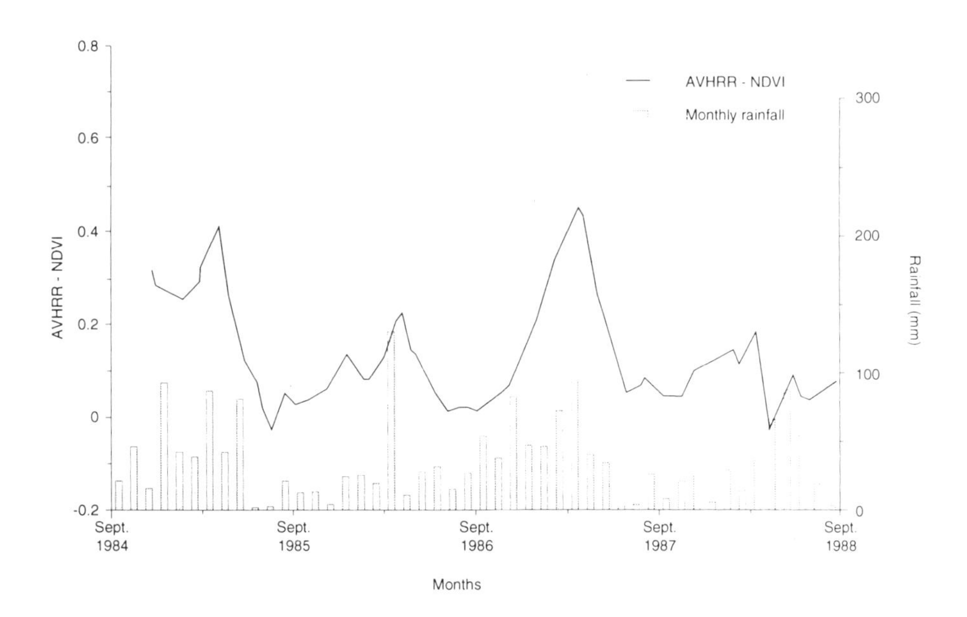

Figure 3.3 Thala (central Tunisia): mean AVHRR-NDVI profile for agricultural years 1984/5–1987/8 (starting in November 1984), and mean monthly rainfall totals, September 1984–August 1988.

coccifera and *Cistus monspeliensis* which provide grazing for sheep, goats and cattle. Figure 3.2 also indicates the monthly rainfall over the same period for the Bizerte synoptic station located 12 km from the site.

Bizerte experiences a Mediterranean climate, with mild, wet winters and hot, dry summers and the vegetation phenology clearly responds to this seasonal pattern. The AVHRR-NDVI response is similar for all years except 1984/5. The vegetation commences growth in November, achieving a peak AVHRR-NDVI value of around 0.7 by late March to early April. In 1984/5 peak AVHRR-NDVI also occurs in early April but only attains a value of 0.55. The trough in AVHRR-NDVI occurs between June and October in all years. Rain falls during the autumn and winter, and the peak rainfall is generally four to eight weeks before peak AVHRR-NDVI. Sometimes, however, large rainfall events occur during the summer (e.g. July 1985), but these have little effect on vegetation growth because plants are generally dormant by this time, although they could affect AVHRR-NDVI because of changes in soil moisture.

The site at Thala consists of steppic grasses and shrubs such as *Artemisia herba-alba* and *Stipa tenacissima* (alfa grass). The vegetation is grazed and the alfa grass is also harvested for paper production. Variations in the AVHRR-NDVI phenology over different growing seasons are visible in Figure 3.3. As at Bizerte, the growing season commences in November and peak AVHRR-NDVI occurs around the beginning of April. In the years 1984/5 and 1986/7 peak AVHRR-NDVI is around 0.4, while in 1985/6 and 1987/8 AVHRR-NDVI does not exceed 0.2. In 1985/6 and 1987/8 the trough in AVHRR-NDVI lasts from May through to October, while in 1984/5 and 1986/7 it is a month shorter and starts in June. The annual trends in the AVHRR-NDVI are reflected in the monthly rainfall for the same period recorded at the meteorological station in Thala, located 1 km from the site.

Data such as these display strong correlations with rainfall and moisture availability. For example, empirical relationships between the AVHRR-NDVI and soil moisture availability, similar to those outlined above, have also been found by Henrickson and Durkin (1986) in Ethiopia, Malo and Nicholson (1990) in Sahelian West Africa and Kerr *et al.* (1989) in Senegal.

However, in many dryland areas vegetation production is related to the supply of irrigation water, rather than rainfall. In such areas the AVHRR-NDVI curves for irrigated and rainfed areas can be markedly different. Figures 3.4 and 3.5 show temporal profiles of AVHRR-NDVI derived from NOAA Global Area Coverage (GAC) data for two 3 × 3 pixel areas in southern Pakistan (Figure 3.1(b)). Figure 3.4 is for a rainfed agricultural area in the semi-arid zone of south east Sind, and Figure 3.5 is for an irrigated area in southern Punjab.

The first four years in Figure 3.4 show a unimodal peak in AVHRR-NDVI (which varies from 0.19 to 0.28) in July and August which corresponds to vegetation growth that was stimulated by the monsoon rains. AVHRR-NDVI values decline as the rains stop and reach their lowest values, around 0.01, between April and June. In the years 1985/6 and 1986/7 the narrow AVHRR-NDVI peak is replaced by a broad, variable peak with individual monthly AVHRR-NDVI values ranging from 0.08 to 0.12. This peak lasts from July to May in 1985/6 and July to March in 1986/7. From 1981/2 to 1984/5 there is a close match between rainfall amount and timing and the AVHRR-NDVI

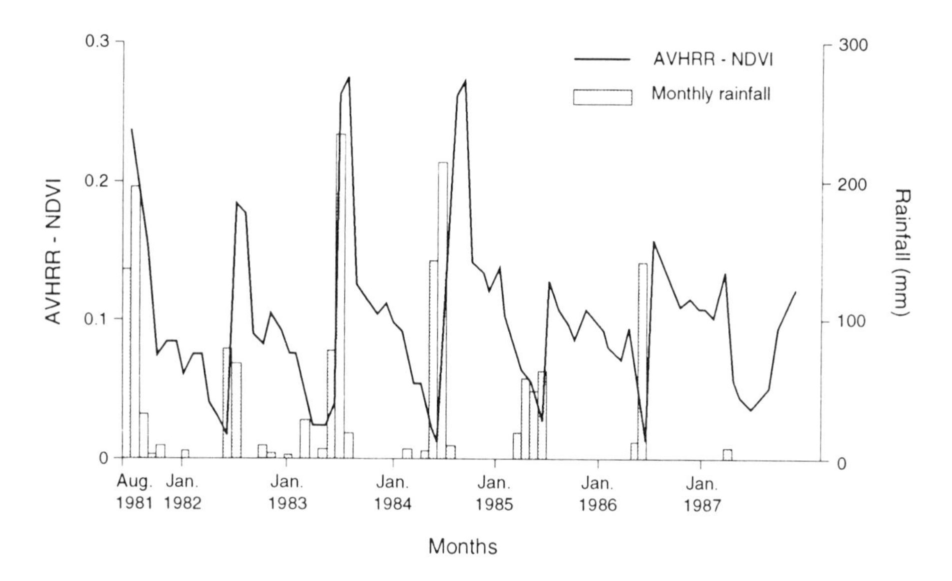

Figure 3.4 South-East Sind (Pakistan): mean AVHRR-NDVI profile for agricultural years 1981/2–1987/8, and mean monthly rainfall totals for Badin, July 1981–December 1988.

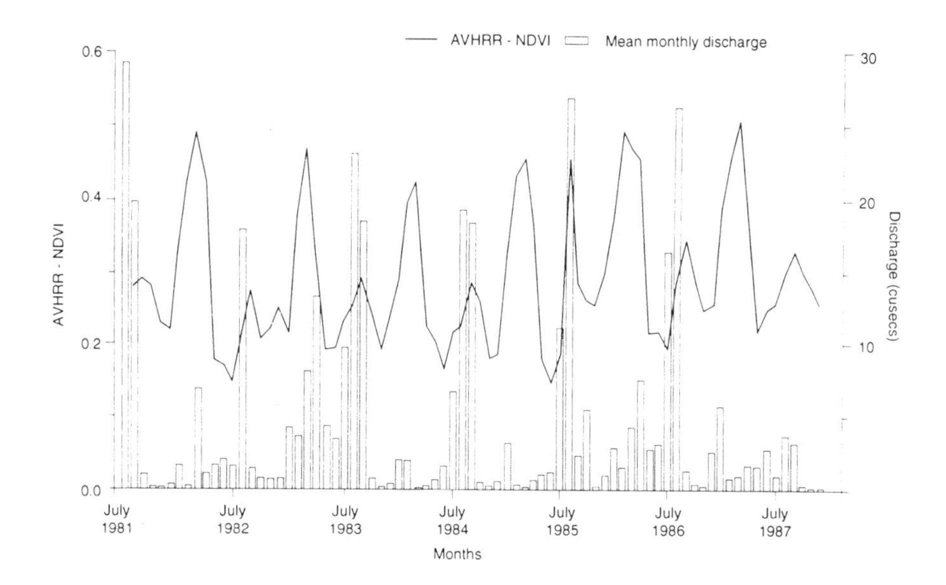

Figure 3.5 Okara (Pakistani Punjab): mean AVHRR-NDVI profile for agricultural years 1981/2–1987/8, and mean total discharge for the Balloki Headworks, River Ravi, July 1981–December 1988.

peaks. The situation changes from 1985/6 onwards: there is a marked decline in rainfall amount and distribution, and this is clearly shown in the different AVHRR-NDVI response in these two years. The 1985/6 and 1986/7 growing seasons were severely affected by a drought in this area and this is reflected in the pattern of vegetation growth.

Figure 3.5 shows the AVHRR-NDVI response of an irrigated area in Okara, fed by the Lower Bari-Doab Canal which gets its water from the River Ravi at Balloki. The AVHRR-NDVI behaviour for each agricultural year, beginning in August and ending in July, is similar for all years except 1985/6. The crops growing during the rabi (northern hemisphere summer) cropping season show a peak AVHRR-NDVI of about 0.3 in September, the peak AVHRR-NDVI values for the kharif (northern hemisphere winter) cropping season are much higher (0.4–0.5) and occur in February or March. The resulting curve is bimodal with two AVHRR-NDVI peaks, the first being lower than the second. The intervening troughs represent the bare fields between November and December, and May and July. However in 1985/6 this typical curve was replaced by one in which the rabi cropping season peak reached approximately 0.44 giving a bimodal curve with two almost equal peaks. The relationships between AVHRR-NDVI and discharge at the Balloki Headworks is not as clear as the examples of AVHRR-NDVI and rainfall. This is mainly because of the distance (about 50 km) between the headworks and the 3 × 3 pixel area in this example. Nonetheless, major fluctuations in canal discharge do correlate with the timing of the AVHRR-NDVI peaks. The amplitude of the AVHRR-NDVI peak is, however, more a function of crop type than the amount of available irrigation water. The first peaks in each cycle, which equate to the rabi season, are dominated by cotton cultivation, whereas the higher amplitude peaks in the kharif season reflect mainly the growth of wheat. The differences in amplitude are due to the fact that wheat develops a far more extensive green canopy than cotton. Similar reasoning has been used to explain the differences in amplitiudes in AVHRR-NDVI peaks of rice and wheat growth in the Indian Punjab (Malingreau, 1986).

MODELLING VEGETATION GROWTH AND AVHRR-NDVI RESPONSES

The examples analysed above show a close correspondence between AVHRR-NDVI and rainfall or irrigation canal run-off data. They provide an approximation of the amount of moisture available for use by vegetation, but two major drawbacks become apparent when trying to explain the AVHRR-NDVI behaviour using such data. First, comparing rainfall or run-off data with the AVHRR-NDVI in this way fails to take into account the fact that only a small percentage of the available water is actually utilised by vegetation (i.e., that transpired during photosynthesis). The majority of the rainfall is lost through direct evaporation from the soil surface, run-off and drainage. Second, because this type of analysis is often done using monthly or ten-day composite AVHRR-NDVI data, the rainfall and run-off data are usually of a similar time scale. Consequently, they do not account for the distribution of water over the time interval.

To overcome the first drawback it is necessary to explore the rainfall and AVHRR-NDVI relationships in much more detail than has been done previously. In particular, there is a need to examine the relationship between AVHRR-NDVI and transpiration, because transpiration provides a measure of the water that is actually used by vegetation. It is more difficult, however, to overcome the second drawback.

A relationship between transpiration and plant production was noted as early as 1913 in Briggs and Schantz' classic study of crop growth in the Great Plains (Briggs and Schantz, 1913). Further research by de Wit (1958) concluded that in climates with a large proportion of bright sunshine hours (which includes arid and semi-arid areas), and where water is non-limiting (as a result of irrigation), a relation:

$$M = mWE_o^{-1} \quad [1]$$

exists between total dry matter yield (M), total transpiration during growth (W), and free water evaporation (E_o), where m is a constant dependent on crop type.

Hanks (1974) used de Wit's relationship to develop the following equation relating relative dry matter yield to relative seasonal transpiration:

$$\frac{Y}{Y_p} = \frac{T}{T_p} \quad [2]$$

where Y is actual yield, T is actual transpiration, T_p is potential transpiration, (i.e., when soil water is not limiting), and Y_p is potential yield when transpiration occurs at the potential rate.

Hank's equation has subsequently been used in a number of models which predict crop or rangeland production (e.g., Wight and Hanks, 1981; Wight *et al.*, 1984). The ratio T/T_p is an index of the growing season's climatic effect on yield; it accounts for the amount of precipitation, its distribution, associated evaporative demand and the soil water content.

A model based the work of Hanks (1974), Wight and Hanks (1981) and Van Keulen (1975), which estimates T/T_p values for Tunisia from climatic and soils data and vegetation parameters, was developed to examine the relationships between AVHRR-NDVI and transpiration. The basis of this model is the water balance equation:

$$T = P + S - E - R - D \quad [3]$$

where T is actual transpiration, P is precipitation, S is the change in soil moisture, E is actual evaporation from the soil surface, R is run-off and D is drainage.

Potential transpiration (T_p) and potential evaporation (E_p) are calculated from an estimate of potential evapotranspiration (ET_o).

$$T_p = b.Kc.ET_o \quad [4]$$

$$E_p = b.Kc.ET_o - T_p \quad [5]$$

where b is the percentage vegetative ground cover and Kc is the crop coefficient, and actual transpiration is calculated as follows:

$$T = T_p \ if \ \frac{SWC}{AW} > C \qquad [6]$$

$$T = \frac{T_p}{C} \times \frac{SWC}{AW} \ if \ \frac{SWC}{AW} < C \qquad [7]$$

where SWC (soil water content) is the amount of water available in the soil, AW is the available soil water and C is the proportion of available water required before transpiration can proceed at the potential rate.

Water enters the soil in the form of precipitation and is removed either by transpiration, evaporation or drainage below the potential rooting zone of the vegetation. The field capacity (FC) and permanent wilting point (WP) of each soil horizon have been used to calculate the available water capacity of the whole soil profile. Should precipitation exceed this, it contributes to drainage. The removal of water by evaporation is a function of potential evaporation (E_p) and the time since the soil was last wet (Hanks, 1974; Rasmussen and Hanks, 1978; Wight and Hanks, 1981) and is estimated by:

$$E = \frac{E_m}{t^{1/2}} C \qquad [8]$$

where E is actual evaporation, E_m is potential evaporation, and t is the number of days since the soil was last wet (i.e., since the last rainfall event).

Evaporation is also limited to those layers which lie within 30 cm of the soil surface (Hanks, 1985). Furthermore, it is assumed to cease when the wilting point is reached, although in reality evaporation removes water to lower potentials than plant roots do.

The model calculates T_p, T, E_p, E and SWC at ten-day intervals. These values are then used to determine T/T_p for each agricultural year (September–August).

The inputs to the model, for the study in Tunisia, are ten-day precipitation data (obtained from the Institut National de la Météorologie (INM), Tunis). However, Class-A pan potential evapotranspiration data from the INM is incomplete (gaps of ten months are not uncommon), so potential evapotranspiration was estimated using the Food and Agriculture Organisation of the United Nations' (FAO) version of Makkink's (1957) radiation method (Doorenbos and Pruitt, 1977). Information on soils were derived from the Afrika-Kartenwerk 1:1 000 000 scale soils map of North Africa. Typical profiles for these soil types were obtained from the accompanying volume (Brechtel and Rohmer, 1980). The field capacity, wilting point and hence available water capacity of each textural class were estimated from curves produced by Hall *et al.* (1977). The available water content of both the profile and the upper 30 cm, which corresponds to the evaporation limit, were calculated.

Most of the coefficients employed in this model have been estimated from literature. These are better documented for agricultural than semi-natural vegetation, except for parts of North America, Israel and Australia.

The crop coefficient (Kc) converts ET_o (the potential evapotranspiration of an extensive surface of 8–15 cm tall green grass cover) to the potential evapotranspiration of the vegetation of interest. Wight and Hanks (1981) determined Kc values of 0.7–0.9 for rangelands in Montana. They used the growing season average of 0.85 and found that this gave accurate yield and soil water estimates. The vegetation at this site consisted mainly of prairie grasses such as *Agropyron smithii*, *Boutelouda gracilis*, *Stipa comata*, *Carex filifolia* and *Carex eleocharis*. These species are similar to those found in the central Tunisian steppes and, therefore, a value of 0.85 was used in the model for Tunisia.

A number of different values for C have been used to estimate the relationship between T, T_p and soil moisture. Hanks (1974) and Rasmussen and Hanks (1978) used a value of 0.5 for modelling cereal crops. De Jong and MacDonald (1975) used a value of 0.7 for native range in Canada. Work by Wight and Hanks (1981) on rangeland production in Montana and North Dakota, suggested that the models were not sensitive to changing values of C and they therefore used a value of 1. Because no further information was available concerning transpiration and soil moisture requirements and, as a C-value of 1 appeared to have performed satisfactorily in models of North American rangelands, this value was used initially in the Tunisian model. However, subsequent investigation of the sensitivity of the model output to variations in the C-value in Tunisia contradict the findings of Wight and Hanks (1981).

The values of percentage ground-cover (b) used in the model are based on measurements made at each field site in Tunisia. They are therefore specific to the date and the locations at which they were collected. In most cases this was around the period of peak standing biomass.

Sensitivity analysis was carried out in order to determine how variations in these coefficients affected the output of the model. The model was run 28 times for each location using the ten-day rainfall data corresponding to a single agricultural year. Each time one of the coefficients was varied by a small amount while the remaining coefficients were held constant. The resulting change in output (T/T_p) was noted for each run. This method was repeated at least twice for each meteorological station, using the rainfall data from the wettest and driest years.

The sensitivity function for the different values of the coefficients were calculated using the formula:

$$S(Fi) = \frac{\delta Fo}{\delta Fi} \qquad [9]$$

where $S(Fi)$ = sensitivity function of factor Fi, Fo = model output and Fi = model input coefficient. The results are presented in Table 3.1.

Table 3.1 indicates that the model is most sensitive to variations in the percentage ground cover (b). At all the locations, the greatest response was found when the simulated ground cover was increased from 90–100 per cent.

Table 3.1 Sensitivity of the model to changing values of coefficients.

Coefficient tested	Range of values examined	Sensitivity function Minimum	Maximum
Crop coefficient (*Kc*)	0.5-1.2	0.035	0.00
% Ground cover (*b*)	10-100	0.236	0.00
C-value	0.1-1.0	−0.096	0.00

However, field measurements of ground cover made during this research in Tunisia have not exceeded 60 per cent. In the range 0–60 per cent cover the model has lower sensitivity to changing values of ground cover. The region of the crop coefficient response curve over which the model is most sensitive varies between locations and years. In general the sub-humid sites are most sensitive to changes at the higher end of the range (0.9–1.2), while the more arid sites are sensitive at lower values (0.5–0.8). The sensitivity response curve for the *C*-value (the fraction of soil moisture required before potential transpiration commences) depends on both the rainfall and the location, such that no two years or sites are equally sensitive at the same point. Figure 3.6 shows the response curve (modelled output (T/T_p) versus *C*-value) for a location in the Saharan bioclimatic zone. This curve indicates that the model

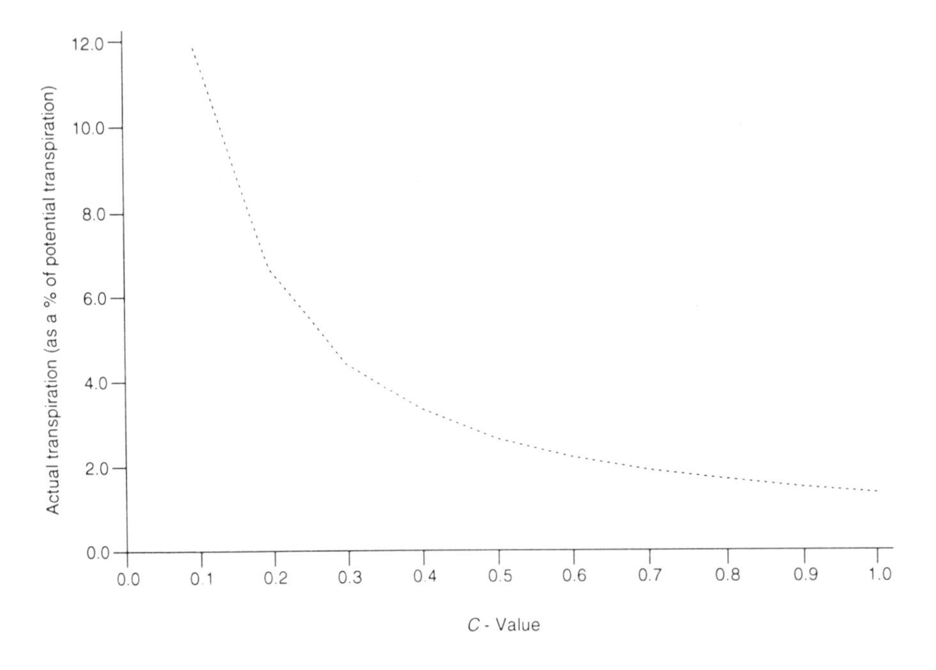

Figure 3.6 Sensitivity of model to variations in the *C*-value: Tozeur (southern Tunisia), agricultural year 1983/4.

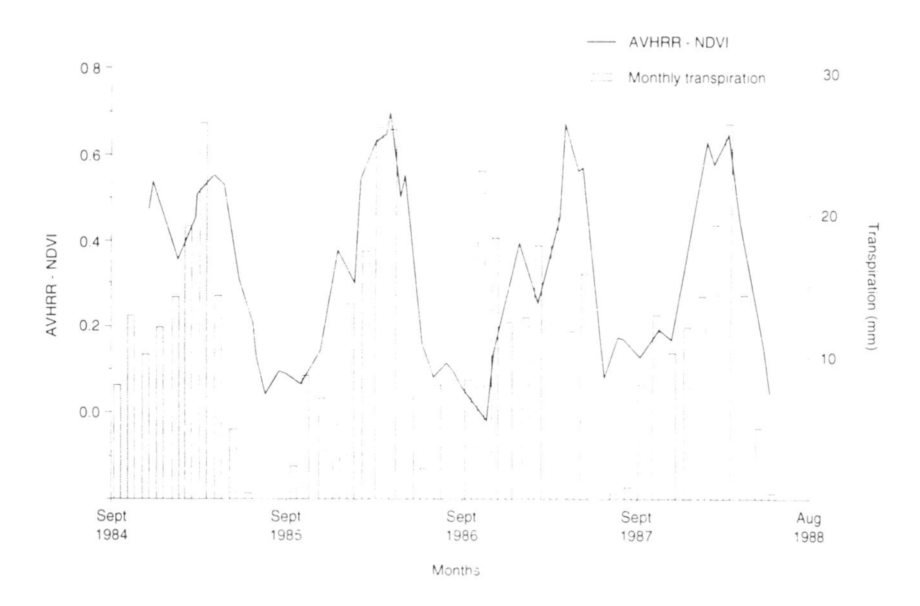

Figure 3.7 Bizerte (North-East Tunisia): mean AVHRR-NDVI profile for agricultural years 1984/5–1987/8 (starting in November 1984), and estimated monthly actual transpiration from the vegetation production model, September 1984–August 1988.

is highly sensitive to variations in the *C*-value over the range 0.1 to 0.4 but that variations in the value towards the higher end of the range have little effect upon the estimated yield. There are very little data available concerning the relationship between transpiration and soil moisture levels, which makes the determination of the *C*-value difficult.

Ten-day averages of estimated actual transpiration (T) calculated throughout the agricultural year were aggregated to give monthly values. These were then compared with temporal profiles of AVHRR-NDVI. The AVHRR-NDVI and transpiration curves for Bizerte are shown in Figure 3.7. If the actual transpiration data are compared to the corresponding rainfall data (Figure 3.2) it can be seen that there is a difference in magnitude between them. Actual transpiration represents approximately one-fifth of the annual rainfall. The relationship between the actual transpiration and AVHRR-NDVI profiles is stronger than that between rainfall and AVHRR-NDVI. Peak transpiration and peak AVHRR-NDVI occur simultaneously, as would be anticipated, from late March to early April. Moisture availability is also distributed more regularly than rainfall through the period of vegetation growth. The modelling approach also incorporates rainfall from single, large events into the soil moisture term in equation 3.

The model-derived yield estimates (T/T_p) for each agricultural year have also been compared with integrated AVHRR-NDVI values at each site. At present these results are tentative because only four years of data were available for comparison. Nonetheless, for some years the relationship between T/T_p and integrated AVHRR-NDVI is strong, whilst in other years it is not. T/T_p and integrated AVHRR-NDVI values appear to correspond in years when rainfall occurs as several large events. But, in years when there are many small (< 5 cm) rainfall events throughout the growing season, the relationship between T/T_p and integrated AVHRR-NDVI is weak. The reasons for this are not known exactly, but it is probable that they result from the lack of a run-off coefficient in the model used.

A further explanation of this pattern may lie with the number of images used to calculate integrated AVHRR-NDVI and the distribution of these images throughout the year. The imagery used in this study is spaced at approximately monthly intervals and was selected on the basis that there was minimal cloud cover over Tunisia. However, each agricultural year contains a different number of images ranging from ten in 1987/88 to 15 in 1985/6. Moreover, because of the lack of suitable imagery during December 1987 there is a ten-week interval between images in the early part of the 1987/8 growing season.

DISCUSSION

Comparison of estimates of actual transpiration over a growing season are more closely related to temporal plots of AVHRR-NDVI than rainfall data over the same period. This is because actual transpiration data provides a more meaningful estimate of the amount of moisture available for use by vegetation. Actual transpiration data can also be used to provide an estimate of the growing season's yield using Hanks' relationship between relative transpiration and relative yield (equation 2). However, more years of AVHRR-NDVI imagery and modelled yield data have to be examined before the statistical validity of such a comparison can be determined.

Further investigation of the coefficients used in the model also needs to be undertaken. In particular, a better estimate of the C-value for each soil type is required. Run-off coefficients also need to be incorporated into the model, as this is an important factor in determining how much moisture is available for use by vegetation in drylands.

Determining how the number of images used to calculate the integrated AVHRR-NDVI and their distribution over a growing season affect its value, is more difficult. A bias towards stressed vegetation will be apparent because of the increased probability of obtaining cloud-free imagery during dry spells, and this will be accentuated in the ten-day composite imagery. A comparison of integrated AVHRR-NDVI values calculated from a series of individual dates with that from ten-day composite data would illustrate this bias.

Finally, the model assumes that the only factor limiting biomass production is moisture availability. This assumption is probably valid in those areas with annual rainfall totals of less than 250–300 mm (Breman and de Wit, 1983). However, in wetter semi-arid and sub-humid regions, soil nutrient shortages, particularly of phosphorous and nitrogen, may be a greater constraint on production. Such a cut-off point is not absolute; animal activity, run-off and local soil conditions can all affect nutrient and water availability.

.... SPATIAL ANALYSIS OF MULTI-TEMPORAL AVHRR DATA

It has been noted that a major problem encountered when mapping land cover in drylands is the fluctuation in vegetation cover between years. Although in many areas this is a function of rainfall, other factors also contribute to the dynamic nature of dryland vegetation cover (Table 3.2).

Table 3.2 Environmental and socio-economic factors which affect plant growth and cause land cover variations, with particular reference to Pakistan.

Main environmental divisions (and types of agroecosystems)	Causal factors
Arid and semi-arid areas (Rain-fed cultivation and pastoralism)	Rainfall variations Land degradation processes (soil erosion, dune encroachment woodland clearance, plant cover destruction) Impact of agricultural projects (especially installation of irrigation systems) Crop pests Crop diseases
Irrigated areas (Irrigated cultivation)	Fluctuations in water discharge Land degradation (secondary salinization, waterlogging) Crop pests Crop diseases Expansion of irrigation systems Reclamation of waterlogged and salt-affected land (SCARP Projects)
Montane regions (Rain-fed and Irrigated cultivation)	Climatic variations (rainfall, temperature and snowfall variations) Variations in valley (snow-melt) flooding Deforestation Crop pests Crop diseases

While describing and explaining these fluctuations is the key task of monitoring, they do hinder traditional forms of land cover mapping using remotely sensed data. This is because such changes limit the usefulness of land cover maps derived from a single year of remotely sensed data. However, by explicitly incorporating land cover dynamics into land cover classification, the accuracy of land cover maps can be increased. Consequently, a robust land cover zonation for a dryland area should be derived from data characterised by the following attributes:

(i) the data for the entire region of interest should be contemporary to allow field verification;
(ii) the spatial resolution of the data must be fine enough to allow identification of the principal land cover variations, but not so fine as to result in an unmanageable quantity of data;
(iii) the data must describe both inter-seasonal and inter-annual land cover variations to make use of the land cover implications of phenology;
(iv) the dataset must be comprehensive in that it captures the major parameters which determine land cover; and

(v) the dataset should be multi-purpose in that it can be used in several different types of analysis.

The requirements for the data to be at an appropriate spatial resolution, dynamic and contemporary imply the use of NOAA AVHRR imagery. Furthermore, the need to have data which are comprehensive and multi-purpose suggests the combined use of the AVHRR-NDVI and digital environmental data in a geographical information system (GIS).

The land cover mapping methodology presented here develops the concept of the *core zone* to reconcile the dynamics of land cover and the need to produce a map for a particular time interval. A core zone is defined as the minimum spatial extent of a land cover class during the period of monitoring. In core zones the inter-annual variations in land cover are low, although inter-seasonal variations may be high, and there is a strong likelihood that any area within the zone will exhibit the same land cover from one year to the next. Conversely, outside the core zone the likelihood of change is high. Therefore, core zones and non-core zones represent the minimum and maximum extents of land cover classes in unfavourable and favourable years respectively (Figure 3.8).

DEVELOPMENT AND IMPLEMENTATION OF THE CORE ZONE CONCEPT IN PAKISTAN

The core zone methodology for land cover mapping was developed and applied in Pakistan using GAC-derived AVHRR-NDVI data acquired between 1981 and 1987 as part of a biomass fuel survey (Saull *et al.*, 1991; Millington *et al.*, in prep.). It was found to be particularly useful in Pakistan because the dryland environments which comprise most of the country are very susceptible to climatic and human-induced changes in land cover (Table 3.2). In such environments traditional land cover mapping techniques cannot produce maps which account for these variations; neither do they remain valid for long due to land cover fluctuations.

Two types of data were used in this application. Remotely sensed data were used to divide the country into spectrally-distinct zones and various digital environmental data were used help to translate these zones into land cover classes (Table 3.3).

Image processing involved two stages. First, the data were inspected and preliminary image inspection and analysis was carried out prior to field checking. After this, the data were analysed using an approach which combined digital image data and digital environmental data in a GIS and allowed the production of a series of maps describing land cover and its dynamics. The steps in the image processing and GIS pathway are described below and represented in Figure 3.9; the details of which can be found in Saull *et al.* (1991) and Millington *et al.* (in prep.).

Initially the imagery was inspected in 12-month sequences which began in August 1981. The time period from August to July represents the *agricultural year* over much of Pakistan. Spatial and temporal patterns of AVHRR-NDVI from areas of natural vegetation and farmland were compared to maps of vegetation and agricultural activities (Survey of Pakistan, 1986) and published

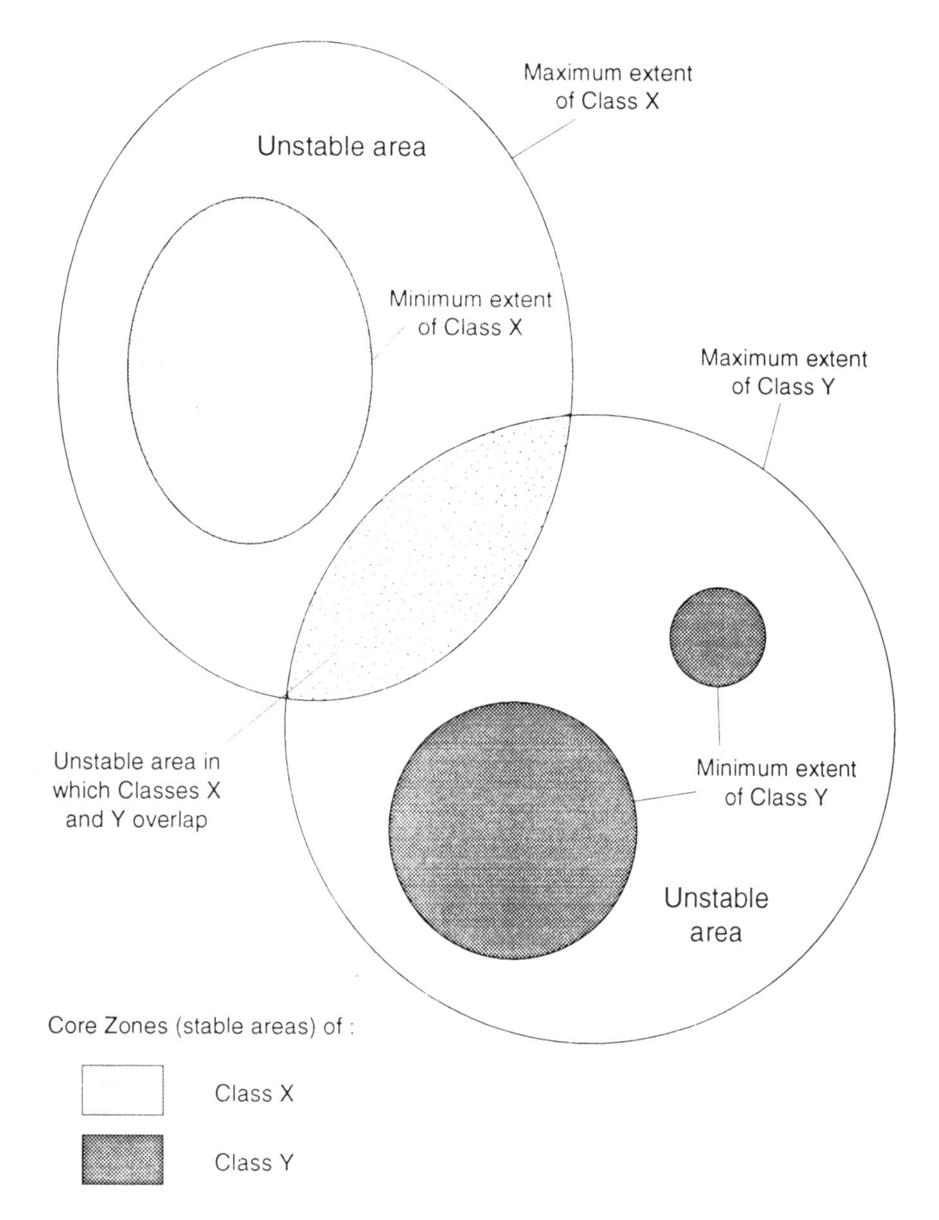

Figure 3.8 The core zone concept.

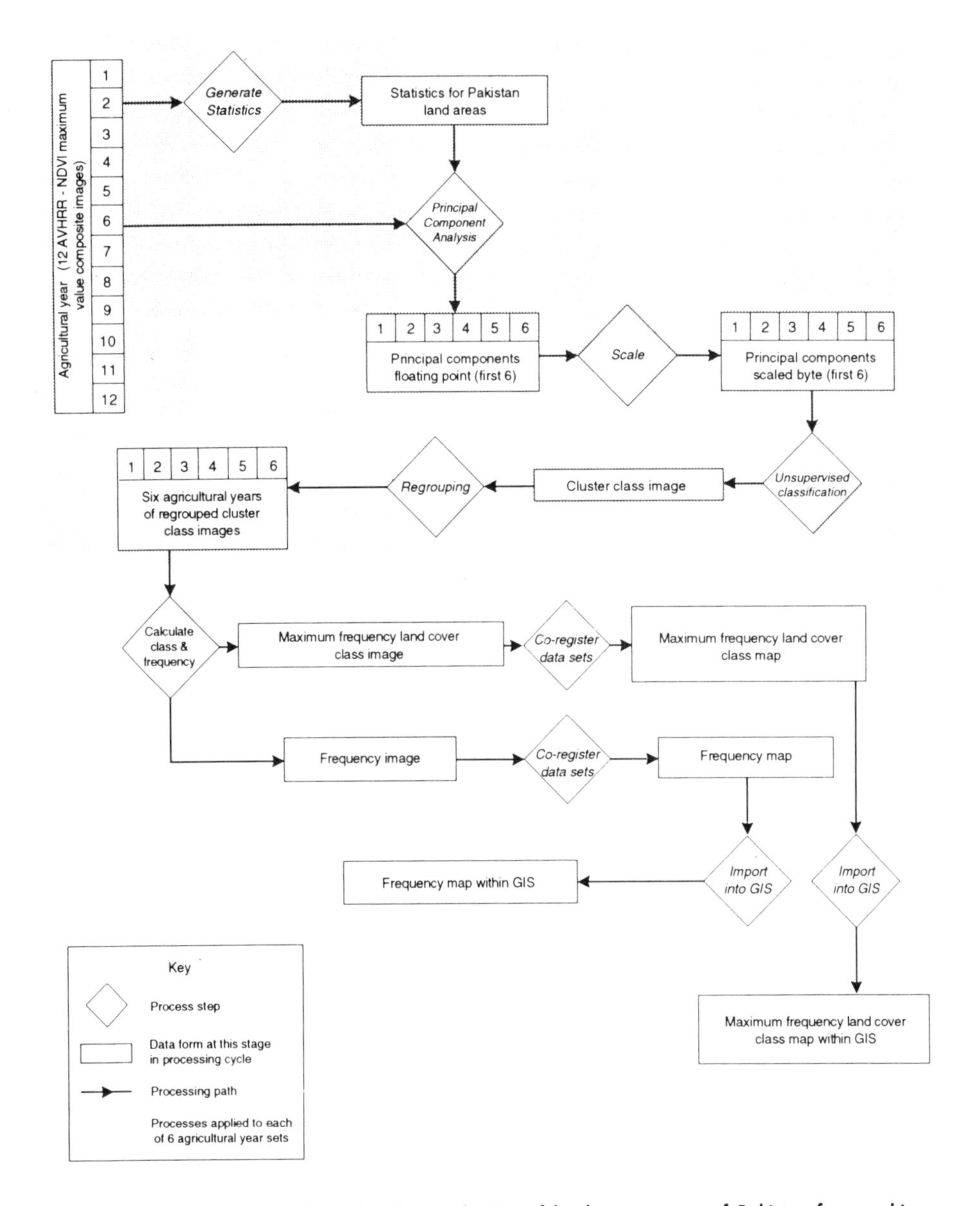

Figure 3.9 Processing pathway for the production of land cover maps of Pakistan from multi-temporal AVHRR-NDVI imagery.

Table 3.3 Sources of data used for land cover mapping of Pakistan.

Data	Description	Original format	Source
Digital elevation data	USGS[1] 5′(ETOPO5) DEM[2] of the world	CCT[3]	UNEP/GRID
Rainfall	Monthly means for main PMD[4] station, 1980–89	Tabular	PMD
River and canal discharge	10-day discharge measurements for main river and canal head gauging stations, 1980–89	Tabular	Water & Power Development Authority
Irrigation canals	Map of main canals and extent of irrigated area	Paper maps	Atlas of Pakistan
Soils	Main soil types	Paper maps	Atlas of Pakistan[5]
Climate	Mean extent of monsoon	Paper maps	Atlas of Pakistan
Remotely sensed data	Monthly AVHRR-NDVI (July 1981–Dec 1987)	CCT	NASA/GSFC[6]

[1] United States Geological Survey
[2] Digital Elevation Model
[3] Computer Compatible Tape
[4] Pakistan Meteorological Department
[5] Original data are from the Soil Survey of Pakistan
[6] NASA/GSFC – NASA Goddard Space Flight Center, Maryland.

crop calendars. Temporal profiles of AVHRR-NDVI were obtained for twenty-five 3 × 3 pixel areas, which were selected because of their proximity to meteorological stations with reliable data. In addition, for each of the agricultural years unsupervised classifications were performed on six, alternate monthly images.

Fieldwork designed to incorporate checking of (i) the seasonal and inter-annual trends in AVHRR-NDVI temporal profiles, (ii) the broad patterns of land cover classes and their boundaries (from the unsupervised classification images), and (iii) a qualitative evaluation of the levels of between-class heterogeneity and within-class homogeneity was undertaken in the spring of 1990. The results of the preliminary unsupervised classifications were found to be relatively accurate during field verification, although it is not possible to quantify the accuracy.

In previous work of this type (Tucker *et al.*, 1985a; Townshend *et al.*, 1987; Millington *et al.*, 1989; 1993) supervised classification has been used to generate the final land cover class maps. However, a different approach was adopted in this study because it differs from these other studies in three ways. First, AVHRR data for more than one year were available, allowing inter-year comparisons. Second, significant amounts of relatively detailed environmental supporting data were available. Third, the implementation of the core zone concept requires that land cover classes to be mapped independently in each agricultural year. Comparisons are then made between years, to quantify the stability of the land cover classes over time.

The implementation of the core zone concept was achieved through the use of unsupervised classification and principal components analysis. Unsupervised classification was preferred to supervised classification because the latter technique would not be able to take into account variations in the location of different land cover classes in different years with the current ground sampling scheme. This is because the location of a training site for a particular land cover class on one year's imagery may not refer to the same land cover class in subsequent years and would, therefore, require six years of training site data. The high levels of autocorrelation between monthly AVHRR-NDVI imagery questions the statistical accuracy of attempting an unsupervised classification of 12 sequential images. Therefore, for each agricultural year a principal components analysis was performed on the 12 images and the resulting principal components examined. In each year the maps of individual principal components images showed similar spatial distributions and it was, therefore, assumed that each principal component could be explained by the same factors each year. The first six principal components accounted for 94.6 to 97.1 per cent of the variation in the six agricultural years analysed. These six principal components were then classified to create cluster class images for each year.

The cluster class maps for each agricultural year had a varying number of classes. However, when inspected, the cluster classes formed eight distinct groups (e.g., a single cluster class in one year would form two classes in a subsequent year). Such variation occurs because of the nature of the unsupervised classification algorithm. To overcome this, the six cluster class maps were examined in conjunction with field observations and ancillary map and meteorological data. The cluster classes were then reclassified into eight groups. The digital numbers of each of the original cluster classes were rewritten according to this grouping.

The variation in the extent of each regrouped cluster class was analysed by producing observed frequency and most frequent class maps based on the behaviour of land cover in the six regrouped cluster class maps. The number of occurrences of each of the eight regrouped cluster classes for a given pixel were recorded; and the observed frequency of the pixels were calculated thus:

$$\text{Observed frequency} = \frac{\text{Number of occurrences in most frequent class}}{\text{Number of agricultural years}}$$

The class number for a given pixel was then set to the class number with the highest observed frequency. For pixels where the observed frequency was < 50 per cent this value can be ambiguous as other classes may have equal observed frequencies. In such cases, the first class encountered was entered into the most frequent class map.

The observed frequency and most frequent class maps were transferred to a GIS so that they could be combined with other environmental data to produce the final land cover maps. The image maps were registered to the polyconic projection used in the GIS using a nearest neighbour algorithm to avoid corrupting the class values.

The most frequent class map is made up of classes that have similar AVHRR-NDVI phenologies over an agricultural year, therefore they are based entirely on the behaviour of the reflectance spectra. Although reasonably accurate, such maps do not include other information that can further explain land cover divisions. In Pakistan, for example, many of the areas of evergreen forest flanking the southern Himalayas are spectrally similar to high productivity irrigated land in the adjacent Punjab Plains, and appear in the same spectral class in each year. Therefore, although the classification is spectrally correct, it may be divided on the basis of other factors which do not affect the reflectance spectra (e.g., elevation, extent of irrigation) to provide a more meaningful land cover zonation. Therefore, the provision of such a land cover map requires other data to convert the most frequent land cover class map to a series of more accurate land cover classes.

This process was performed in a GIS and termed environmental constraint splitting. To split the regrouped cluster classes, all of the relevant digital data (see Table 3.3) were combined in a map of environmental constraints (see Plate 1(a)). A table was constructed in the GIS to define the way in which the environmental constraint map was used to split the regrouped cluster classes in the most frequent class map. After environmental constraint splitting, 14 land cover classes were identified. The observed frequencies of these classes for each pixel were calculated and a maximum frequency class map was produced (see Plate 1(b)) – this is equivalent to a traditional land cover map. In addition a map of the observed frequency of the most frequent land cover class for each pixel was produced. The temporal stability of these classes was examined by intersecting the maximum frequency class map and the observed frequency map. By adopting a frequency threshold of 0.83 (i.e., at least five out of six years in the same class) a map of core areas was produced (see Plate 1(c)).

RESULTS

The 14 land cover classes naturally form four groups (see Plate 1(b)) in terms of the external factors which affect vegetation growth. The two arid classes are mainly restricted to west Baluchistan and the western Thar Desert in East Sind and South East Punjab. The four semi-arid to sub-humid classes occur along the mountain ranges that extend from the Afghanistan border south of 34° N to the west of Karachi; in South East Sind; in the Thal Desert; and on the Pothwar Plateau which flanks the Himalayan Foothills. The four montane classes are restricted to northern Pakistan. Finally, there are three classes in the Indus Irrigated Zone.

It is clear from Table 3.4 and Plate 1(b) that the proportion of the maximum extent of each land cover class between 1981 and 1987 occupied by its core zones varies considerably. Some classes (e.g., High mountain valleys and Marginal irrigated) have proportionately small core zones; whereas other classes (e.g., the two arid classes, the wetter semi-arid and sub-humid classes, the Temperate and sub-tropical Himalayan Foothills and the

Table 3.4 Total areas of land cover classes and *the equivalent* core zone areas.

Class	Total area (km^2)	Proportion of country (%)	Stable area 1981–1987 (km^2)	Class area that is stable as a proportion of total class area (%)
ARID CLASSES				
1. Hot, hyperarid desert	58,062	5.9	29,640	50.7
2. Arid desert	195,950	20.0	93,835	47.9
SEMI-ARID and SUB-HUMID CLASSES				
3. Transitional arid/semi-arid	110,445	11.3	40,031	36.2
4. Moderate-productivity semi-arid	68,291	7.0	28,066	41.1
5. High-productivity semi-arid	46,960	4.8	29,453	62.9
6. Sub-tropical rain-fed agriculture	24,496	2.5	14,788	60.4
MONTANE CLASSES				
7. Permanent snow	57,976	5.9	11,131	19.2
8. High mountain valleys	42,909	4.4	1,430	3.4
9. Alpine and temperate scrub and forest	135,014	13.8	32,944	24.4
10. Temperate and Sub-tropical Himalayan Forest	27,647	2.8	18,443	66.7
IRRIGATED CLASSES				
11. Marginal irrigated	10,435	1.1	629	6.0
12. Medium-productivity irrigated	84,409	8.6	48,838	57.9
13. High-productivity irrigated	111,186	11.4	64,978	58.4
14. Indus delta swamps	5,071	0.5	829	16.3

Medium and High-productivity irrigated classes) exhibit core zones that cover over 50 per cent of their maximum area (Table 3.4).

VERIFICATION

The use of the core zone approach was verified by analysing temporal profiles of AVHRR-NDVI for a stratified random sample of core zones and non-core zones from Plate 1(c). Wherever possible, sites were selected close to meteorological stations or, in the irrigated zone, with reference to canal and river discharge records. Mean AVHRR-NDVI values were extracted for 3 × 3 pixel areas and converted to latitude and longitude coordinates to facilitate accurate geo-referencing.

The observed frequency threshold for core zones was set at 0.83 in this study. Therefore, it would be expected that an AVHRR-NDVI temporal profile would be similar in at least five years out of six in a core zone. In non-core zones the curve shapes should show a greater variation between years, with at least two of the six years being different. Mean AVHRR-NDVI temporal profiles for four locations are shown in Figure 3.10. These represent two core zone and two non-core zone locations in both rain-fed and irrigated land cover types. It is clear that the AVHRR-NDVI profiles for the two core zones (Figures 3.10(a) and 3.10(b)) fit the criterion outlined above. In one case (Figure 3.10(a)) five out of six years have a similar curve, and in the other profile all six years are similar (Figure 3.10(b)). The AVHRR-NDVI profiles of the two non-core zones (Figures 3.10(c) and 3.10(d)) show far greater variation between years than in the core zones. In the case of the irrigated area (Figure 3.10(c)) inter-annual fluctuations in AVHRR-NDVI have been related to variations in canal run-off (Millington *et al.*, in prep.). In the rainfed area the variations are related to rainfall and have already been noted in Figure 3.4.

DISCUSSION

It is clear that the 14 land cover types identified in Pakistan show a wide variation in stability over the six agricultural years, ranging from only 3.4 per cent of the High mountain valleys class to 66.7 per cent of the Temperate and Sub-tropical Himalayan forests class (see Table 3.4). If nothing else, this result demonstrates the shortcomings of land cover mapping based on data from a single year. An understanding of why such fluctuations occur in different land cover classes is important because it provides both information on the type of land cover change that may lead to environmental degradation and also on the limitations of monitoring dryland land cover using AVHRR data.

If the land cover classes are examined in the groups outlined in Table 3.4 it can be seen that the range of the percentage of stable land cover in the two arid classes is 47.9 to 50.7 per cent, for the semi-arid and sub-humid classes it is 36.2 to 62.9 per cent, for the montane classes it is 3.4 to 66.7 per cent, and for the irrigated classes it is 6.0 to 58.4 per cent. Perhaps the most straightforward of these groups is the semi-arid and sub-humid group in which there is a positive trend between increasing rainfall and increasing

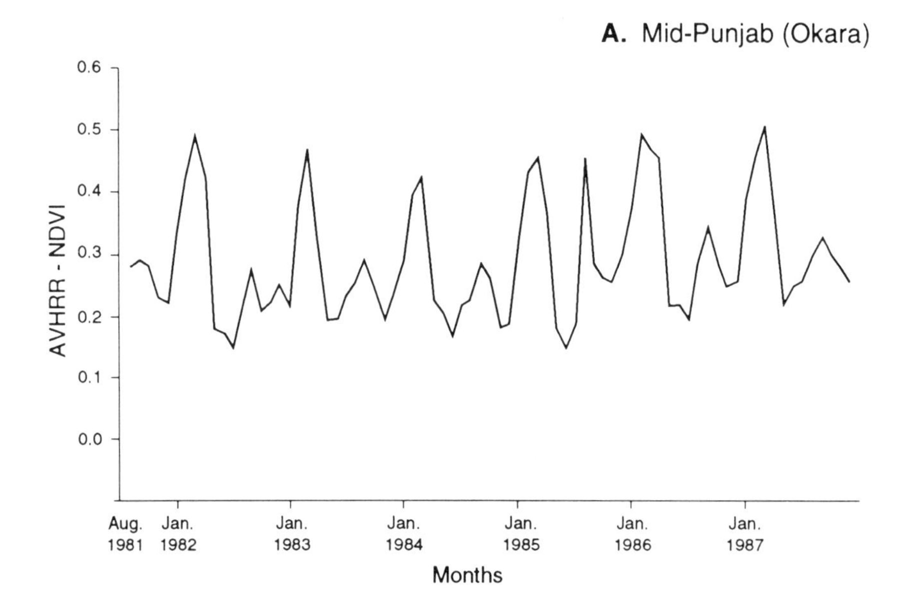
A. Mid-Punjab (Okara)
0.6
0.5
0.4
0.3
0.2
0.1
0.0
AVHRR - NDVI
Aug. 1981
Jan. 1982
Jan. 1983
Jan. 1984
Jan. 1985
Jan. 1986
Jan. 1987
Months

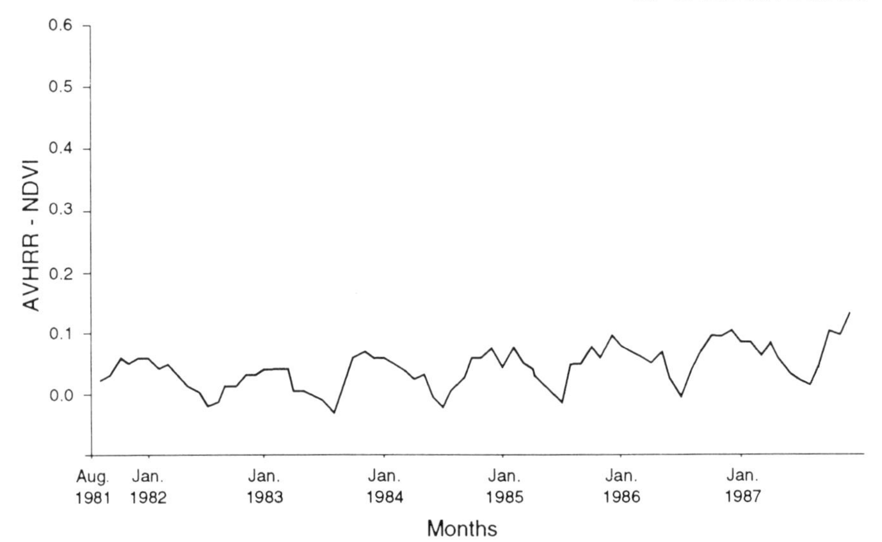
B. Cholistan Desert
0.6
0.5
0.4
0.3
0.2
0.1
0.0
AVHRR - NDVI
Aug. 1981
Jan. 1982
Jan. 1983
Jan. 1984
Jan. 1985
Jan. 1986
Jan. 1987
Months

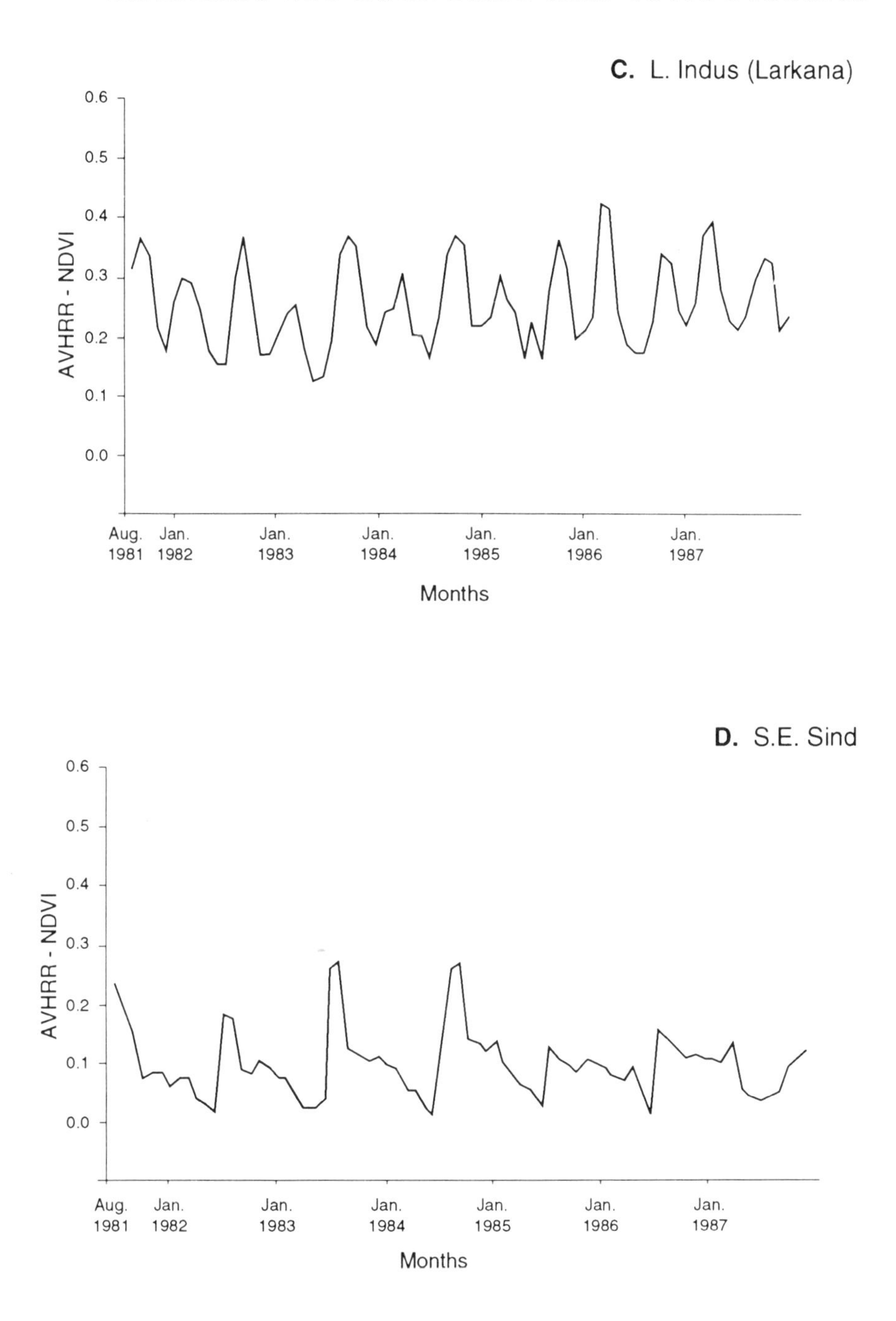

Figure 3.10 Verification of the application of the core zone by the use of mean AVHRR-NDVI profiles for agricultural years 1981/2–1986/7: (a) Okara (Irrigated core zone), (b) Cholistan Desert (Semi-arid core zone), (c) Larkana (Irrigated non-core zone), (d) South-East Sind (Semi-arid non-core zone).

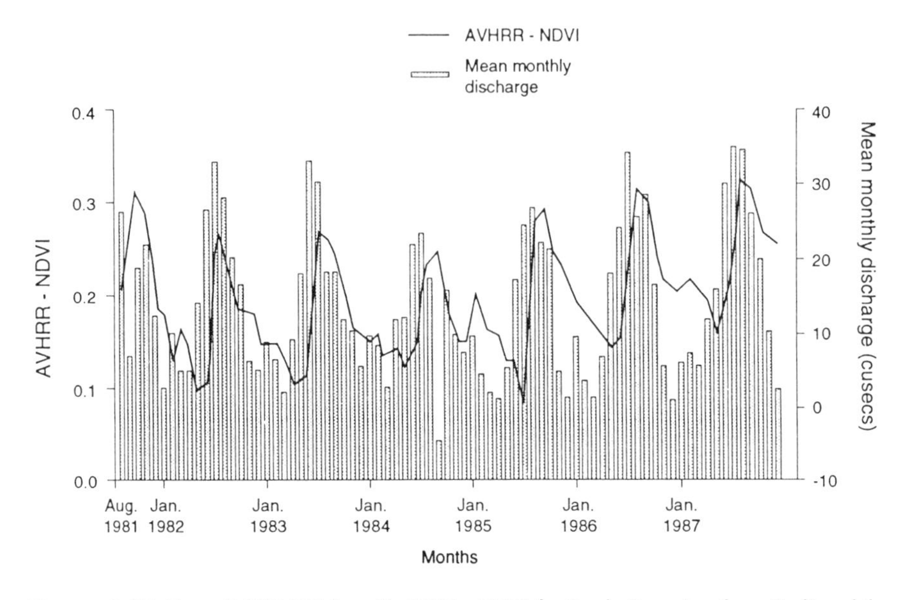

Figure 3.11 Mean AVHRR-NDVI profile (1981–1987) for Tando Bago (southern Sind) and the equivalent canal draw-down discharges for the Gulam Mohammed Barrage at Hyderabad.

proportion of the core zone. The trend is probably a function of rainfall variability which is known to increase with decreasing mean annual rainfall. The two arid classes have similar core zone proportions and there is much overlap between these two classes. In all of these classes most of the variation can be attributed to fluctuations in rainfall (e.g., Figure 3.4).

In the irrigated zone the fluctuations in land cover were far greater than anticipated. Here, the fluctuations are usually between the irrigated land cover classes; although some, especially in the Marginal irrigated class, may be artifacts of image misregistration. However, its low productivity is also a reflection of these areas being fed by the canals at the extremities of the water distribution network. Such canals often do not receive water if supplies are low as the water is extracted before it reaches the area, this leads to crop failures and increased salinization, both of which change surface reflectance properties. A combination of natural and anthropogenic causes leads to high variability in this case. In the irrigated zones the fluctuations in AVHRR-NDVI profiles are best related to canal and river discharge records. This can be seen clearly in the area around Tando Bago in lower Sind which is irrigated by water drawn off from the Gulam Mohammed Barrage on the Indus at Hyderabad (Figure 3.11). In particular, the suspension of water supplies in September 1984 was reflected in a significant decline in AVHRR-NDVI during November and December 1984.

.... CONCLUSIONS

While relationships between the AVHRR-NDVI and soil moisture availability (particularly parameters such as rainfall and effective rainfall) are attractive, they fail to account for the complexities of water use by vegetation during growth. Furthermore, in many dryland areas such growth is related to irrigation supplies, which are not considered if only the AVHRR-NDVI-rainfall relationships are examined.

The latter point indicates that data other than rainfall may need to be used when explaining AVHRR-NDVI behaviour in drylands. Whichever data are used, it is apparent that modelling of vegetation growth can improve the estimate of moisture used by vegetation. This estimate can, in turn, be related to AVHRR-NDVI to obtain a significantly stronger relationship between AVHRR-NDVI and moisture availability than has been available so far.

A further problem surrounds land cover mapping in drylands when imagery for only one year is utilised. Here there is a potential conflict between the demands of monitoring (which describes and explains land cover fluctuations) and traditional land cover mapping (which has the implicit assumption of land cover stability built into it). It is clear from the Pakistani case study that all land cover types in drylands have high inter-annual fluctuations in their areal extent; but by incorporating this dynamic element into land cover maps their reliability can be improved. Fundamental to such mapping, and of great importance in studies of desertification, is the identification of core zones of stable land cover.

The potential of AVHRR data has been underexploited in the study of desertification, despite its advantages of high temporal resolution and empirical and theoretical relationships between AVHRR-NDVI and vegetation growth. Nevertheless, in such studies it is important that the relationships between AVHRR-NDVI and vegetation growth, and their variations over time, are fully understood for specific land cover types before spatial analysis is attempted. In this respect, vegetation growth modelling holds great promise.

.... ACKNOWLEDGEMENTS

The Tunisian work was supported by a NERC Training Award (GT4/89/TLS/49) to JW. The meteorological data set for Tunisia was freely supplied by the Institut National de la Météorologie, Tunis. The work in Pakistan was carried out for the World Bank/UNDP Energy Sector Management Assistance Programme and, in this context we acknowledge the support of Mike Crosetti, Gary Archer, Azeddine Querghi and Mohammed Gul Naqvi. NUTIS is supported under NERC contract S60/G6/12.

.... REFERENCES

Brechtel, R., Rohmer, W., 1980, *Bodenkunde-Nordafrika (Tunesien, Algerien) Afrika-Kartenwerk N4*, Gebrüder Borntraeger, Berlin, 67pp.

Bremen, H., de Wit, C.T., 1983, 'Rangeland productivity and exploitation in the Sahel', *Science*, **221**: 1341–7

Briggs, L.J., Schantz, H.L., 1913, *The water requirement of plants*, U.S. Department of Agriculture Bureau Plant Industries Bulletin, 285 de Jong, E., Macdonald, K.B., 1975, 'The soil moisture regime under native grassland', *Geoderma*, **14**: 207–21

de Wit, C.T., 1958, *Transpiration and crop yields*, Verslangen Van Landbouwkundige Onderzoekingen, 64, Wageningen, The Netherlands

Diallo, O., Diouf, A., Hanan, N.P., Ndiaye, A., Prevost, Y., 1991, 'AVHRR monitoring of savanna primary production in Senegal, west Africa, 1987–1988', *International Journal of Remote Sensing*, **12**: 1259–80

Doorenbos, J., Pruitt., W.O., 1977, *Guidelines for predicting crop water requirements*, FAO Irrigation and Drainage Paper, 24, Food and Agriculture Organisation, Rome, 144pp.

Eidenshink, J.C., Haas, R.H., 1992, 'Analysing vegetation dynamics of land systems with satellite data', *Geocarto International*, **7**: 53–61

Franklin, J., Hiernaux, P.H.Y., 1991, 'Estimating foliage and woody biomass in Sahelian and Sudanian woodlands using a remote sensing model', *International Journal of Remote Sensing*, **12**: 1369–86

Hall, D.G.M., Reeve, M.J., Thomassen, A.J., Wright, W.F., 1977, *Water retention, porosity and density of field soils*, Technical Monograph 9, Soil Survey of England and Wales, Harpenden, 75pp.

Hanks, R.J., 1974, 'Model for predicting plant yield as influenced by water use', *Agronomy Journal*, **66**: 660–5

Hanks, R.J., 1985, 'Crop coefficients for transpiration', in *Advances in evapotranspiration*, Proceedings National Conference on Advances in Evapotranspiration, ASAE Publ. 14–85, American Society of Agricultural Engineers, St. Joseph, Michigan: 431–8

Henrickson, B.L., Durkin, J.W., 1986, 'Growing period and drought early warning in Africa using satellite data', *International Journal of Remote Sensing*, **7**: 1583–1608

Justice, C.O., Hiernaux, P.H.Y., 1986, 'Monitoring the grasslands of the Sahel using NOAA AVHRR data: Niger', *International Journal of Remote Sensing*, **7**: 1475–97

Kennedy, P.J., 1989, 'Monitoring the phenology of Tunisian grazing lands', *International Journal of Remote Sensing*, **10**: 835–45

Kerr, Y.H., Imbernon, J., Dedieu, G., Hautecoeur, O., Lagouarde, J.P., Seguin, B., 1989, 'NOAA AVHRR and its uses for rainfall and evapotranspiration monitoring', *International Journal of Remote Sensing*, **10**: 847–54

Makkink, G.G., 1957, 'Testing the Penman formula by means of lysimeters', *Journal Institute of Water Engineers*, **11**: 277–88

Malingreau, J-P., 1986, 'Monitoring tropical wetland rice production systems: a test for orbital remote sensing', in Eden, M.J., Parry, J.T. (editors), *Remote sensing and tropical land management*, Wiley, Chichester: 279–305

Malo, A.R., Nicholson, S.E., 1990, 'A study of rainfall and vegetation dynamics in the African Sahel using normalised difference vegetation index', *Journal of Arid Environments*, **19**: 1–24

Millington, A.C., Townshend, J.R.G., Kennedy, P.A., Saull, R., Prince, S.D., Madams, R., 1989, *Biomass Assessment, Woody Biomass in the SADCC Region*, Earthscan Publications, London

Millington, A.C., Critchley, R.W., Douglas, T.D., Ryan, P., 1993, *Woody biomass assessment in sub-Saharan Africa*, World Bank Publications, Washington D.C.

Millington, A.C., Crosetti, M., Saull, R.J., in prep., 'Mapping land cover in changing environments using AVHRR-NDVI data: The development and implementation of a methodology in Pakistan', *Remote Sensing of Environment* (submitted).

Prince, S.D., Tucker, C.J., 1986, 'Satellite remote sensing of rangelands in

Botswana II: NOAA AVHRR and herbaceous vegetation', *International Journal of Remote Sensing*, 7: 1555–70

Prince, S.D., 1991, 'Satellite remote sensing of primary production: comparison of results for Sahelian grasslands 1981–1988', *International Journal of Remote Sensing*, **12**: 1301–11

Rasmussen, V.P., Hanks, R.J., 1978, 'Spring wheat yield model for limited moisture conditions', *Agronomy Journal*, **70**: 940–4

Robinson, T.P., 1991, *Modelling the seasonal distribution of habitat suitability for armyworm population development in East Africa using GIS and remote sensing techniques*, unpublished Ph.D. Thesis, University of Reading, Reading

Saull, R.J., Millington, A.C., Crosetti, M., 1991, 'Pakistan: A land cover zonation from multitemporal AVHRR and environmental data using GIS techniques', *Proceedings 5th AVHRR Data Users Meeting*, EUMETSAT, Tromso: 217–21

Schneider, S.R., McGinnis, D.F., Gatlin, J.A., 1985, 'Monitoring Africa's Lake Chad Basin with Landsat and NOAA satellite data', *International Journal of Remote Sensing*, **6**: 59–73

Survey of Pakistan, 1986, *Atlas of Pakistan*, Survey of Pakistan, Rawalpindi

Townshend, J.R.G., Justice, C.O., Kalb, V.T., 1987, 'Characterization and classification of South American land cover types using satellite data', *International Journal of Remote Sensing*, **8**: 1189–1207

Tucker, C.J., Gatlin, J.A., Schneider, S.R., Kuchinos, A., 1984, 'Monitoring vegetation in the Nile Valley with NOAA-6 and NOAA-7 AVHRR', *Photogrammetric Engineering and Remote Sensing*, **50**: 53–61

Tucker, C.J., Townshend, J.R.G., Goff, T.E., 1985a, 'African land cover classification using satellite data', *Science*, **227**: 233–50

Tucker, C.J., Vanpraet, C.L., Sharman, M.J., Van Ittersum, G., 1985b, 'Satellite remote sensing of total herbaceous biomass production in the Senegalese Sahel: 1980–1984', *Remote Sensing of Environment*, **17**: 223–49

Tucker, C.J., Sellers, P., 1986, 'Satellite remote sensing of primary production', *International Journal of Remote Sensing*, **7**: 1395–1416

Tucker, C.J., Justice, C.O., Prince, S.D., 1986, 'Monitoring the grasslands of the Sahel 1984–1985', *International Journal of Remote Sensing*, **7**: 1571–83

Van Keulen, H., 1975, *Simulation of water use and herbage growth in arid regions*, Pudoc, Wageningen

Wight, J.R., Hanks, R.J., 1981, 'A water-balance, climate model for range herbage production', *Journal of Range Management*, **34**: 307–11

Wight, J.R., Hanson, C.L., Whitmer, D., 1984, 'Using weather records with a range forage production model to forecast range forage production', *Journal of Range Management*, **37**: 3–6

Wylie, B.K., Harrington, J.A., Prince, S.D., Denda, I., 1991, 'Satellite and ground-based pasture production assessment in Niger: 1986–88', *International Journal of Remote Sensing*, **12**: 1281–1300

4. The Use of Remote Sensing to Characterise the Regenerative States of Tropical Forests

Paul J. Curran and Giles M. Foody

.... TROPICAL FORESTS AND THE GLOBAL CARBON CYCLE

The current increase in atmospheric CO_2 of about 3.2 gigatonnes (Gt) per year is, in its simplest terms, a result of an increase in the contribution of CO_2 sources in relation to CO_2 sinks (IGBP, 1990; Jarvis and Moncrieff, 1992). The balance between sources and sinks of CO_2 is poorly understood (Enting and Mansbridge, 1989; 1991; Fan *et al.*, 1990; Tans *et al.*, 1990) (Table 4.1) with particular uncertainty over the rôle of terrestrial vegetation in maintaining that balance.

> The primary reason that there is still controversy over whether terrestrial vegetation is a source or a sink of CO_2 in the carbon cycle ... is that we are unable to quantify seasonal CO_2 balances over large areas. (Roughgarden *et al.*, 1991, p. 1919.)

Mature tropical forests are in carbon balance, absorbing as much CO_2 as is released (Shugart *et al.*, 1986; Aber and Melillo, 1991). These mature forests may become a major source of CO_2 on deforestation or a major sink of CO_2 on regeneration (Detwiler and Hall, 1988; Myers, 1989; Brookfield, 1992). Remotely sensed data have been used to investigate or illustrate the notion of tropical forests as a source rather than as a sink of CO_2 (Woodwell, 1984; Eden, 1986; Houghton *et al.*, 1985; 1987; 1991a; 1991b; Houghton and Skole, 1990; Houghton, 1991). Theoretically, this is because the tropical forest source of CO_2 is an order of magnitude larger than the tropical forest sink of CO_2 (Houghton, 1990). Practically, this is because it is much easier to use remotely sensed data to estimate the presence of forest (as in studies of deforestation) than it is to estimate the characteristics of that forest (as in studies of regeneration).

REMOTE SENSING OF DEFORESTATION IN THE TROPICS

The negative impact of deforestation in the tropics on atmospheric carbon, species diversity and climate stability have been well documented (Myers,

Table 4.1 The annual global CO_2 balance sheet in 1990, with uncertain entries marked with a question mark.

Sources	Gt yr^{-1}	Sinks	Gt yr^{-1}
Fossil fuels	5.7	Atmosphere	3.2
Tropical deforestation	2.1?	Oceans	1.0?
CO_2 and CH_4 from burning vegetation and soil changes	0.7?	Temperate and boreal forests	1.8?
		Tropical forests and grasslands	2.5?
Total sources	8.5?	Total sinks	8.5?

Source: Modified from Jarvis and Moncrieff (1992).

1980, 1991; Henderson-Sellers *et al.*, 1988; Wood, 1990; Proctor, 1990; 1991; Soussan and Millington, 1992). The potential of Landsat sensors for monitoring this deforestation was demonstrated in pioneering studies by Morain and Klankamsorn (1978), Lanley (1982) and Woodwell *et al.* (1984). Despite early over-enthusiasm (Baltaxe, 1980), the Multispectral Scanning Systems (MSS) and later the Thematic Mappers (TM) on Landsat satellites (Table 4.2) have proved to be valuable tools for measuring and monitoring deforestation. For instance, the work by Woodwell *et al.* (1987) in which a small but changing area of Amazonian forest was mapped in 1976, 1978 and 1981 is a powerful illustration of the ability of Landsat MSS to monitor deforestation. Today Landsat MSS imagery is used to study deforestation at a range of scales from the local to the national (Nelson and Holben, 1986; Nelson *et al.*, 1987; Westman *et al.*, 1989; Tardin and Cunha, 1990; Changchui and Chaudhury, 1991). However, while Landsat MSS and also Landsat TM and SPOT HRV data are available at a suitable spatial resolution (Table 4.2), they are recorded too infrequently and are too expensive for the study of deforestation at the regional scale (Woodwell, 1984; Campbell and Bowder, 1992; Pereira, 1992). During the 1980s, considerable advances were made in the use of NOAA AVHRR data for regional scale studies (Tucker *et al.*, 1985). These data have a coarse spatial resolution but are available daily at a low cost. Several local to national scale studies have demonstrated the value of AVHRR data for the study of deforestation and have gone further to point to the potential of these data for the monitoring of deforestation at regional to global scales. Typically AVHRR imagery have been used on their own (Päivinen and Witt, 1988; Malingreau and Tucker, 1988; Malingreau *et al.*, 1989), in a temporal sequence (Achard and Blasco, 1990; Malingreau, 1991) or in unison with finer spatial resolution imagery (e.g., SPOT HRV, Landsat TM) (Jeanjean and Husson, 1991; Malingreau and Reichert, 1991; Amaral, 1992; Malingreau, 1992; Malingreau *et al.*, 1992). The NOAA AVHRR sensor was not designed for terrestrial applications (Hastings and Emery, 1992) and its use in the study of tropical forests poses many problems (Millington *et al.*, 1989; Goward *et al.*, 1991); not least is its coarse spatial resolution. To minimise the effect of a coarse spatial resolution on the accuracy of forest cover estimates, a number of researchers have tried to determine the

Table 4.2 Characteristics of the twelve satellite sensors of most value for the remote sensing of tropical forests.

		Wavelengths sensed							
Satellite	Sensor	visible	near infrared	middle infrared	thermal infrared	microwave waveband	Spatial resolution (m)	Swath (km)	Amount of cloud-free imagery available of tropical forests in 1992 (estimate) (1 = small, 5 = large)
Landsat	MSS	✓	✓				80	185	3
Landsat	TM	✓	✓	✓	✓		30	185	2
SPOT	HRV/XS	✓	✓				20	60	1
SPOT	HRV/Pan	✓					10	60	1
MOMS	MESSR	✓	✓				50	100	1
IRS	LISS	✓	✓				40	70	1
NOAA	AVHRR	✓	✓		✓		1,100	2,700	4
Seasat	SAR					L	25	100	2
Space Shuttle	SIR A/B					L	40	100	2
Almaz	SAR					S	15	30	2
ERS-1	SAR					C	35	100	2
Fuyo-1	SAR	✓	✓	✓		L	18	75	2

Note that the Landsat TM band 6 (thermal infrared) has a spatial resolution of 120 m and Almaz SAR ceased operation in mid 1992. For further details on these satellite sensors refer to CEOS (1992).

proportion of forest within a pixel (e.g., Cross *et al.*, 1991). Such an ability will be vital if AVHRR data are to be used to monitor changes in forest cover as a result of deforestation (or forest characteristics as a result of regeneration) at time scales shorter than a decade (Townshend, 1992).

The tropical forests are cloud-covered for most of the year and this restricts severely the amount and quality of remotely sensed data available from optical sensor systems. Synthetic aperture radar (SAR) systems, by contrast, operate in microwave wavelengths which can penetrate cloud cover without significant attenuation. Since the early 1970s airborne SAR sensors have been used successfully to map forest cover in the tropics (Sicco-Smit, 1978; 1992; Hammond, 1977; Parry and Trevett, 1979; Furley, 1986; Campbell *et al.*, 1992). In addition, local coverage by SAR sensors onboard Seasat and the Space Shuttle have shown promise for the study of deforestation (Sader *et al.*, 1990; Churchill and Sieber, 1991). However, SAR data have only become available for parts of the tropics on a routine basis following the launch of ERS-1 in 1991 and Fuyo-1 (formerly named JERS-1) in 1992. Initial evaluations of the data collected from these spaceborne SARs report them to be well suited to the mapping of deforestation (Suzuki and Shimada, 1992; de Groof *et al.*, 1992).

In a period of a few years remotely sensed data have become the major tool for monitoring deforestation at scales from the local to the national (Myers, 1988; Sader and Joyce, 1988; Gilruth *et al.*, 1990; Blasco and Achard, 1990; Green and Sussman, 1990; Hall *et al.*, 1991; Campbell and Bowder, 1992). Developments in the use of AVHRR and SAR data have made possible the monitoring of deforestation at regional to global scales. To ensure that plans for regional to global scale studies of deforestation are implemented requires political will (Kummer, 1992) and the type of programmes outlined below.

MAJOR PROGRAMMES INVOLVED WITH THE REMOTE SENSING OF DEFORESTATION

There are several initiatives aimed at investigating the use of remotely sensed data for the study of deforestation at regional to global scales (e.g., Chomentowski and Lawrence, 1992). Perhaps the most important have been based on the development of five, existing or proposed, NOAA AVHRR data bases. These data bases are (i) the global vegetation index (GVI) data at 15–20 km spatial resolution from the NOAA; (ii) the regional 'global' area coverage (GAC) data at 4 km spatial resolution from the NASA Goddard Space Flight Center; (iii) the regional 'local' area coverage (LAC) data at 1.1 km spatial resolution from the individual AVHRR receiving stations; (iv) the proposed global 'Pathfinder' data set at 9 km spatial resolution from the NOAA/NASA and (v) the proposed global data set at 1 km spatial resolution from the International Geosphere Biosphere Programme (IGBP) (Townshend, 1992).

These data bases are made available for the study of deforestation but their use has not been well co-ordinated. Interestingly, only the proposed IGBP database would provide data of adequate quality, spatial resolution and coverage for the study of regeneration stages at a global scale.

In addition to the AVHRR data bases there are three established and well co-ordinated programmes that are using AVHRR and other remotely sensed data to map, monitor and model tropical deforestation.

(i) THE GLOBAL RESOURCE INFORMATION DATA BASE (GRID) OF THE UNITED NATIONS The United Nations Environmental Programme (UNEP) is to provide a globally comprehensive database of current tropical forest cover, derived from the analysis of AVHRR imagery. The analysis of the imagery and the construction of the data base is being undertaken in collaboration with twelve research/educational organisations in Brazil, Europe and the USA.

The data base should be completed during 1993. However, limited access is already possible to the data on forest cover in both West Africa and Amazonia (Jaakkola, 1990; Johnson, 1991).

(ii) TROPICAL FOREST MONITORING PROGRAMME OF THE FOOD AND AGRICULTURE ORGANISATION (FAO) Researchers within the Food and Agriculture Organisation (FAO) of the United Nations have been mapping and monitoring tropical forest cover since the 1970s. They have used a 10 per cent areal sample of forests for 87 tropical countries every 10 years in order to produce their influential deforestation assessments (Table 4.3). In 1989 the more ambitious 'Forest Resource Assessment 1990' project was launched. This involves mapping tropical forest cover for 1980, 1990 and beyond using AVHRR imagery augmented with other remotely sensed imagery (e.g., Landsat MSS). The aim will be to produce decadal estimates of deforestation areas and rates with an accuracy of 10 per cent (95 per cent confidence level) (Singh, 1990; 1992).

(iii) TROPICAL ECOSYSTEM ENVIRONMENT OBSERVATIONS BY SATELLITE (TREES) PROGRAMME OF THE EUROPEAN COMMUNITY (EC) AND EUROPEAN SPACE AGENCY (ESA) The Tropical Ecosystem Environment Observations by Satellites (TREES) programme is a joint European Community and European Space Agency venture that was launched formally in 1990. It is now one of the five remote sensing research projects organised by the Joint Research Centre at Ispra (Achard, 1991; JRC, 1991). The aim of the TREES programme is to produce a three-class inventory of tropical forests (100 per cent cover, < 70 per cent cover and secondary forest) using NOAA AVHRR and ERS-1 SAR imagery and checked using Landsat TM and SPOT HRV imagery (ESA, 1990). This map will be updated and used to model deforestation as a route to understanding the links between deforestation, climate change, hydrology and biomass burning (Verstraete *et al.*, 1990a). Current research within the TREES programme is focused on forest mapping using AVHRR imagery (Malingreau *et al.*, 1992) and ERS-1 SAR data (de Groof *et al.*, 1992).

REMOTE SENSING OF TROPICAL FOREST REGENERATION

Forest clearance results in a complex mosaic of forest types with each area representing a point in a successional sequence from clear-cut to mature tropical forest. The successional sequences followed by any particular area are determined by the local environment and also the area's initial condition or 'starting point', as some clearances involve complete denudation while others leave patches of primary forest (Whitmore, 1990; Aber and Melillo, 1991). However, successional sequences are characterised typically by increases in

Table 4.3 Preliminary estimates by FAO of forest area and rate of deforestation for eighty-seven countries in the tropics.

Sub-Region	Number of countries studied	Total land area	Forest area 1980	Forest area 1990	Area deforested annually 1981–90	Rate of change in forest area, 1981–90
		(	——— 10^3 ha ———		)	% yr^{-1}
LATIN AMERICA	32	1,675,700	923,000	839,000	8,300	−0.9
Central America & Mexico	7	245,300	77,000	63,500	1,400	−1.8
Caribbean Sub-Region	18	69,500	48,800	47,100	200	−0.4
Tropical South America	7	1,360,800	797,100	729,300	6,800	−0.8
ASIA	15	896,600	310,800	274,900	3,600	−1.2
South Asia	6	445,600	70,600	66,200	400	−0.6
Continental South East Asia	5	192,900	83,200	69,700	1,300	−1.6
Insular South East Asia	4	258,100	157,000	138,900	1,800	−1.2
AFRICA	40	2,243,400	650,300	600,100	5,000	−0.8
West Sahelian Africa	8	528,000	41,900	38,000	400	−0.9
East Sahelian Africa	6	489,600	92,300	85,300	700	−0.8
West Africa	8	203,200	55,200	43,400	1,200	−2.1
Central Africa	7	406,400	230,100	215,400	1,500	−0.6
Tropical Southern Africa	10	557,900	217,700	206,300	1,100	−0.5
Insular Africa	1	58,200	13,200	11,700	200	−1.2
TOTAL	87	4,815,700	1,884,100	1,714,800	16,900	−0.9

Source: Modified from Singh (1991).

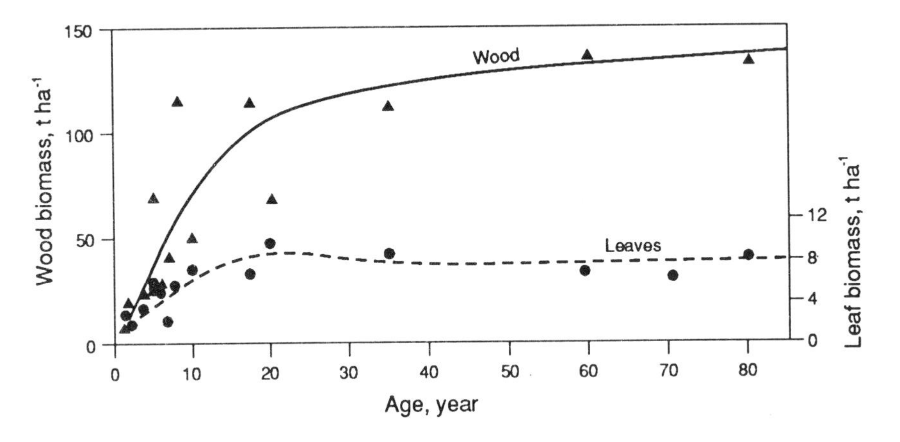

Figure 4.1 Biomass of leaves and wood (twigs, branches and stems) of different ages of regenerating tropical forest, with lines illustrating general trends (modified from Brown and Lugo, 1990).

leaf biomass, wood biomass and canopy roughness that can be sensed remotely. These three variables tend to increase rapidly after clearance (Fölster *et al.*, 1976; Uhl and Jordon, 1984; Uhl, 1987) with, for example, half of the potential maximum leaf biomass being reached after about 14 years of growth (Saldarriaga *et al.*, 1988). The amount of leaf biomass and wood biomass will approach a maximum after around 20 and 40 years respectively (Brown and Lugo, 1990) (Figure 4.1). Before these maxima are reached there is generally a floristic change from a few fast-growing pioneer species of similar age to many slower growing late pioneer and mature forest species of various ages (Swaine and Hall, 1983; Saldarriaga *et al.*, 1988). This often results in a gradual increase in canopy roughness, only reaching a maximum when a mature tertiary canopy is formed (Whitmore, 1990). Leaf biomass, wood biomass and canopy roughness provide indications of how far regeneration has progressed. The carbon flux between the forest and the atmosphere is also changing as the forest regenerates, with the maximum amount of carbon being fixed some 5 and 30 years after clearance (Uhl, 1987; Sader *et al.*, 1989). Therefore, regeneration during this 25-year period will be an important one to discriminate remotely. Throughout the period of increasing and then decreasing carbon fixation, measurements of the three key variables related to forest regeneration, leaf biomass, wood biomass and canopy roughness are required. These may be made remotely by one or more of three approaches. First, the use of reflected red and near infrared radiation to estimate leaf biomass; second, microwave sensors to estimate leaf biomass, wood biomass and canopy roughness and third, multiple view angle optical sensors to estimate canopy roughness.

ESTIMATING LEAF BIOMASS WITH RED AND NEAR INFRARED RADIATION

Remotely sensed radiation in red (R $\simeq$ 630–690 nm) and near infrared

(NIR $\simeq$ 760–900 nm) wavebands has potential for use in estimating leaf biomass (Peterson and Running, 1989). In R wavelengths absorption by leaf pigments is large and, therefore, there is little reflection, while in NIR wavelengths within-leaf scattering is large and, therefore, reflection from the canopy is also large (Jensen, 1983). Shadow decreases canopy reflectance in both R and NIR wavelengths. As a consequence of shadow, leaf biomass would be expected to exhibit a negative relationship with R radiation, a no or positive relationship with NIR radiation and a positive relationship to the difference between R and NIR radiation, at least up to the reflectance asymptote of the canopy. A number of environmental factors (solar angle, understorey vegetation, atmospheric conditions, phenology etc.) cause variability in these relationships and are reviewed in Curran (1983). For temperate forests strong correlations have been observed between measures of leaf biomass, R radiation and the normalised difference vegetation index (NDVI = (NIR − R)/(NIR + R)) (Peterson *et al.*, 1987; Curran *et al.*, 1992; Danson and Curran, 1993). Far fewer studies have been undertaken on the remote sensing of leaf biomass in tropical forests. However, researchers have reported that the reflectance of tropical forest canopies is generally small as a result of shadow and that while R radiation varies much more than NIR radiation across a tropical forest canopy it is the more affected by atmospheric haze (Singh, 1987). The correlations between measures of leaf biomass and remotely sensed radiation have, in general, been based on small sample sizes and ground data of questionable validity. Nevertheless, occasionally small correlations have been observed between measures related to leaf biomass and R and NIR radiation (Box *et al.*, 1989; Cook *et al.*, 1989) that may (Vanclay and Preston, 1990) or may not (Sader *et al.*, 1989) be increased by taking measures related to the difference of the two (e.g., NDVI).

The estimation of leaf biomass with R and NIR radiation for a range of vegetation types is well developed. However, its application to tropical forests is in its infancy with very few investigators managing to bring together appropriate ground data and both cloud-free and haze-free imagery for the same time and place.

ESTIMATING LEAF BIOMASS, WOOD BIOMASS AND CANOPY ROUGHNESS WITH SAR

Microwave backscatter measured by synthetic aperture radars (SAR) can be used to estimate the leaf biomass, wood biomass and canopy roughness of forests (Ford and Wickland, 1985; Westman and Paris, 1987; Wu, 1987; 1990). The penetration of microwaves into the forest is dependent primarily upon wavelength. Short wavelength (e.g., X band $\simeq$ 3 cm) microwaves interact with the surface of the canopy and provide information on canopy roughness; medium wavelength (e.g., C band $\simeq$ 6 cm) microwaves interact with the volume of the canopy and provide information on leaf biomass, while long wavelength (e.g., L band $\simeq$ 22 cm) microwaves penetrate the canopy, interact with the tree stems and provide information on wood biomass (Carver, 1988; Kasischke *et al.*, 1991). The depth of microwave penetration into the canopy is also dependent on the angle of incidence of the sensor and canopy openness (Ulaby *et al.*, 1982; Churchill and Sieber, 1991).

Many SARs can both transmit and receive microwaves at two polarisations and this provides yet further information on surface roughness and geometric regularities in the forest stand (Durden *et al.*, 1989; Aschbacher, 1991). Even more information may be derived from polarimetric radars which may soon be providing remotely sensed data of tropical regions (Carver, 1988).

Forest age is correlated positively to total biomass which in turn is correlated positively to backscatter; as a result there is a strong non-causal positive correlation between forest age and backscatter (Stone *et al.*, 1989; Durden *et al.*, 1991; Baker *et al.*, 1993; Briggs, 1993). Unfortunately, in C band the negative relationship between leaf biomass and backscatter can mask the positive relationship between wood biomass and backscatter, and vice versa (Kasischke and Christensen, 1990). Nevertheless, SAR imagery, especially when recorded in a number of wavebands, has the potential to differentiate between those differences in leaf biomass, wood biomass and canopy roughness that characterise a regenerating forest canopy.

ESTIMATING CANOPY ROUGHNESS WITH MULTI-ANGLE OPTICAL SENSORS

Tropical forests have a very rough canopy over the distance of a few metres but a very smooth canopy over the distance of a few kilometres. This relationship between canopy roughness and distance is non-linear with roughness decreasing to a minima at the distance of a few tree crowns. To estimate this roughness remotely we need to measure either (i) the variability between pixels, or (ii) the effect of illumination/observation geometry and polarisation within pixels.

(i) RECORDING THE VARIABILITY BETWEEN PIXELS The variability between pixels can be measured using a first or second order statistic on a block of at least 3×3 pixels (e.g., standard deviation) (Curran, 1985; Cohen and Spies, 1992). Non-isotropic variability can be determined by calculating the semi-variance for transects of pixels orientated at various directions across the imagery (Curran, 1988).

(ii) RECORDING THE EFFECT OF ILLUMINATION/OBSERVATION GEOMETRY AND POLARISATION WITHIN PIXELS In optical wavelengths the remotely sensed radiation (R) is a function of the five variables of location (x), time (t), wavelength (λ), geometry of illumination and observation (θ) and polarisation (p) (Gerstl, 1990; Verstraete and Pinty, 1992; Barale *et al.*, 1993) where:

$$R = f(x, t, \lambda, \theta, p). \qquad [1]$$

The remote sensing of tropical forests has concentrated upon the first three variables to locate, monitor and to identify forests with little attention being paid to either the geometry of illumination and observation or polarisation. Both of these variables provide information on the roughness of the canopy (Curran, 1982; Pinty *et al.*, 1990; Verstraete *et al.*, 1990b; Vanderbilt *et al.*, 1991;) and both can be measured. For instance, the influence of the geometry of illumination and observation on radiation can be determined by recording the radiation at a point on a stable forest canopy, at various positions within

AVHRR swaths or at various look angles of the SPOT HRV sensor (Barnsley and Muller, 1991; Roujean *et al.*, 1992). The influence of polarisation on radiation can be determined by measuring the difference in tone between polarised and unpolarised photographs taken, for example, from a camera carried on the Space Shuttle.

While the use of broadband optical sensors to estimate canopy roughness and thereby regenerative stage is the most speculative of the three techniques presented in this section, it does offer the potential of recording change during the latter stages of regeneration.

AIM OF THE RESEARCH REPORTED IN THIS CHAPTER

Three approaches have been identified for the remote sensing of regeneration in tropical forests. The most feasible, for the moment at least, is the use of R and NIR radiation. However, three areas of uncertainty have been identified in the application of this approach: (i) the availability of satellite sensor data for tropical forests, (ii) the relationships between remotely sensed data in R and NIR wavelengths and those forest measures that relate to leaf biomass and (iii) the feasibility of estimating the areal coverage of forest within a pixel (Curran and Foody, 1992).

The aim of the research reported in this chapter was to investigate these three areas of uncertainty in the context of one country in the forested tropics.

.... STUDY AREA

The forests of West Africa are undergoing the most rapid rate of change in the world (Gornitz, 1985; Whitmore, 1990) (Table 4.3), thus providing us with large areas of regenerating forest. The study area chosen for this research was the tropical forest region of south-west Ghana centred on 2°W and 7°N (Taylor, 1952) (Figures 4.2 and 4.3). This region is covered by approximately 1.7×10^6 ha of forest of which 96.8 per cent belongs to the four forest classes of wet evergreen, moist evergreen, moist semi-deciduous and dry semi-deciduous forest (Swaine and Hall, 1976; Gornitz, 1985; Grainger, 1986). These four classes are arranged on a strong moisture gradient with over 1,750 mm yr^{-1} of rain falling on the wet evergreen forest in the south and around 1,000 mm yr^{-1} of rain falling on the dry semi-deciduous forest in the north (Varley and White, 1958; FAO, 1981; Sayer *et al.,* 1992). There are large environmental and associated biotic differences along this south to north environmental gradient, notably a decrease in tree size, tree density and species numbers (Hall and Swaine, 1981).

This region was chosen for four reasons. First, it had a slightly dryer season during November to March (Gibbs and Leston, 1970) and the anticipated decrease in cloud cover during this dryer season is likely to increase the availability of remotely sensed imagery in optical wavelengths. Second, the region contains some 200 forest reserves (Boateng, 1966) and around a third of these approximate a 'natural' structure as a result of controlled logging (World Bank, 1988). Third, relative to other countries in West Africa the rate of deforestation has slowed in recent years giving some stability to the

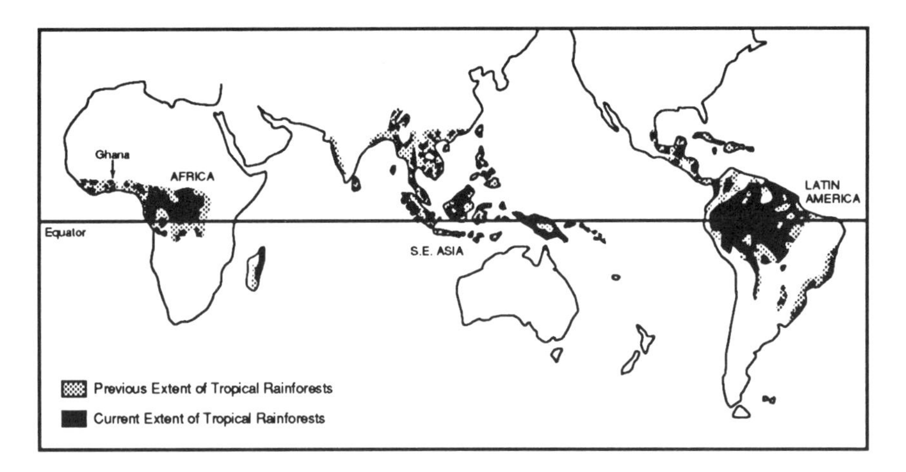

Figure 4.2 Global coverage of tropical rainforests both prior to the current period of intensive logging and in the late 1980s (modified from DOE, 1990).

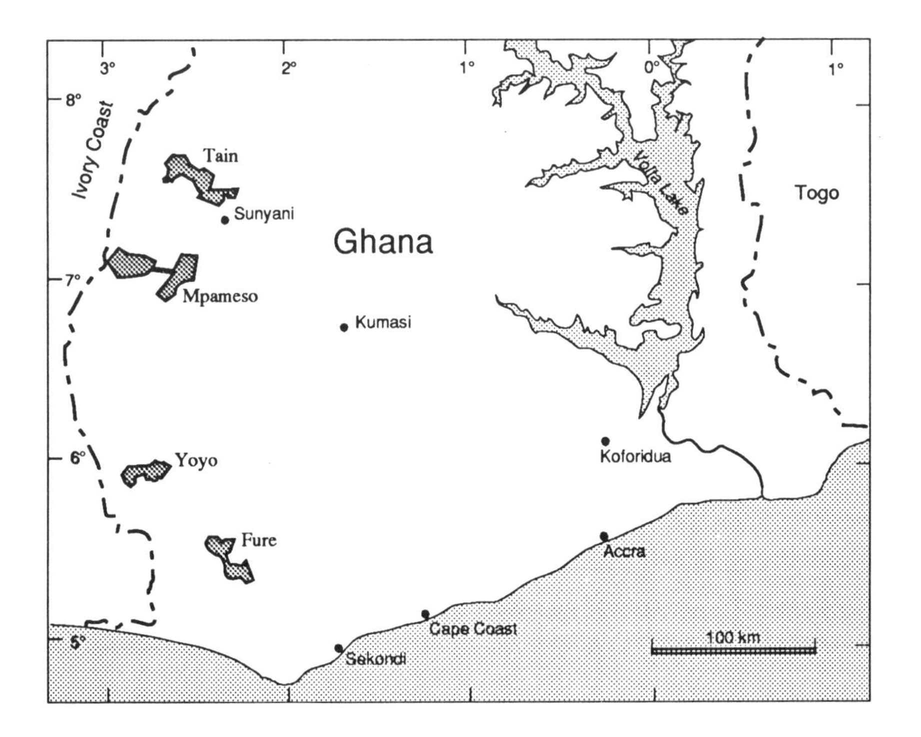

Figure 4.3 Southern Ghana showing the location of the four forest reserves from the dryer north to the wetter south.

Table 4.4 The four study sites ordered from north to south.

Forest reserve	Forest type	Latitude (N)	Longitude (W)	Area (km^2)
Tain	Dry semi-deciduous	7°35′	2°30′	509
Mpameso	Moist semi-deciduous	7°05′	2°53′	323
Yoyo	Moist evergreen	5°55′	2°48′	236
Fure	Wet evergreen	5°28′	2°20′	328

landscape (FAO, 1981; Gornitz, 1985). Fourth, several of the larger forest reserves on which logging is controlled are managed as part of a £4 million programme by the Overseas Development Agency (ODA) (DOE, 1990) and have within them permanent sample plots (PSPs) for which a range of biophysical and floristic details are recorded.

One forest reserve or group of forest reserves were selected from each forest class on the anticipated availability of remotely sensed and ground data (Figure 4.3). These forest reserves were, from north to south, the dry semi-deciduous reserve of Tain Tributaries II (hereinafter referred to as Tain), the moist semi-deciduous reserves of Mpameso, Anhwaiso North and Bia Tano (hereinafter referred to as Mpameso), the moist evergreen reserve of Yoyo River (hereinafter referred to as Yoyo) and the wet evergreen reserves of Fure Headwaters and Fure River (hereinafter referred to as Fure) (Table 4.4). PSP records were not available for Mpameso and so this forest reserve was used only for the study of forest/non-forest classification reported later. The other three reserves of Tain, Yoyo and Fure were very different to each other (Hall and Swaine, 1981), and it was hoped that this would ensure a large range in leaf biomass and thereby remotely sensed radiance. The differences between the forest reserves were the result of rainfall but also (i) topography (Yoyo is hillier than the other sites), (ii) the canopy (Fure has a traditional three layer tropical forest canopy whereas Tain has only two layers and its larger trees are deciduous during the dry season) and (iii) logging that has affected all forest reserves (Tain and Fure have been logged most heavily) (Table 4.5).

.... REMOTELY SENSED DATA

Searches were undertaken for remotely sensed data coverage of the four forest reserves (Table 4.4). The optical imagery investigated were from NOAA AVHRR, Landsat MSS, Landsat TM, multispectral SPOT HRV, Panchromatic SPOT HRV, MOMS 1 MESSR and the Space Shuttle Metric Camera. The microwave imagery investigated were from SIR-A/B, ERS-1, Fuyo-1 and Almaz-1 SARs.

REMOTELY SENSED DATA IN OPTICAL WAVELENGTHS

There are few cloud-free remotely sensed images of south-western Ghana. Analysis of the NOAA AVHRR archive imagery for 1989/1990 revealed that

Table 4.5 An illustration of some of the major differences between the three forest reserves for which ground data are available.

					Mean basal area		Density of trees	
Forest reserve	Topography	Canopy form	Deciduousness (% of total trees)	Logging intensity	All trees	Large trees	All trees	Large trees
Tain	Fairly flat	Continuous	> 20 ($\simeq 100\%$ in larger trees)	Medium	Small	Small	Low	Low
Yoyo	Hilly	Discontinuous	< 20	Light	Medium	Medium	Medium	Medium
Fure	Fairly flat	Discontinuous and undulating	< 20	Medium	Medium	Small	High	High

Sources: The information is derived from the Survey of Ghana (1969), Hall and Swaine (1981), FAO (1981), Lieberman (1982), ground data, notes supplied by the ODA and field observation.

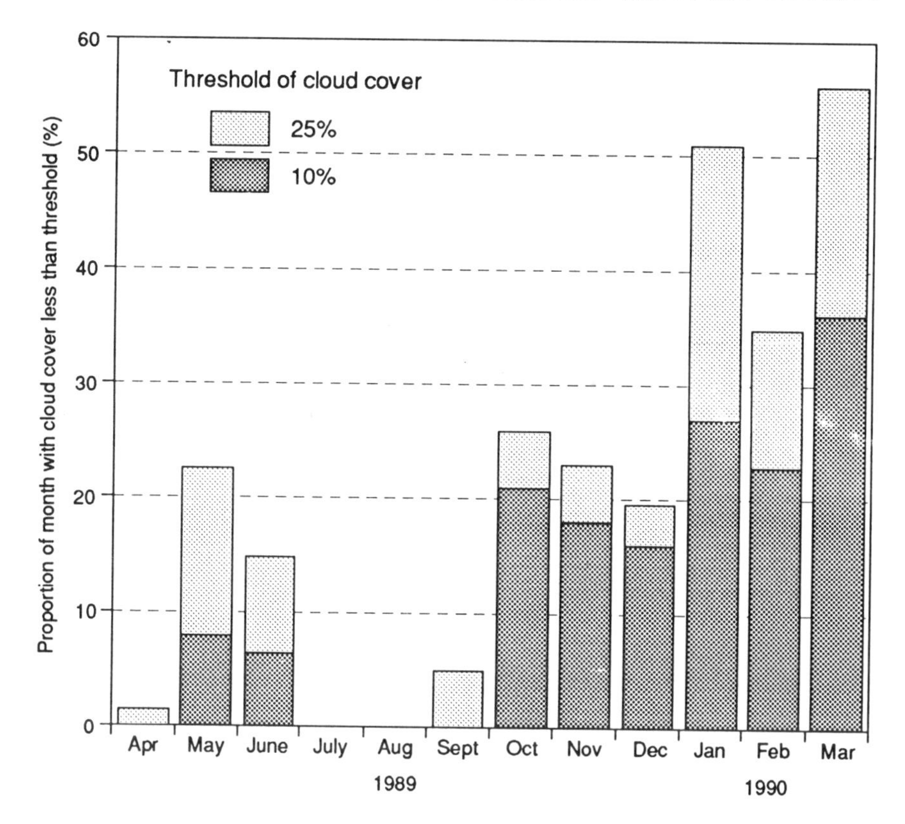

Figure 4.4 Cloud cover over Ghana derived from NOAA AVHRR archive imagery for 1989/90. There was no cloud-free imagery in July and August but around a quarter of the images recorded in the dryer months of January, February and March had less than ten per cent cloud cover. Note: (i) the cloud cover is greatest in the south and over forest reserves and (ii) there are only a few haze-free days a year.

cloud cover at the time of the AVHRR overpass was 100 per cent during the two rainy season months of July and August, but decreased to less than 25 per cent cloud cover for nearly half of the time during the dryer months of January, February and March (Figure 4.4). Unfortunately, cloud cover is much greater in the forest zone than in the non-forest zone (Gibbs and Leston, 1970; Hall and Swaine, 1981) and, as a result, the probability of obtaining AVHRR imagery of the forest reserves with less than 25 per cent cloud cover was small. From the 1989/1990 archive imagery and with the aid of binomal expansion (Gregory, 1978) this probability was estimated to be 0.01–0.02 (95 per cent confidence level) for the nine wetter months and 0.07–0.11 (95 per cent confidence level) for the three dryer months. Therefore, 9–15 AVHRR images a year could be expected (95 per cent confidence level) in which at least one of the four forest reserves was visible and 6–10 of these images would be recorded during the three dryer months. AVHRR imagery archived during 1991 supported this estimate. Closer inspection of the visible

(a)

Figure 4.5 Subscenes of Landsat MSS images recorded in band 4 (near infrared). Note: images in other bands had less contrast as a result of haze and are not shown, and both subscenes are displayed at full spatial resolution for an area of approximately 100 × 10^3 km^2. Image (a) was recorded on 25 November 1973 and covers the forest reserves of Mpameso and Tain, and image (b) was recorded on 21 December 1986 and covers the forest reserve of Yoyo.

waveband of cloud-free AVHRR images revealed that spatial detail was suppressed severely by haze within forest reserves (Singh, 1987). It was only on those days when the Harmattan wind (FAO, 1981; Goudie, 1985) was blowing that both cloud-free and haze-free imagery were recorded.

It had been hoped to record the reflectance properties of the four forest reserves at several points on the AVHRR swath. Unfortunately, no suitable sets of imagery were available from either 1989 or 1990. Therefore, just one cloud-free and haze-free AVHRR image recorded in February 1990 was selected for this study (Table 4.6 and Plate 2).

Imagery from the Landsat and SPOT sensors are recorded much less frequently than imagery from the NOAA AVHRR. Consequently, few cloud-free images of Ghana have been recorded by these sensors. A search of the archives for Landsat sensor and SPOT sensor imagery revealed that between launch and May 1991 2.6 per cent of the Landsat sensor images had less than 30 per cent cloud cover and 3.5 per cent of the SPOT sensor images had less

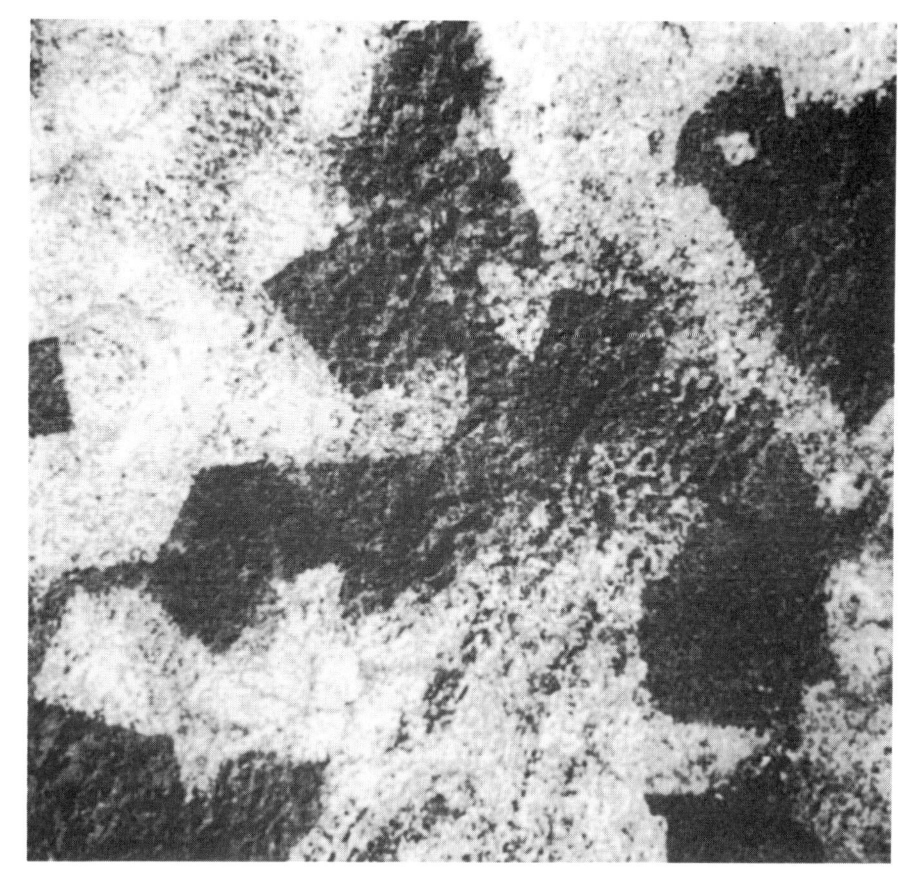

(b)

than 25 per cent cloud cover. This cloud cover tended to occur over forest reserves and those images with very little cloud cover were often hazy. Three Landsat sensor images (Figure 4.5) but no SPOT sensor images were found that provided cloud-free and haze-free coverage of at least one of the four forest reserves (Table 4.6). Interestingly, the image with the least cloud contamination was the Landsat TM image for December 1986 and this had also been selected by Päivinen and Witt (1988) for a forest mapping project. The archives of MOMS-1 MESSR and Space Shuttle Metric Camera imagery are much less extensive than those for NOAA AVHRR, Landsat sensor or SPOT sensor imagery. However, one almost cloud-free image from each of these sensors was available (MOMS-1 MESSR 109/110, Space Shuttle Metric Camera V1-0672) and both covered the northern two forest reserves. Unfortunately, these two images were too hazy to be useful.

The four remotely sensed images used in this study were atmospherically corrected and then the Landsat MSS and TM imagery were radiometrically corrected (Chavez, 1988; 1989). As a result the Landsat MSS and TM data were in absolute values of radiance (Robinove, 1982) whereas the AVHRR data were in relative values of DN (Holben *et al.*, 1990; Goward *et al.*, 1991).

Table 4.6 Remotely sensed imagery discussed in this chapter. Note, the limited amount of remotely sensed data and the time lag between the collection of ground data and the recording of both Landsat MSS and Landsat TM imagery.

Remotely sensed imagery	Date of image acquisition	Years before the collection of ground data (approximate)	Coverage of forest reserves			
			Fure	Yoyo	Mpameso	Tain
NOAA AVHRR	27/02/90	0	Yes	Yes	Yes	Yes
Landsat MSS	21/12/86	4		Yes		
Landsat MSS	25/11/73	17			Yes	Yes
Landsat TM	21/12/86	4	Yes			

Geometric correction using a nearest neighbour procedure (Mather, 1987) was attempted on all four images but was abandoned in the absence of sufficient ground control points to achieve an accurate transformation.

REMOTELY SENSED DATA IN MICROWAVE WAVELENGTHS

No imagery recorded using microwave wavelengths was available for Ghana. The SIR-A/B image swaths missed the four forest reserves. Although ERS-1 SAR data have been recorded for most of the world since July 1991, there is, unfortunately, no receiving station that covers West Africa and no ERS-1 SAR data will be available for Ghana until one is built (ESA, 1991; Rapley, 1992). The Fuyo-1 SAR recorded its first images in March 1992 but the imagery of West Africa has yet to be made routinely available. The Almaz-1 SAR did not record data for Ghana and ceased to transmit data from mid 1992.

.... GROUND DATA

Two sources of 'ground' data were used; the first was a simple forest/non-forest cover map of the Mpameso reserve and the second were biophysical data for Tain, Fure and Yoyo forest reserves.

FOREST/NON-FOREST MAP

A Landsat MSS band 7 (NIR) image of the Mpameso forest reserve area (see Table 4.4) was classified into forest/non-forest by means of a density slice. This two-class land cover map was used to determine the accuracy with which the coverage of forest within a coarse spatial resolution image pixel could be estimated (Foody and Cox, 1991).

BIOPHYSICAL DATA

Detailed biophysical and floristic records for 1990 were available for 29, 1 ha permanent sample plots (PSP) (Alder and Synnott, 1992) in the three forest reserves of Tain (n = 9), Fure (n = 10) and Yoyo (n = 10) (Figure 4.6). From these records six variables were derived that between them characterised the differences in total biomass and tree density that would, it is hoped, have expression in leaf biomass. These variables were mean basal area (m^2 ha^{-1}) for all trees with diameters larger than 10 cm, 30 cm and 50 cm and density of trees (trees ha^{-1}) for all trees, again with diameters larger than 10 cm, 30 cm and 50 cm (Figure 4.7 and Table 4.7). The range of each of these variables was much larger than expected (Hall and Swaine, 1981) and may have resulted from the logging of larger trees, particularly in the dryer northern forest reserve of Tain. The magnitude of this range is illustrated by the difference between both the mean basal area and density (for all trees with diameters larger than 10 cm) on the wet evergreen forest reserve (Fure) and the dry semi-deciduous forest reserve (Tain). By comparison with data recorded in the 1970s (Hall and Swaine, 1981), the range of these two variables had increased three-fold and seven-fold respectively.

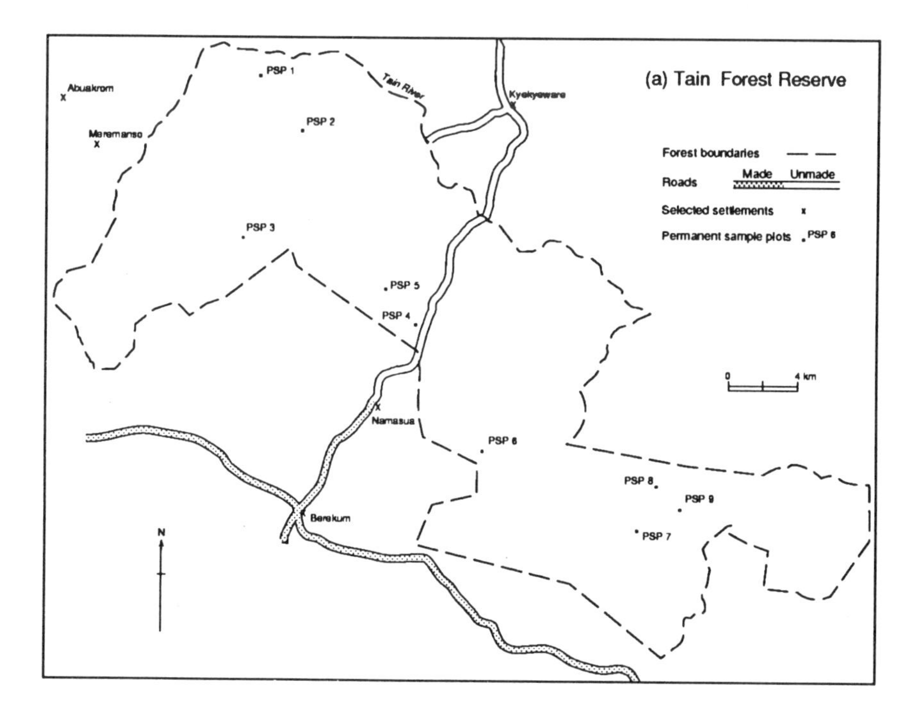

Figure 4.6 The three forest reserves for which ground data were available. From the north to the south of the study region there are (a) Tain (full name, Tain Tributaries II) with nine permanent sample plots (PSPs), (b) (*overleaf*) Yoyo (full name, Yoyo River) with ten PSPs, and (c) (*overleaf*) Fure (full name, Fure Headwaters and Fure River) with ten PSPs. A PSP is a permanent demarked 1 ha area of forest that is remeasured periodically (Alder and Synnott, 1992).

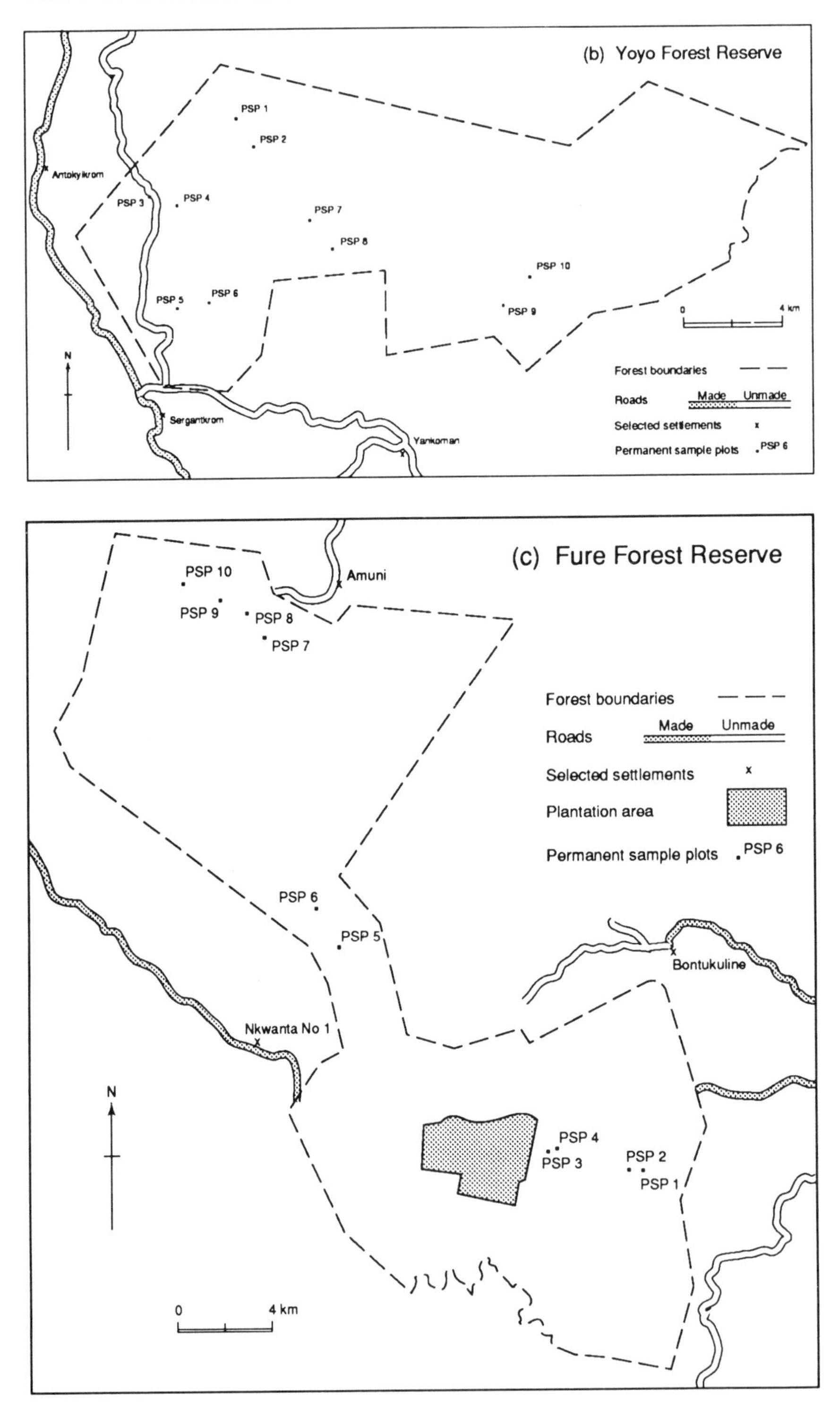
(b) Yoyo Forest Reserve
PSP 1
PSP 2
PSP 3
PSP 4
PSP 5
PSP 6
PSP 7
PSP 8
PSP 9
PSP 10
Antokyikrom
Sergantkrom
Yankoman
N
0
4 km
Forest boundaries
Roads
Made
Unmade
Selected settlements
Permanent sample plots
PSP 6
(c) Fure Forest Reserve
Amuni
PSP 10
PSP 9
PSP 8
PSP 7
PSP 6
PSP 5
PSP 4
PSP 3
PSP 2
PSP 1
Nkwanta No 1
Bontukuline
N
0
4 km
Forest boundaries
Roads
Made
Unmade
Selected settlements
Plantation area
Permanent sample plots
PSP 6

Figure 4.7 A tree in a permanent sample plot at the Yoyo forest reserve. Note the line for girth measurement from which tree diameter was estimated and the sample number tag.

Table 4.7 Summary statistics for the three forest reserves for which ground data were available. Note, these reserves represent points on a moisture gradient from the dryer north (Tain) to the wetter south (Fure) and the two measures with the greatest coefficient of variation ($\sigma/\bar{x}$) are the density of trees with diameters larger than 10 cm and the number of trees with diameters larger than 50 cm.

Forest reserve	Statistic	Mean basal area (m^2ha^{-1}) for three diameter ranges of tree			Density (trees ha^{-1}) for three diameter ranges of tree		
		>10 cm	>30 cm	>50 cm	>10 cm	>30 cm	>50 cm
Tain	$\bar{x}$	18.1	7.2	4.3	309	17	6
	σ	4.3	3.8	2.7	45	10	4
Yoyo	$\bar{x}$	27.1	14.0	9.2	402	33	14
	σ	7.8	7.0	4.6	43	18	7
Fure	$\bar{x}$	27.3	9.6	5.9	517	22	7
	σ	5.2	3.5	2.8	59	8	3
Tain, Yoyo and Fure	$\bar{x}$	24.4	10.4	6.5	443	24	9
	σ	7.2	5.6	3.9	98	14	6

Five of the six variables were intercorrelated ($r > 0.7$) at the 95 per cent confidence level on all three forest reserves. The variable that was uncorrelated with the other five, at the 95 per cent confidence level, was the density of trees with diameters greater than 10 cm. This variable had the largest range and variability, is a measure of the number of younger as well as older trees and is a potential indicator of regenerative stage (Hartshorn, 1980).

None of the six variables appeared to have a normal distribution, although the sample size at each reserve was too small for a reliable assessment. However, the variables for trees with diameters larger than 30 cm had a binomial distribution (young trees/old trees) that was not normalised readily, and the variables for trees with diameters larger than 30 cm and 50 cm had such a minor negative skew that normalisation was considered unnecessary. The range and distribution of the ground data for the forest reserves collectively and singularly are illustrated in Figure 4.8.

.... CORRELATION BETWEEN REMOTELY SENSED DATA AND GROUND DATA

It was hypothesised that the mean basal area and tree density were related positively to leaf biomass and would therefore exhibit relationships that were negative with R radiation, non existent to slightly positive with NIR radiation and positive with NDVI.

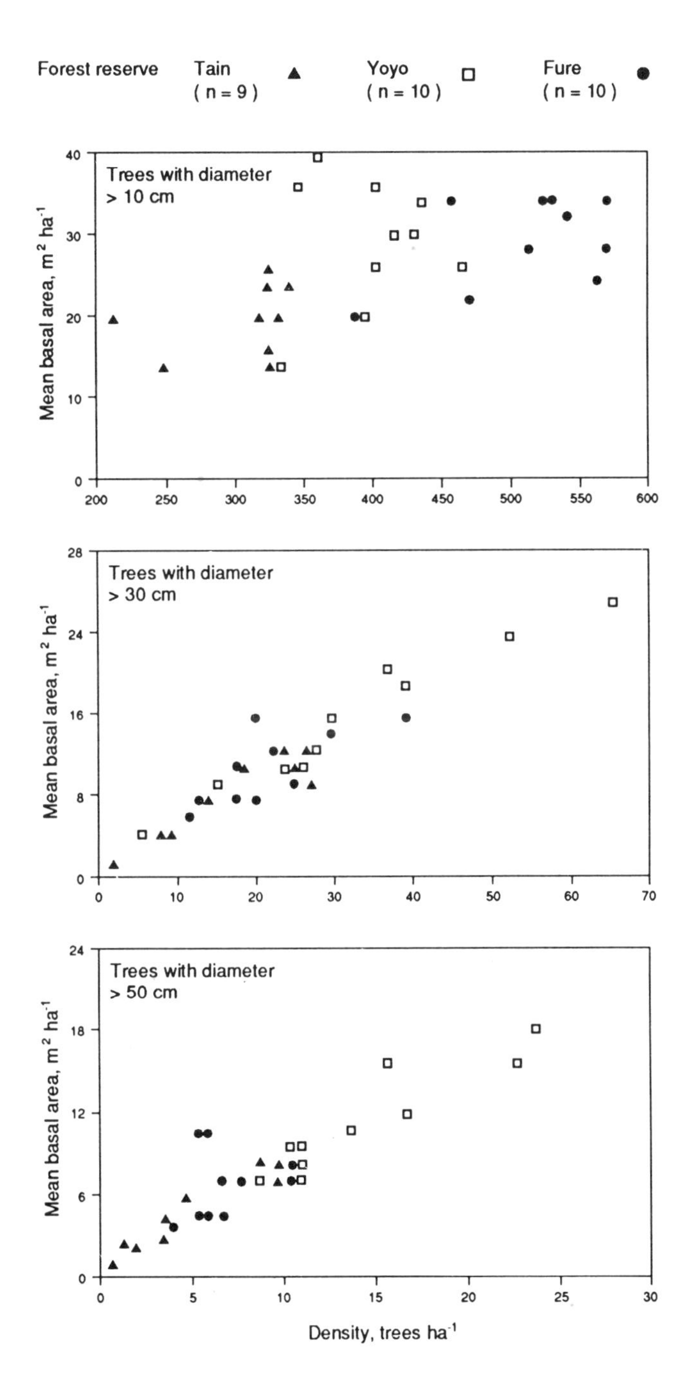

Figure 4.8 The relationship between mean basal area and density for three tree diameter ranges (› 10 cm, › 30 cm and › 50 cm) at three forest reserves (Tain, Yoyo, Fure).

Table 4.8 Correlations, significant at the 90 per cent confidence level or greater, between radiation in red (R) wavelengths, near infrared (NIR) wavelengths, NDVI and the six ground data variables. The remotely sensed data sets are NOAA AVHRR for all of the 29 PSPs (A), AVHRR for the 10 PSPs in Fure (AF), AVHRR for the 9 PSPs in Tain (AT) and Landsat TM (smoothed) for the 10 PSPs in Fure (TMF). The numbers in bold indicate a correlation significant at the 95 per cent confidence level.

		Remotely sensed radiation (radiance, DN)		
Ground data	Tree diameter (cm)	R	NIR	NDVI
	>10	**−0.80(A)** −0.59(AF)	**+0.79(TMF)**	**+0.62(A)** **+0.70(AF)**
Density (trees ha^{-1})	>30	+0.66(AT)		
	>50	**−0.72(TMF)**	+0.60(AF) −0.58(AT)	−0.64(AT)
	>10	−0.32(A)		
Mean basal area (m^2 ha^{-1})	>30			
	>50			

NOAA AVHRR DATA

The hypothesised relationships above were, in general, observed between AVHRR data and ground data at all of the forest reserves and at Fure alone, but not at Tain and Yoyo alone (Table 4.8 and Figure 4.9). Tain is a semi-deciduous reserve and at the time of imaging the larger trees would have been leafless (Hall and Swaine, 1981), thus modifying the relationship. It was assumed that at the PSPs with a large number of large trees the reflectance of R radiation could have been increased by the light-toned bark and reduced leaf area, while reflectance of NIR radiation may have been reduced by shadow. The result of this was a strong negative relationship between the number of the larger trees and NDVI. Yoyo by contrast is hilly and it was the terrain that dominated the remotely sensed response with a pattern of topographically reduced DN in valleys and on west-facing slopes, an effect that has been well documented in other hilly areas (Smith *et al.*, 1980; Hugli and Frei, 1983; Sader *et al.*, 1989; Cohen and Spies, 1992). As a result of this spatially variable illumination, no significant relationship between R DN, NIR DN or NDVI and any of the ground data was observed at Yoyo.

LANDSAT MSS DATA

There was no cloud-free Landsat MSS data of Fure and there were no relationships between Landsat MSS radiance and ground data at either Tain or Yoyo. The lack of a relationship at Tain was attributed to atmospheric haze, the leafless upper canopy and the seventeen-year time lag between the collection of the remotely sensed and ground data (Table 4.6). The lack of a relationship at Yoyo was again attributed to the effects of atmospheric haze and the terrain (Figure 4.5). To investigate the effect of terrain further, the R and NIR radiance was corrected topographically (Teillet, 1986) using a digital terrain model generated in the 'Idrisi' geographical information system (Curran and Foody, 1992). The topographical data were on a 17 m × 17 m raster grid and were digitised at a scale of 1:25,000 from a 1:50,000 scale Ghanaian Forestry Commission map. This DTM was used to perform a 'cosine correction' of the Landsat MSS radiance for Yoyo (Teillet *et al.*, 1982). To a first approximation it was assumed that slopes facing the Sun had a larger radiance than those facing away from the Sun,

$$Lh = Lt \cos Z / \cos i \qquad [2]$$

where, Lh is the radiance corrected to that for a horizontal surface, Lt the radiance from an inclined surface, Z the solar zenith angle and i the incident angle with respect to surface normal. Lt is the recorded radiance, Z is recorded with the image and

$$\cos i = (\cos Z \cos s) + (\sin Z \sin s \cos (A-a)) \qquad [3]$$

where, A is the solar azimuth, s the terrain slope angle and a the terrain slope aspect. A is recorded with the image while s and a are the mean of the 18 DTM pixels that cover 1 Landsat MSS pixel (Downward, 1991). The topographically corrected Landsat MSS radiance values were appreciably different to the original radiance values. For example, the NIR radiance at PSP 3 increased by a factor of 2.44 while at PSP 7 it decreased by a factor of 0.62. However, this topographic correction did not result in a large change in the degree of correlation between the ground and remotely sensed data. While topography clearly alters the radiance recorded by the Landsat MSS (Figure 4.5) it is also related to the density and basal area of the trees, such that the sloping sites have more and larger trees. For example, trees with a diameter larger than 10 cm had a mean density of 419 trees ha^{-1} and a mean basal area of 30.9 m^2 ha^{-1} on the five PSPs located on steeper slopes, while on the five flatter PSPs the mean density was 386 trees ha^{-1} and the mean basal area was 23.4 m^2 ha^{-1}. In addition, the degree of interrelationship between topography, tree density and tree basal area varied spatially as a result of increased logging nearer to the one main road (Figure 4.6). For example, trees with a diameter larger than 10 cm had a mean density of 372 trees ha^{-1} and mean a basal area of 25.9 m^2 ha^{-1} at the five PSPs nearest to the road (mean distance to road of 2.5 km). However, the same sizes of tree had a mean density of 431 trees ha^{-1} and a mean basal area of 28.2 m^2 ha^{-1} at the five PSPs furthest from the road (mean distance to road of 9.4 km).

Given the effect of atmospheric haze in visible wavelengths and the presence of a complex relationship between remotely sensed radiance, ground data and topography it was considered unwise to pursue further the topographic correction of remotely sensed radiance with the available data.

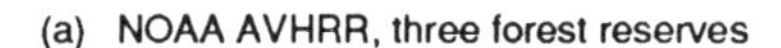
(a) NOAA AVHRR, three forest reserves

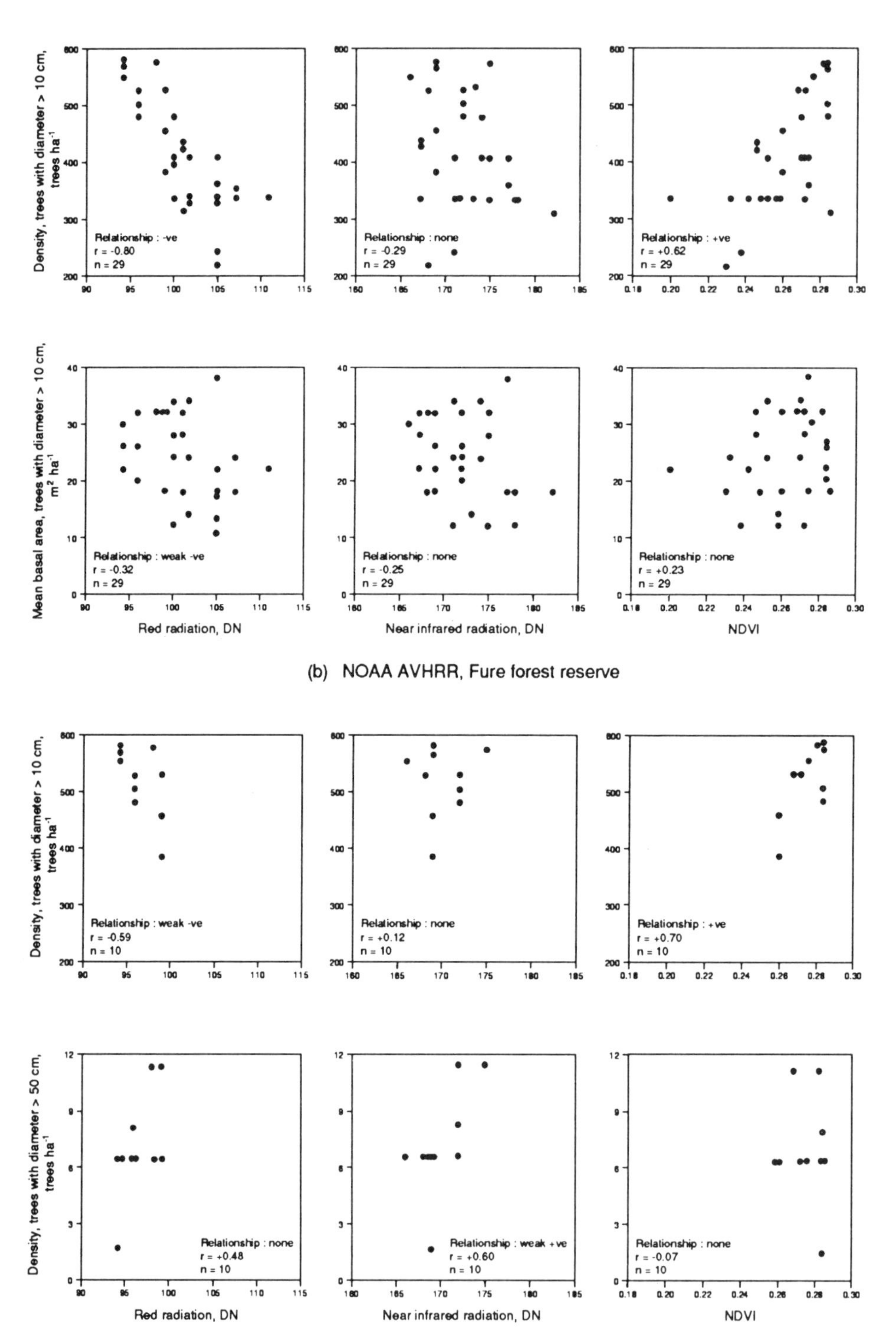
Density, trees with diameter > 10 cm, trees ha-1
Relationship : -ve
r = -0.80
n = 29
Relationship : none
r = -0.29
n = 29
Relationship : +ve
r = +0.62
n = 29
Mean basal area, trees with diameter > 10 cm, m2 ha-1
Relationship : weak -ve
r = -0.32
n = 29
Relationship : none
r = -0.25
n = 29
Relationship : none
r = +0.23
n = 29
Red radiation, DN
Near infrared radiation, DN
NDVI
(b) NOAA AVHRR, Fure forest reserve
Density, trees with diameter > 10 cm, trees ha-1
Relationship : weak -ve
r = -0.59
n = 10
Relationship : none
r = +0.12
n = 10
Relationship : +ve
r = +0.70
n = 10
Density, trees with diameter > 50 cm, trees ha-1
Relationship : none
r = +0.48
n = 10
Relationship : weak +ve
r = +0.60
n = 10
Relationship : none
r = -0.07
n = 10
Red radiation, DN
Near infrared radiation, DN
NDVI

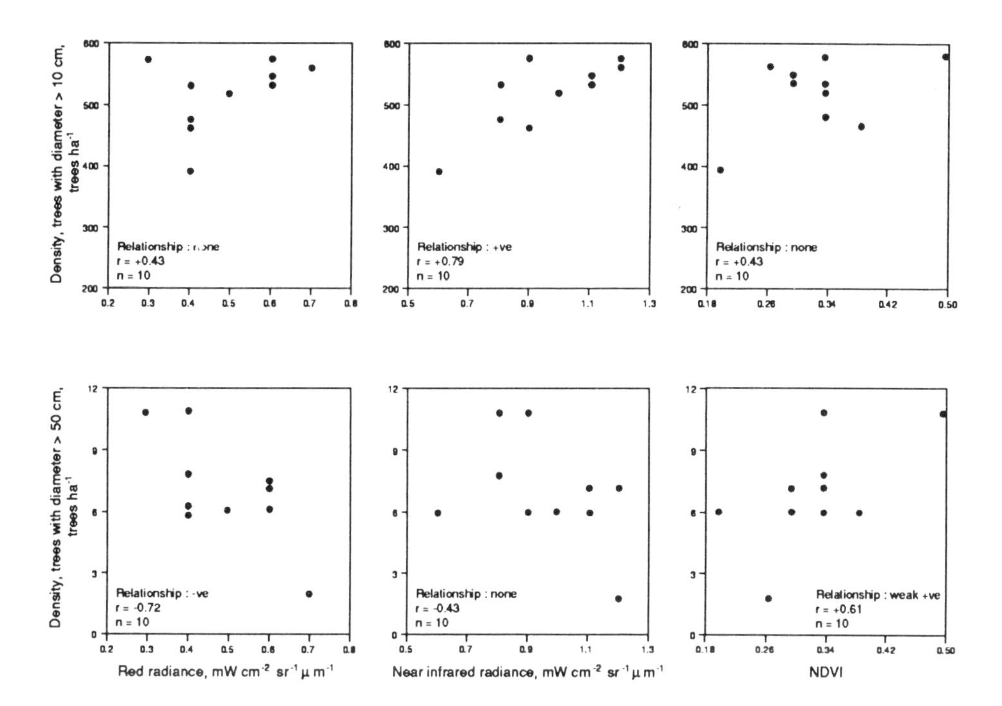

Figure 4.9 Illustrative relationships between remotely sensed data and ground data: (a) AVHRR DN and both mean basal area and density for all three forest reserves, (b) AVHRR DN and two measures of density for Fure forest reserve, and (c) TM radiance (smoothed) and two measures of density for Fure forest reserve.

LANDSAT TM DATA

There was no cloud-free Landsat TM data of Tain and Yoyo and there was no relationship between Landsat TM radiance and ground data at Fure. The lack of a relationship was attributed to the difficulty in linking a PSP with a specific pixel (Vanclay and Preston, 1990). Therefore, the Landsat TM imagery of Fure was smoothed with a 5 × 5 pixel low-pass filter (Curran, 1985) to increase the pixel size and hence the probability of linking each PSP with its relevant pixel. As a result of this image smoothing there was a negative relationship between the density of trees with a diameter greater than 50 cm and R radiance, a positive relationship between the density of trees with a diameter greater than 10 cm and NIR radiance, but no relationship between any measure of ground data and NDVI (Table 4.8 and Figure 4.9).

Table 4.9 The range at and standard error of tree density associated with a given value of AVHRR-NDVI.

AVHRR NDVI	Range (1 standard error) of density for trees with diameters >10 cm (trees ha^{-1}) estimated using AVHRR NDVI	
	Equation 1 (three reserves)	Equation 2 (Fure alone)
0.20	113–269	182–270
0.22	199–355	243–331
0.24	284–440	319–407
0.26	370–526	364–452
0.28	455–611	424–512
0.30	541–697	484–572

THE POSSIBILITY OF ESTIMATING GROUND DATA WITH REMOTELY SENSED DATA

Illustrations of the relationships between remotely sensed data and ground data for all three forest reserves with (AVHRR data) and Fure with (AVHRR and smoothed TM data) are given in Figure 4.9. These plots draw attention to variability in the direction of the relationships, the small range of remotely sensed and ground data at Fure and the apparent unsuitability of AVHRR-NDVI for the estimation of any of the ground variables. For example, when using AVHRR-NDVI data to estimate the density of all trees with a diameter greater than 10 cm (D_{10}) for all reserves (equation 4) and Fure alone (equation 5) the coefficients are unstable and less than half of the variance was accounted for.

$$D_{10} = -664 \ (4275 \ \text{NDVI}), \ n = 29, \ SE = 78, \ R^2 = 38.4 \ \% \qquad [4]$$

$$D_{10} = -378 \ (3021 \ \text{NDVI}), \ n = 29, \ SE = 44, \ R^2 = 49.0 \ \% \qquad [5]$$

However, the AVHRR-NDVI may have a rôle in estimating very broad classes of tree density and thereby regeneration stage. For example, using the above two equations it would be possible to use AVHRR-NDVI data to estimate three classes of low density (AVHRR-NDVI < 0.23), medium density (AVHRR-NDVI 0.23–0.27) and high density (AVHRR-NDVI > 0.27) forest (Table 4.9). Future research should, therefore, focus on the rôle that AVHRR-NDVI could play in estimating readily measured ground variables that are related to leaf biomass and thereby regeneration stage (Swaine and Hall, 1983).

.... CLASSIFYING FOREST COVER WITH COARSE SPATIAL RESOLUTION IMAGERY

A prerequisite to the identification of regenerative stages is the discrimination of forest from non-forest land cover. As the spatial resolution of an image decreases, so a greater proportion of pixels will have a partial forest cover and so the accuracy of a forest/non-forest classification will decrease. To overcome this problem there is a need to spectrally unmix the forest from the non-forest proportions within individual pixels and this was attempted using four images of Mpameso forest reserve (Figure 4.3). Three images had spatial resolutions of 300 m, 600 m and 1,200 m, were spatially degraded versions of a band 7 (NIR) Landsat MSS image (Table 4.6), and were derived using the filtering method presented by Justice *et al.* (1989). The fourth image had a spatial resolution of 1,100 m and comprised waveband 1 (R) and 2 (NIR) of the AVHRR scene listed in Table 4.6. Two AVHRR wavebands were used because a discriminant analysis revealed that for this particular haze-free image the use of the R waveband in conjunction with the NIR waveband provided slightly greater discrimination between forest and non-forest than the NIR waveband alone.

The prediction of sub-pixel forest cover in the coarse spatial resolution imagery was achieved by defining a regression relationship between the fuzzy membership function for forest against the observed forest proportion in 33 pixels for each of the spatially degraded Landsat MSS images and in 23 pixels for the AVHRR image. This regression was then used to estimate the forest cover in a further 32 of the spatially degraded Landsat MSS pixels and 19 AVHRR pixels via the fuzzy membership functions (Figure 4.10). The algorithm used to generate these fuzzy membership functions was a supervised version of that presented by Bezdek *et al.* (1984). This is based on a fuzzy *c*-means partition (equation 6)

$$M = \{U{:}u_{ik}\in[0,1];\ \sum_{k=1}^{n} u_{ik} > 0,\ i = 1...c;\ \sum_{i=1}^{c} u_{ik} = 1,\ k = 1...n\} \qquad [6]$$

where U is a fuzzy c-partition of n observations and c fuzzy groups, u_{ik} is an element of U and represents the membership of an observation x_k to the ith class, where x_k is a vector the length of which is the number of attributes, p, used. The optimal fuzzy c-partition is identified through the minimisation of the generalised least-squared errors functional J_m,

$$J_m(U,V) = \sum_{k=1}^{n} \sum_{i=1}^{c} (u_{ik})^m (d_{ik})^2 \qquad [7]$$

where, V is a c by p matrix whose elements, v_{ik}, represent the mean of the kth attribute in the ith class, m is a weighting component, $1<m<\propto$, and d_{ik} is a measure of dissimilarity based on the distance between an observation and a class centroid which can be determined from,

$$(d_{ik})^2 = (x_k - v_i)^{\mathrm{T}} A (x_k - v_i) \qquad [8]$$

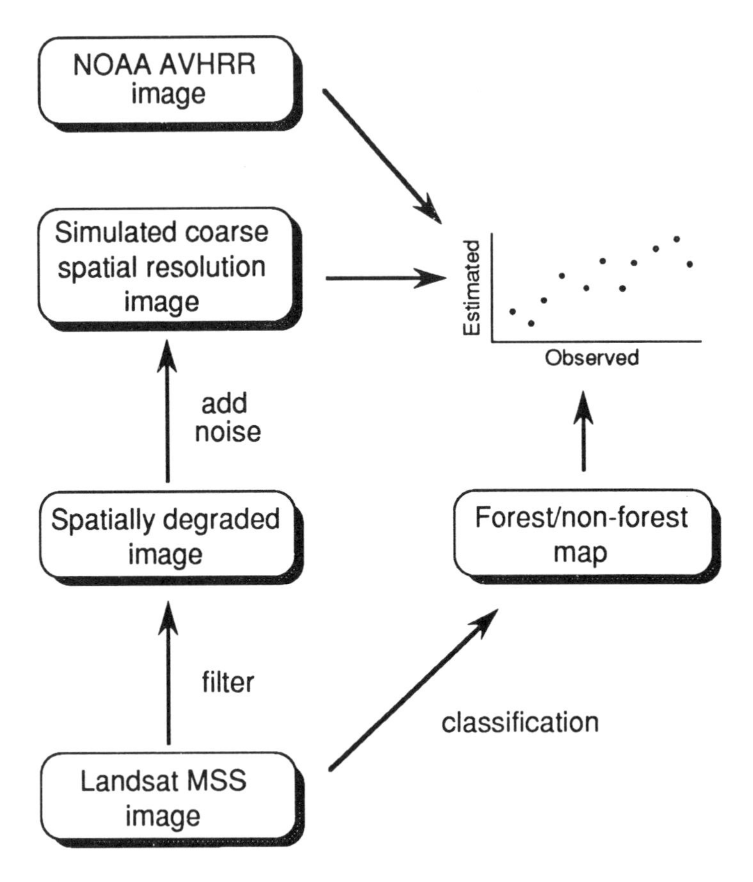

Figure 4.10 Summary of the method used to determine the accuracy with which remotely sensed data with a coarse spatial resolution could be used to estimate forest cover within a pixel.

in which A is a weight matrix which determines the norm to be used (Bezdek, 1981; Bezdek *et al.*, 1984). Only the Mahalanobis norm was used and attention focused on the fuzzy membership functions, u_{ik}, output with parameter $m = 2.0$. The latter variable is related positively to the degree of 'fuzziness', with a conventional classification of the observations obtained from an analysis with $m = 1.0$. The fuzzy membership functions lie on a scale between 0 and 1 and sum to 1 for each observation. They are therefore similar in some respects to posterior probabilities of class membership and their magnitude will be related to the proportion of a particular class within a pixel (Fisher and Pathirana, 1990; Foody, 1992).

There were relatively small root-mean-square (r.m.s.) errors of forest cover estimation and statistically significant correlations between the estimated and observed forested proportion of pixels in each of the four images (Figure 4.11). For the spatially degraded MSS imagery alone the r.m.s. error and the correlation between estimated and observed forested proportion of the pixels was related positively to spatial resolution. However, for areal error the

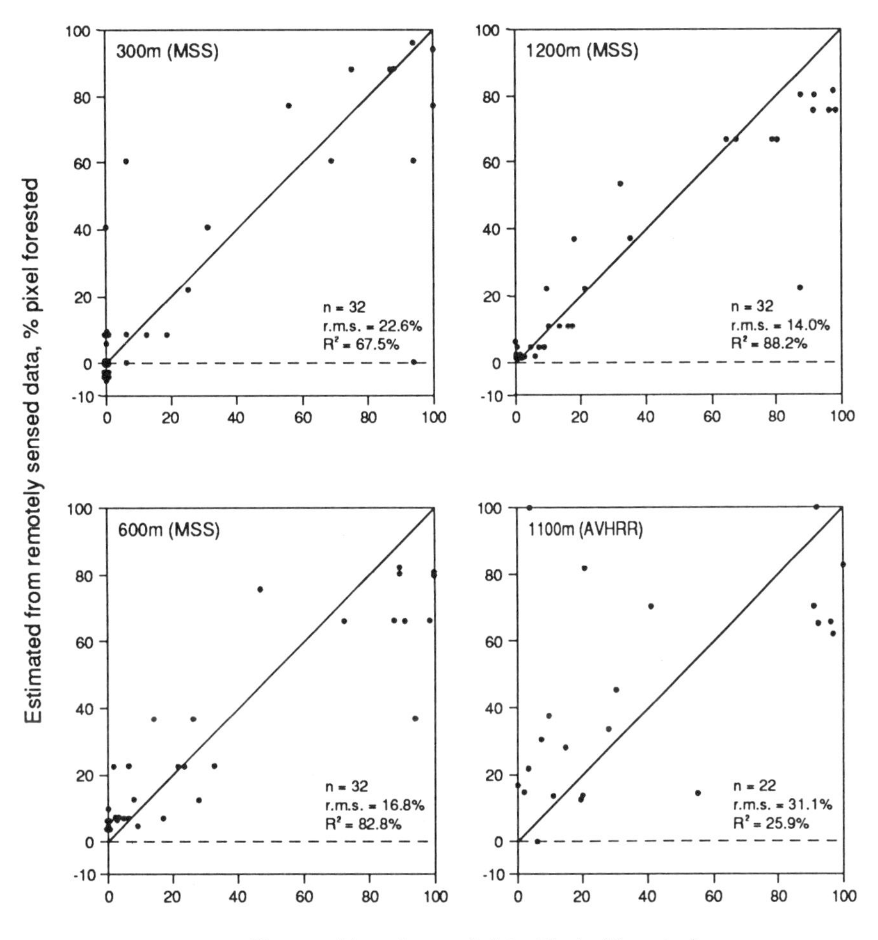

Figure 4.11 Comparison between the observed and estimated forest coverage at Mpameso forest reserve, expressed as a percentage of pixel forested. The remotely sensed data comprise Landsat MSS at three degraded spatial resolutions and NOAA AVHRR data at one spatial resolution and the 'ground' data are derived from a spatially undegraded Landsat MSS image (Figure 4.5).

relationship was reversed with areal errors of 0.02 km^2, 0.061 km^2 and 0.2 km^2 for spatial resolutions of 300 m, 600 m and 1,200 m respectively. The r.m.s. error was larger and R^2 values were smaller than expected for the AVHRR data. This may simply be the result of a smoke plume that covered about one quarter of the forest-non-forest boundary (Plate 2) and imprecise registration of the AVHRR data and forest/non-forest classification. Perhaps most importantly, the r.m.s. error of forest cover estimation was less than would be derived from a classification of all four images (Table 4.10).

Estimating the areal extent of tropical forest may be achieved more appropriately after the application of a spectral mixture model than directly

Table 4.10 Comparison of the root-mean-square (r.m.s.) errors, expressed as a percentage of pixel, observed for estimates derived from a classification and after pixel unmixing using the regression relationship between the fuzzy membership functions and forest proportions.

Remotely sensed imagery	Spatial resolution (m)	r.m.s. error (%) classification	unmixed
Landsat MSS	300	27.2	22.6
Landsat MSS	600	23.2	16.8
Landsat MSS	1,200	23.3	14.0
NOAA AVHRR	1,100	31.1	25.9

from an image classification. These results concur with those of others who have applied different mixture modelling approaches in attempts to increase the accuracy of the estimation of tropical forest extent from coarse spatial resolution imagery (e.g., Cross *et al.*, 1991).

.... CONCLUSIONS AND RECOMMENDATIONS

Two key conclusions can be drawn. First, the three hypothesised relationships

(i) R radiation is related negatively to tree density and basal area,
(ii) NIR radiation has no, or a positive relationship, with tree density and basal area, and
(iii) NDVI has a positive relationship with tree density and basal area

were supported broadly, especially when using the NOAA AVHRR data. Instances where these relationships were not observed may be accounted for by complicating factors such as topography, haze and the problem of colocating ground and remotely sensed data. Second, estimation of the areal extent of tropical forests may be more accurately made after unmixing pixel composition.

Further work to refine the relationships between remotely sensed radiation and biophysical properties related to regenerative state would be valuable. Similarly, there seems to be a potential to estimate the extent of the forest regenerative classes identified through the use of unmixed coarse spatial resolution data (Curran and Foody, 1992).

It was recommended that future research on the remote sensing of regenerative stage in tropical forests should use the NDVI derived from NOAA AVHRR data and also evaluate the potential of microwave radar and multi-angle optical sensors.

.... ACKNOWLEDGEMENTS

We wish to thank everyone who assisted with the research discussed in this chapter, especially J.H. François (Chief Conservator of Forests, Ghana) for collaboration and access to ground data; N. Bird (formerly of Overseas

Development Agency, Ghana) and E. Sackey (formerly of Forest Inventory and Management Project, Forestry Department, Ghana) for advice, access to ground data and field support; T. Abel (Natural Resources Institute, Chatham), G. D'Souza (Joint Research Centre, Ispra) and staff at Silsoe College for access to Landsat sensor imagery of Ghana; G. Lawson (Institute of Terrestrial Ecology, Bush) and R. Walsh (University College of Swansea) for discussions on tropical forests; R. Davies (Oxford University), S. Downard and M. Honzak (University College of Swansea) for assistance with image processing; J. Hill and R. Lucas (University College of Swansea) for comments on a draft of this chapter; E. Baker and D. Reynolds for word processing an 'evolving' text; the University College of Swansea Research Fund for technical support and the financial support provided by the NERC through its TIGER (Terrestrial Initiative in Global Environment Research) programme, contract 90/105.

.... REFERENCES

Achard, F., Blasco, F., 1990, 'Analysis of vegetation seasonal evolution and mapping of forest cover in West Africa with the use of NOAA AVHRR HRPT data', *Photogrammetric Engineering and Remote Sensing*, **56**: 1359–65

Aber, J.D., Melillo, J.M., 1991, *Terrestrial ecosystems*, Saunders College Publishing, Philadelphia

Achard, F., 1991, 'The methodological approach of the ESA-CEC TREES project in tropical Southeast Asia', *Training in remote sensing applications to deforestation in Southeast Asia*, Centre National d'Etudes Spatiales (CNES): 139–40

Alder, D., Synnott, T.J., 1992, *Permanent sample plot techniques for mixed tropical forest*, Oxford Forestry Institute, Oxford

Amaral, S., 1992, 'Deforestation estimates in AVHRR/NOAA and TM/ Landsat images for a region in central Brazil', *International Archives of Photogrammetry and Remote Sensing, Commission VII*, **29**: 764–7

Aschbacher, J., 1991, 'The complementary nature of SAR and VIR remote sensing for tropical forest monitoring', *Training in remote sensing applications to deforestation in southeast Asia*, Centre National D'Etudes Spatiales (CNES), Paris: 132–7

Baker, J.R., Mitchell, P.L., Cordey, R.A., Groom, G.B., Settle, J.J., Stileman, M.R., 1993, 'Relationships between physical characteristics and polarimetric radar backscatter for Corsican pine stands in Thetford Forest, UK', *International Journal of Remote Sensing* (in press)

Baltaxe, R., 1980, *The application of Landsat data to tropical forest surveys*, Food and Agriculture Organisation, Rome

Barale, V., Curran, P.J., Deschamps, P.T., Fischer, J., Grassl, H., Malingreau, J.P., Morel, A., Verstraete, M., 1993, *The Medium Resolution Imaging Spectrometer (MERIS)*, European Space Agency, Paris, 43pp.

Barnsley, M.J., Muller, J.P., 1991, 'Measurement, simulation and analysis of the directional reflectance properties of Earth surface materials', *Proceedings, 5th International Colloquium, Physical Measurements and Signatures in Remote Sensing*, European Space Agency, SP-319, Noordwijk: 375–82

Bezdek, J.C., 1981, *Pattern recognition with fuzzy objective function algorithms*, Plenum, New York

Bezdek, J.C., Ehrlich, R., Full, W., 1984, 'FCM: The fuzzy *c*-means clustering algorithm', *Computers and Geosciences*, **10**: 191–203
Blasco, F., Achard, F., 1990, 'Analysis of vegetation changes using satellite data', in Bouwman, A.F. (editor), *Soils and the greenhouse effect*, Wiley, New York: 303–10
Boateng, E.A., 1966, *A geography of Ghana* (Second Edition), Cambridge University Press, Cambridge
Box, E.O., Holben, B.N., Kalb, V., 1989, 'Accuracy of the AVHRR vegetation index as a predictor of biomass, primary production and net CO_2 flux', *Vegetatio*, **80**: 71–89
Briggs, S.A., 1993, 'Remote sensing and terrestrial global environmental research – the TIGER programme', in Foody, G.M., Curran, P.J. (editors), *Environmental remote sensing from regional to global scales*, Belhaven Press, London (this volume)
Brookfield, H., 1992, '"Environmental colonism", tropical deforestation and concerns other than global warming', *Global Environmental Change*, **21**: 93–6
Brown, S., Lugo, A.E., 1990, 'Tropical secondary forests', *Journal of Tropical Forestry*, **6**: 1–32
Campbell, F.H.A., Ahern, F.J., Raney, R.K., 1992, 'Canada's tropical forestry initiative in Latin America – an airborne SAR program', *International Archives of Photogrammetry and Remote Sensing, Commission VII*, **29**: 768–70
Campbell, J.B., Bowder, J.O., 1992, 'SPOT survey of agricultural land uses in the Brazilian Amazon', *International Archives of Photogrammetry and Remote Sensing, Commission VII*, **29**: 159–65
Carver, K., 1988, *SAR Synthetic Aperture Radar – Earth Observing System*, Instrument Panel Report Vol. IIf, National Aeronautics and Space Administration, Washington D.C.
CEOS, 1992, *The relevance of satellite missions to the study of the global environment*, Committee on Earth Observation Satellites, British National Space Centre, London
Changchui, H., Chadhury, M.U., 1991, 'The deforestation process in South East Asia', *Training in remote sensing applications to deforestation in south east Asia*, Centre National D'Etudes Spatiales (CNES), Paris: 19–26
Chavez, P.S., 1988, 'An improved dark-object subtraction technique for atmospheric scattering correction of multispectral data', *Remote Sensing of Environment*, **24**: 459–79
Chavez, P.S., 1989, 'Radiometric calibration of Landsat Thematic Mapper multispectral images', *Photogrammetric Engineering and Remote Sensing*, **55**: 1289–94
Chomentowski, W., and Lawrence, W., 1992, 'NASA project studies tropical deforestation', *Monitor*, **4**: 9
Churchill, P.N., Sieber, A.J., 1991, 'The current status of ERS-1 and the role of radar remote sensing for the management of natural resources in developing countries', in Belward, A.S., Valenzuela, C.R. (editors), *Remote sensing and geographical information systems for resource management in developing countries*, Kluwer Academic, Dordrecht: 111–43
Cohen, W.B., Spies, T.A., 1992, 'Estimating structural attributes of Douglas-Fir/Western Hemlock forest stands from Landsat and SPOT imagery', *Remote Sensing of Environment*, **41**: 1–17
Cook, E.A., Iverson, L.R., Graham, R.L., 1989, 'Estimating forest productivity with Thematic Mapper and biogeographical data', *Remote Sensing of Environment*, **28**: 131–41
Cross, A.M., Settle, J., Drake, N.A., Päivinen, R.T.M., 1991, 'Subpixel

measurement of tropical forest cover using AVHRR data', *International Journal of Remote Sensing*, **12**: 1119–29

Curran, P.J., 1982, 'Polarized visible light as an aid to vegetation classification', *Remote Sensing of Environment*, **12**: 491–9

Curran, P.J., 1983, 'Multispectral remote sensing for the estimation of green leaf area index', *Philosophical Transactions of the Royal Society London Series A*, **309**: 257–70

Curran, P.J., 1985, *Principles of remote sensing*, Longman, Harlow

Curran, P.J., 1988, 'The semi-variogram in remote sensing: an introduction', *Remote Sensing of Environment*, **24**: 483–507

Curran, P.J., Dungan, J.L., Gholz, H.L., 1992, 'Seasonal LAI in slash pine estimated with Landsat TM', *Remote Sensing of Environment*, **39**: 3–13

Curran, P.J., Foody, G.M., 1992, *The use of remote sensing to characterise the regenerative states of felled tropical forests*, Report to the Natural Environment Research Council on TIGER contract 90/105, University College of Swansea, Swansea, 79pp.

Danson, F.M., Curran, P.J., 1993, 'Factors affecting the remotely sensed response of coniferous forest plantations', *Remote Sensing of Environment*, **43**: 55–65

de Groof, H.D., Malingreau, J.P., Sokeland, A., Achard, F., 1992, A first look at ERS-1 data over the tropical forest of West Africa', *World Forest Watch conference on global forest monitoring*, INPE, São Paulo: 19

Detwiler, R.P., Hall, C.A.S., 1988, 'Tropical forests and the global carbon cycle', *Science*, **239**: 42–7

DOE, 1990, *This common inheritance: UK environmental strategy*, Department of Environment, HMSO, London

Downward, S., 1991, *An assessment of the use of a GIS digital terrain model for the topographic correction of Landsat MSS radiance values, Yoyo National Forest Reserve, Ghana*, unpublished Diploma in Topographic Science Thesis, University College of Swansea, Swansea, 56pp.

Durden, S.L., van Zyl, J.J., Zebker, H.A., 1989, 'Modelling and observation of the radar polarization signature of forested areas', *IEEE Transactions on Geoscience and Remote Sensing*, **27**: 290–301

Durden, S., Freeman, A., Klein, J., Vane, G., Zebker, H., Zimmermann, R., Oren, R., 1991, 'Polarimetric radar measurements of a tropical forest', in van Zyl, J.J. (editor), *Proceedings, Third Airborne Synthetic Aperture Radar (AIRSAR) Workshop*, JPL Workshop, JPL Publications 91-30. Jet Propulsion Laboratory, Pasadena: 223–9

Eden, M.J., 1986, 'The management of renewable resources in the tropics: the use of remote sensing', in Eden, M.J., Parry, J.T. (editors), *Remote sensing and tropical land management*, Wiley, New York: 3–15

Enting, I.G., Mansbridge, J.V., 1989, 'Seasonal sources and sinks of atmospheric CO_2: direct inversion of filtered data', *Tellus*, **41B**: 111–26

Enting, I.G., Mansbridge, J.V., 1991, 'Latitudinal distribution of sources and sinks of CO_2: results of an inversion study', *Tellus*, **43B**: 156–70

ESA, 1990, *Cooperation with the EEC: TREES*, European Space Agency Earth Observation Programme Board ESA/PB-EO(90)91, European Space Agency, Paris

ESA, 1991, *ERS-1*, European Space Agency, Noordwijk

FAO, 1981, *Tropical forest resources assessment project: Forest resources of Tropical Africa. Part II: Country briefs*, Food and Agriculture Organisation of the UN, Rome

Fan, S.M., Wofsy, S.C., Bakwin, P.S., Jacob, D.J., 1990, 'Atmosphere-biosphere exchange of CO_2 and O_3 in the Central Amazon Forest', *Journal of Geophysical Research*, **95**: 16851–64

Fisher, P.F., Pathirana, S., 1990, 'The evaluation of fuzzy membership of land cover classes in the suburban zone', *Remote Sensing of Environment*, **34**: 121–32

Fölster, H., de Lassalas, G., Khana, P., 1976, 'A tropical evergreen forest with perched water table, Magdalena Valley, Colombia', *Oceologica Plantarum*, **11**: 297–320

Foody, G.M., 1992, 'A fuzzy sets approach to the representation of vegetation continua from remotely sensed data: an example from lowland heath', *Photogrammetric Engineering and Remote Sensing*, **58**: 221–5

Foody, G.M., Cox, D.P., 1991, 'Estimation of sub-pixel land cover composition from spectral mixture models', *Spatial Data 2000*, 'Remote Sensing Society, Nottingham: 186–95

Ford, J.P., Wickland, D.E., 1985, 'Forest discrimination with multi-polarization imaging radar', *International Geoscience and Remote Sensing Symposium*, IGARSS, New York, **1**: 462–5

Furley, P.A., 1986, 'Radar surveys for resource evaluation in Brazil: an illustration from Rondônia', in Eden, M.J., Parry, J.T. (editors), *Remote sensing and tropical land management*, Wiley, New York: 79–99

Gerstl, S.A.W., 1990, 'Physics concepts of optical and radar reflectance signatures. A summary review', *International Journal of Remote Sensing*, **11**: 1109–17

Gibbs, D.G., Leston, D., 1970, 'Insect phenology in a forest farm locality in West Africa', *Journal of Applied Ecology*, **7**: 519–48

Gilruth, P.T., Hutchinson, C.F., Barry, B., 1990, 'Assessing deforestation in the Guinea highlands of West Africa using remote sensing', *Photogrammetric Engineering and Remote Sensing*, **56**: 1375–82

Gornitz, V., 1985, 'A survey of anthropogenic vegetation changes in West Africa during the last century – climatic implications', *Climatic Change*, **7**: 285–325

Goudie, A., 1985, *The encyclopaedic dictionary of physical geography*, Blackwell, Oxford

Goward, S.N., Markham, B., Dye, D.G., Dulaney, W., Yang, J., 1991, 'Normalized difference vegetation index measurements from the Advanced Very High Resolution Radiometer', *Remote Sensing of Environment*, **35**: 257–77

Grainger, A., 1986, 'Deforestation and progress in afforestation in Africa', *The International Tree Crops Journal*, **4**: 33–48

Green, G.M., Sussman, R., 1990, 'Deforestation history of the eastern rainforests of Madagascar from satellite images', *Science*, **248**: 212–15

Gregory, S., 1978, *Statistical methods and the geographer* (Fourth Edition), Longman, Harlow

Guo, L.J., 1991, 'Balance contrast enhancement technique and its application in image colour composition', *International Journal of Remote Sensing*, **12**: 2133–51

Hall, J.B., Swaine, M.D., 1981, *Distribution and ecology of vascular plants in a tropical rain forest. Forest vegetation in Ghana*, Dr W. Junk, The Hague

Hall, F.G., Botkin, D.B., Strebel, D.E., Woods, K.D., Goetz, S.J., 1991, 'Large-scale patterns of forest succession as determined by remote sensing', *Ecology*, 72: 628–40

Hammond, A.L., 1977, 'Remote sensing', *Science*, **196**: 511–16

Hartshorn, G.S., 1980, 'Neotropical forest dynamics', *Biotropica*, **12**: 23–30

Hastings, D.A., Emery, W.J., 1992, 'The Advanced Very High Resolution Radiometer (AVHRR): A brief reference guide', *Photogrammetric Engineering and Remote Sensing*, 58: 1183–8

Henderson-Sellers, A., Dickinson, R.E., Wilson, M.F., 1988, 'Tropical deforestation: important processes for climate models', *Climate Change*, **13**: 43–67

Holben, B.N., Kaufman, Y.J., Kendall, J.D., 1990, 'NOAA-11 AVHRR visible and near-IR inflight calibration', *International Journal of Remote Sensing*, **11**: 1511–19

Houghton, R.A., 1990, 'The future role of tropical forests in affecting the carbon dioxide concentration of the atmosphere', *Ambio*, **19**: 204–9

Houghton, R.A., 1991, 'Tropical deforestation and atmospheric carbon dioxide', *Climatic Change*, **19**: 99–118

Houghton, R.A., Boone, R.D., Melillo, J.M., Palm, C.A., Woodwell, G.M., Myers, N., Moore III, B., Skole, D.L., 1985, 'Net flux of carbon dioxide from tropical forests in 1980', *Nature*, **316**: 617–20

Houghton, R.A., Boone, R.D., Fruci, J.R., Hobbie, J.E., Melillo, J.M., Palm, C.A., Peterson, B.J., Shaver, G.R., Woodwell, G.M., Skole, D.L., Myers, N., 1987, 'The flux of carbon from terrestrial ecosystems to the atmosphere in 1980 due to changes in land use', *Tellus*, **39B**: 122–39

Houghton, R.A., Skole, D.C., 1990, 'Carbon', in Turner, B.L. (editor), *The Earth transformed by human action*, Cambridge University Press, Cambridge: 393–408

Houghton, R.A., Lefkowitz, D.S., Skole, D.C., 1991a, 'Changes in the landscape of Latin America between 1850 and 1985 (I). Progressive loss of forests', *Forest Ecology and Management*, **38**: 143–72

Houghton, R.A., Skole, D.L., Lefkowitz, D.S., 1991b, 'Changes in the landscape of Latin America between 1850 and 1985: II a net release of CO_2 to the atmosphere', *Journal of Forest Ecology and Management*, **38**: 173–99

Hughli, H., Frei, W., 1983, 'Understanding anisotropic reflectance in mountainous terrain', *Photogrammetric Engineering and Remote Sensing*, **49**: 671–83

IGBP, 1990, *The International Geosphere-Biosphere Programme: A study of global change. The initial core projects*, International Geosphere Biosphere Programme, Stockholm

Jaakkola, S., 1990, 'Managing data for the monitoring of tropical forest cover: the global resource information database approach', *Photogrammetric Engineering and Remote Sensing*, **56**: 1355–7

Jarvis, P.G., Moncrieff, J.B., 1992, 'Atmosphere-biosphere exchange of CO_2', in Mather, P.M. (editor), *TERRA-1 understanding the terrestrial environment*, Taylor and Francis, London: 85–99

Jeanjean, H., Husson, A., 1991, 'Seameo-France Project: detailed studies with SPOT data and global perception with NOAA data', *Training in remote sensing applications to deforestation in south east Asia*, Centre National D'Etudes Spatiales (CNES), Paris: 34–49

Jensen, J.R., 1983, 'Biophysical remote sensing', *Annals of the Association of American Geographers*, **73**: 111–32

Johnson, G.E., 1991, 'The GRID project in South East Asia', *Training in remote sensing applications to deforestation in South East Asia*, Centre National D'Etudes Spatiales (CNES), Paris: 138

JRC, 1991, *TREES, Tropical Ecosystem Environment Observations by Satellites: Strategy proposal 1991–1993 (AVHRR data collection and analysis)*, Joint Research Centre, Commission of the European Communities, Ispra, 20pp.

Justice, C.O., Markham, B.L., Townshend, J.R.G., Kennard, R.L., 1989, 'Spatial degradation of satellite data', *International Journal of Remote Sensing*, **10**: 1539–61

Kasischke, E.S., Christensen, N.L., 1990, 'Connecting forest ecosystem and microwave backscatter models', *International Journal of Remote Sensing*, **11**: 1277–98

Kasischke, E.S., Bourgeau-Chavez, L.L., Christensen, N.L., Dobson, M.C., 1991, 'The relationship between aboveground biomass and radar backscatter as observed on airborne SAR imagery', in van Zyl, J.J. (editor), *Proceedings, Third Airborne Synthetic Aperture Radar (AIRSAR) Workshop*, JPL Publication 91-30, 'Jet Propulsion Laboratory, Pasadena: 11–21

Kummer, D.M., 1992, 'Remote sensing and tropical deforestation: a cautionary note from the Philippines', *Photogrammetric Engineering and Remote Sensing*, **58**: 1469–71

Lanley, J.P., 1982, *Tropical forest resources*, F.A.O. Forestry Paper 30, United Nations, Food and Agriculture Programme Publications, Rome

Lieberman, D., 1982, 'Seasonality and phenology in a dry tropical forest in Ghana', *Journal of Ecology*, **70**: 791–806

Malingreau, J.P., 1991, 'Remote sensing for tropical forest monitoring: an overview', in Belward, A.S. and Valenzuela, C.R. (editors), *Remote sensing and geographical information systems for resource management in developing countries*, Kluwer Academic, Dordrecht: 253–78

Malingreau, J.P., 1992, 'Satellite based forest monitoring. A review of current issues', *World Forest Watch conference on global forest monitoring*, INPE, São Paulo: 3

Malingreau, J.P., Tucker, C.J., 1988, 'Large scale deforestation in the Southeastern Amazon basin of Brazil', *Ambio*, **17**: 49–55

Malingreau, J.P., Tucker, C.J., Laporte, N., 1989, 'AVHRR for monitoring global tropical deforestation', *International Journal of Remote Sensing*, **10**: 855–67

Malingreau, J.P., Reichert, P., 1991, 'Forestry', *Report of the Earth observation user consultation meeting*, European Space Agency, SP-1143, Noordwijk: 161–72

Malingreau, J.P., Verstraete, M.M., Achard, F., 1992, 'Monitoring tropical forest deforestation: a challenge for remote sensing', in Mather, P.M. (editor), *TERRA-1 understanding the terrestrial environment*, Taylor and Francis, London: 121–31

Mather, P.M., 1987, *Computer processing of remotely-sensed images*, Wiley, Chichester

Millington, A., Townshend, J., Kennedy, P., Saull, R., Prince, S., and Madams, R., 1989, *Biomass assessment: Woody biomass in the SADCC Region*, Earthscan Publications, London

Morain, S.A., Klankamsorn, B., 1978, 'Forest mapping and inventory techniques through visual analysis of Landsat imagery: examples from Thailand', *Proceedings, 12th International Symposium on Remote Sensing of Environment*, University of Michigan, Ann Arbor: 417–26

Myers, N., 1980, *Conservation of tropical moist forests*, National Academy of Sciences, Washington D.C.

Myers, N., 1988, 'Tropical deforestation and remote sensing', *Forest Ecology and Management*, **23**: 215–25

Myers, N. (editor), 1989, *Deforestation rates in tropical forests and their climatic implications*, Friends of the Earth, London

Myers, N., 1991, 'The disappearing forests', in Porritt, J. (editor), *Save the Earth*, Dorling Kindersley, London: 47–56

Nelson, R., Holben, B.N., 1986, 'Identifying deforestation in Brazil using multi-resolution satellite data', *International Journal of Remote Sensing*, **7**: 429–38

Nelson, R., Horning, N., Stone, T.A., 1987, 'Determining the rate of forest conversion in Mato Grosso, Brazil using Landsat MSS and AVHRR data', *International Journal of Remote Sensing*, **8**: 1767–84

Päivinen, R., Witt, R., 1988, 'Application of NOAA/AVHRR data for tropical forest for forest inventory and monitoring', *Department of forest mensuration and research notes*, **21**, University of Helsinki, Helsinki: 163–70

Parry, D.E., Trevett, J.W., 1979, 'Mapping Nigeria's vegetation from radar', *Geographical Journal*, **145**: 265–81

Pereira, M.N., 1992, 'Land use evaluation: a case study in Brazil', *International Archives of Photogrammetry and Remote Sensing, Commission VII*, **29**: 507–10

Peterson, D.L., Running, S.W., 1989, 'Applications in forest science and management', in Asrar, G. (editor), *Theory and applications of optical remote sensing*, Wiley, New York: 429–79

Peterson, D.L., Spanner, M.A., Running, R.W., Teuber, K.B., 1987, 'Relationship of Thematic Mapper Simulator data to leaf area index of temperate coniferous forests', *Remote Sensing of Environment*, **22**: 323–41

Pinty, B., Verstraete, M.M., Dickinson, R.E., 1990, 'A physical model of the bidirectional reflectance of vegetation canopies. 2 Inversion and validation', *Journal of Geophysical Research*, **95**: 11767 –75

Proctor, J., 1990, 'Tropical rain forests', *Progress in Physical Geography*, **14**: 251–69

Proctor, J., 1991, 'Tropical rain forests', *Progress in Physical Geography*, **15**: 291–303

Rapley, C.G., 1992, 'ERS-1 land and ice applications', in Mather, P.M. (editor), *TERRA-1 understanding the terrestrial environment*, Taylor and Francis, London: 147–62

Robinove, C.J., 1982, 'Computation with physical values from Landsat digital data', *Photogrammetric Engineering and Remote Sensing*, **48**: 781–4

Rougharden, J., Running, S.W., Matson, P.A., 1991, 'What does remote sensing do for ecology?', *Ecology*, **76**: 1918–22

Roujean, J.L., Leroy, M., Podaire, A., Deschamps, P.Y., 1992, 'Evidence of surface reflectance bidirectional effects from NOAA/AVHRR multi-temporal data set', *International Journal of Remote Sensing*, **13**: 685–98

Sader, S.A., Joyce, A.T., 1988, 'Deforestation rates and trends in Costa Rica, 1940–1983', *Biotropica*, **20**: 11–19

Sader, S.A., Waide, R.B., Lawrence, W.T., Joyce, A.T., 1989, 'Tropical forest biomass and successional age class relationships to a vegetation index derived from Landsat TM data', *Remote Sensing of Environment*, **28**: 143–56

Sader, S.A., Stone, T.A., Joyce, A.T., 1990, 'Remote sensing of tropical forests: an overview of research and applications using non-photographic sensors', *Photogrammetric Engineering and Remote Sensing*, **56**: 1343–51

Saldarriaga, J.G., West, D.C., Tharp, M.C., Uhl, C., 1988, 'Long term chronosequence of forest succession in the upper Rio Negro of Colombia and Venezuela', *Journal of Ecology*, **76**: 938–58

Sayer, J.A., Harcourt, C.S., Collins, N.M. (editors), 1992, *The conservation atlas of tropical forests, Africa*, Macmillan, Basingstoke

Shugart, H.H., Antonovsky, M. Ya, Jarvis, P.G., Sandford, A.P., 1986, 'CO_2, climatic change and forest ecosystems', in Bolin, B., Döös, B.R., Jager, J., Warrick, R.A. (editors), *The greenhouse effect, climatic change and ecosystems*, Wiley, New York: 475–521

Sicco-Smit, G., 1978, 'SAR for forest type classification in a semi-deciduous tropical region', *ITC Journal*: 385–99

Sicco-Smit, G., 1992, 'A good, a better or the best remote sensing application system for remote sensing in the Amazon region', *International Archives of Photogrammetry and Remote Sensing, Commission VII*, **29**: 760–3

Singh, A., 1987, 'Spectral separability of tropical forest cover classes', *International Journal of Remote Sensing*, **8**: 971–9

Singh, K.D., 1990, 'Design of a global tropical resources assessment', *Photogrammetric Engineering and Remote Sensing*, **56**: 1353–4

Singh, K.D., 1991, 'FAO tropical forest resources monitoring programme', *Training in remote sensing applications to deforestation in southeast Asia*, Centre National D'Etudes Spatiales (CNES), Paris: 141–53

Singh, K.D., 1992, 'FAO tropical forest resources monitoring programme', *World Forest Watch conference on global forest monitoring*, INPE, São Paulo: 9–10

Soussan, J.G., Millington, A.C., 1992, 'Forests, woodlands and deforestation', in Mannion, A.M., Bowlby, S.R. (editors), *Environmental Issues in the 1990s*, Wiley, London: 79–96

Smith, J.A., Lin, T.L., Ranson, K.J., 1980, 'The Lambertian assumption of Landsat data', *Photogrammetric Engineering and Remote Sensing*, **46**: 1183–9

Stone, T.A., Woodwell, G.M., Houghton, R.A., 1989, 'Tropical deforestation in Para Brazil: analysis with Landsat and Shuttle Imaging Radar A', *Proceedings, international geoscience and remote sensing symposium*, IEEE Press, New York, **1**: 192–5

Survey of Ghana, 1969, *Portfolio of Ghana maps*, Survey of Ghana, Accra

Suzuki, T., Shimada, M., 1992, 'Japanese Earth observation satellite program and application of JERS-1 sensors data to forest monitoring', *World Forest Watch conference on global forest monitoring*, INPE, São Paulo: 47

Swaine, M.D., Hall, J.B., 1976, 'An application of ordination to the identification of forest types', *Vegetatio*, **32**: 83–6

Swaine, M.D., Hall, J.B., 1983, 'Early succession in tropical forest', *Journal of Ecology*, **71**: 601–27

Tans, P.P., Fung, I.Y., Takahashi, T., 1990, 'Observational constraints on the global atmospheric CO_2 budget', *Science*, **247**: 1431–8

Tardin, A.T., Cunha, R., 1990, *Evaluation of deforestation in the Legal Amazon using Landsat TM Images*, INPE Publication 5015-RPE/609, INPE, São Paulo

Taylor, C.J., 1952, 'The vegetation zones of the Gold Coast', *Bulletin of the Gold Coast Forestry Department*, **4**: 1–12

Teillet, P.M., 1986, 'Image correction for radiometric effects in remote sensing', *International Journal of Remote Sensing*, **7**: 1637–51

Teillet, P.M., Guidon, B., Goodenough, D.G., 1982, 'On the slope-aspect correction of multispectral scanner data', *Canadian Journal of Remote Sensing*, **8**: 85–105

Townshend, J.R.G. (editor), 1992, *Improved global data for land applications. A proposal for a new high resolution data set*, IGBP Global Change Report No. 20, The International Geosphere-Biosphere Programme, Stockholm

Tucker, C.J., Townshend, J.R.G., Geoff, T.E., 1985, 'Africa's land-cover classification using satellite data', *Science*, **227**: 369–75

Uhl, C., Jordan, C.F., 1984, 'Succession and nutrient dynamics following forest cutting and burning in Amazonia', *Ecology*, **65**: 1476–90

Uhl, C., 1987, 'Factors controlling slash and burn agriculture in Amazonia', *Journal of Ecology*, **75**: 371–407

Ulaby, F.T., Moore, R.K., Fung, A.K., 1982, *Microwave remote sensing active and passive. Volume II radar remote sensing and surface scattering and emission theory*, Addison Wesley Publishing Co., Reading, Massachusetts

Vanclay, J.K., Preston, R.A., 1990, 'Utility of Landsat Thematic Mapper data for mapping site productivity in tropical moist forests', *Photogrammetric Engineering and Remote Sensing*, **56**: 1383–8

Vanderbilt, V.C., Grant, L., Ustin, S.L., 1991, 'Polarization of light by vegetation', in Myeni, R.B., Ross, J. (editors), *Photon-vegetation interactions. Applications in optical remote sensing and plant ecology*, Springer Verlag, Berlin: 191–228

Varley, W.J., White, H.P., 1958, *The geography of Ghana*, Longman, London

Verstraete, M.M., Belward, A.S., Kennedy, P.J., 1990a, 'The Institute for Remote Sensing Applications contribution to the Global Change Research Programme', *Remote Sensing and Global Change*, Remote Sensing Society, Nottingham: vi–xxii

Verstraete, M.M., Pinty, B., Dickinson, R.E., 1990b, 'A physical model of the bidirectional reflectance of vegetation canopies. 1 Theory', *Journal of Geophysical Research*, **95**: 11755–65

Verstraete, M.M., Pinty, B., 1992, 'Extracting surface properties from satellite data in the visible and near-infrared wavelengths', in Mather, P.M. (editor), *TERRA-1 understanding the terrestrial environment*, Taylor and Francis, London: 203–9

Westman, W.E., Paris, J.A., 1987, 'Detecting forest structure and biomass with C-band multipolarization radar: physical model and field tests', *Remote Sensing of Environment*, **22**: 249–69

Westman, W.E., Strong, L.L., Wilcox, B.A., 1989, 'Tropical deforestation and species endangerment: the role of remote sensing', *Landscape Ecology*, **3**: 97–109

Whitmore, T.C., 1990, *An introduction to tropical rain forests*, Clarendon Press, Oxford

Wood, W.B., 1990, 'Tropical deforestation. Balancing regional development demands and global environmental concerns', *Global Environmental Change*, **7**: 23–41

Woodwell, G.M., 1984 (editor), *The role of terrestrial vegetation in the global carbon cycle: Measurement by remote sensing*, Wiley, Chichester

Woodwell, G.M., Hobbie, J.E., Houghton, R.A., Melillo, J.M., Moore, B., Park, A.B., Peterson, B.J., Shaver, G.R., 1984, 'Measurement of changes in vegetation of the Earth by satellite imagery', in Woodwell, G.M. (editor), *The role of terrestrial vegetation in the global carbon cycle*, SCOPE 23, Wiley, New York: 221–40

Woodwell, G.M., Houghton, R.A., Stone, T.A., Nelson, R.F., Kovalick, W., 1987, 'Deforestation in the tropics: New measurements in the Amazon Basin using Landsat and NOAA/AVHRR imagery', *Journal of Geophysical Research*, **92**: 2157–63

World Bank, 1988, *Ghana*, World Bank, Geneva

Wu, S.T., 1987, 'Potential application of multipolarization SAR for pine plantation biomass estimation', *IEEE Transactions on Geoscience and Remote Sensing*, **25**: 403–9

Wu, S.T., 1990, 'Assessment of tropical forest stand characteristics with multi-polarization SAR data acquired over a mountainous region in Costa Rica', *IEEE Transactions on Geoscience and Remote Sensing*, **28**: 752–5

5. Global Land Cover: Comparison of Ground-based Data Sets to Classifications with AVHRR Data

R.S. DeFries and J.R.G. Townshend

.... INTRODUCTION

Quantitative information on the global distribution of vegetation types is required in many arenas of global change research. In climate modelling, for example, the vegetation on the land surface determines fluxes of radiation, water vapour, sensible heat and momentum to the atmosphere (Sellers *et al.*, 1986). Changes in land covers such as forest, wetlands and cropland are important variables in the study of fluxes of greenhouse gases from terrestrial biota to the atmosphere (Watson *et al.*, 1990). In the study of socioeconomic causes of land transformations, information on the extent of land area under different cover types is essential (Richards, 1990).

Each of these lines of enquiry require quantitative and reliable information on global land cover at different spatial and temporal scales (Townshend, 1992). The climate modeller must characterise the land surface at very coarse spatial scales of hundreds of square kilometres and very short temporal scales of less than a day to seasons. Conversely, the scientist studying socioeconomic causes of deforestation might require information at spatial scales as fine as tens of meters but temporal scales as long as years or decades. This variation in scales further challenges abilities to provide reliable, quantitative and up-to-date information on global land cover.

Currently available estimates of the global extent of different land cover types vary widely (Townshend *et al.*, 1991). Such estimates are based on information from separate and diverse sources, many of which may be out-of-date or perhaps erroneous. Even if the information were reliable, integrating such sources from local and regional to global scales is hampered because there is no standard approach to characterising land cover. Some characterisations of vegetative cover are based on physiognomic properties such as growth and life forms, some on floristic characteristics such as dominant species, some on environmental and climatic properties, some on geographical location, and others on various combinations of all of these

(Mueller-Dombois, 1984). In addition, the distinction between natural vegetative cover that could potentially exist given the climatic conditions at a site and the vegetative cover that actually exists at that site is not always clear. The most serious concern, however, is that these estimates cannot provide information on changes in land cover through time because the variability between estimates substantially exceeds that of actual cover changes.

These deficiencies in ground-based sources of global land cover information – namely, problems of internal consistency and inability to reproduce consistent data sets over time – are the primary reasons why efforts have begun to use satellite sensor data to increase our knowledge of global land cover. Koomanoff (1989) compared data from NOAA's Advanced Very High Resolution Radiometer (AVHRR) with a conventional data source for global land cover, and AVHRR data has been used to carry out land cover classifications at continental scales, for Africa (Tucker *et al.*, 1985) and South America (Townshend *et al.*, 1987; 1989).

The purpose of this chapter is to compare several digitised ground-based data sets of global land cover information in order to identify training and testing sites for land cover classifications based on satellite sensor data. Two approaches are used to evaluate whether a given land cover type is present in a particular geographical location; comparison of ground-based data sets to identify geographical areas where they agree with each other, and comparison of each ground-based data set with land cover classifications derived from AVHRR data.

This chapter describes preliminary work directed towards the creation of a global multi-spatial resolution land cover data base. The first part of this work concentrates on the creation of a coarse spatial resolution data set suitable for use in global climate modelling.

.... COMPARISON OF EXISTING GLOBAL LAND COVER DATA SETS

A comparison of existing global land cover data sets was carried out, assuming that the most reliable sites for land cover information are those where ground-based data sets all agree on a given land cover type in a geographical location. The three published, digital data bases of existing land cover used for this comparison were Matthews (1982), Olson *et al.* (1983) and Wilson and Henderson-Sellers (1985). These three data sets are widely available in digitised form to the climate modelling community (Henderson-Sellers *et al.*, 1986).

Each of the three data sets employs a different classification scheme. The data set compiled by Matthews (1982) is based on the UNESCO scheme for classifying vegetation (UNESCO, 1973) and includes 32 categories at a one degree spatial resolution. The 'cultivation' category was derived for this analysis by combining categories 3, 4, and 5 from Matthews' data set of cultivation intensity. Olson *et al.* (1983) includes 40 categories derived for calculating the amount of carbon in live vegetation at a one-half degree spatial resolution. Wilson and Henderson-Sellers (1985) include 53 categories at a one degree spatial resolution, with arrays of both primary and secondary land cover types. Only the data set for primary land type was used in this analysis.

So that the three data sets would be comparable, categories were grouped according to 15 major land cover classes (see Appendix A). The selection of these classes was based on a number of criteria. First, and most importantly, each class must display a unique combination of functional characteristics appropriate to global modelling, such as height of mature vegetation, leaf type, seasonality and fraction of ground surface covered by vegetation. Second, the total number of classes should not exceed what could reasonably be incorporated into a global climate model. Third, each class must cover a significant portion of the global land surface. Fourth, it should be possible to discriminate among classes using currently-available multispectral remotely sensed data.

Consequently, 15 classes based on major growth form were selected. These classes were: broadleaf evergreen trees; broadleaf deciduous trees; needle leaf evergreen trees, closed canopy; needle leaf evergreen trees, open canopy; needle leaf deciduous trees; mixed broadleaf and evergreen trees; succulents and thorn trees and shrubs; shrubs dominant; grass with 10 to 40 per cent woody cover; grass with less than 10 per cent woody cover; moss/lichen with trees/shrubs; shrubs and bare soil; bare; cultivated crops and rice paddy.

While the process of grouping many different cover types into only 15 major classes may misrepresent the original data sets in some cases, there is clearly major disagreement among the three data sets. Figure 5.1 shows the land area derived from each data set for the 15 classes and the land area where the data sets agree that the same land cover type is present. Clearly there are major differences in terms of the area occupied globally by most cover types. Further comparison with several other global compilations is provided in Townshend *et al.* (1989), which demonstrates that the variations between the three data sets considered in this chapter are not atypical.

Note that even when the global area of a given land cover class are similar in all three data sets, the area of agreement can be much smaller because of the absence of correspondence in the geographical coverage of the class. For example, Figure 5.2 shows the land areas in the 'bare' class for the three data sets and Figure 5.3 shows the land area in the 'broadleaf evergreen trees' class. In the former the total areas delineated by Matthews (1982) and Olson *et al.* (1983) are similar for this category, but the correspondence between the two data sets is poor both in Asia and southern Africa. In Figure 5.3, broadleaf evergreen forest is depicted with similar distribution throughout the world, but in south east Asia and central South America there are substantial areas of disagreement. It is conceivable that the data sets would agree because the same source of information was used rather than because there is true independent agreement. Examination of the sources used – such as atlases and land resource maps – and the general lack of agreement among the data sets suggest that the data sets are not based predominantly on the same sources of information in most instances.

Figure 5.1 indicates the area where there is agreement between the three data sets by a solid bar for each class. Clearly there is less agreement than disagreement. Overall, the land area where the data sets agree, for the 15 classes, covers only 26 per cent of the total land surface area. While these areas of agreement can be used for training and testing sites, the use of satellite sensor data to verify the ground-based data sets might provide additional information.

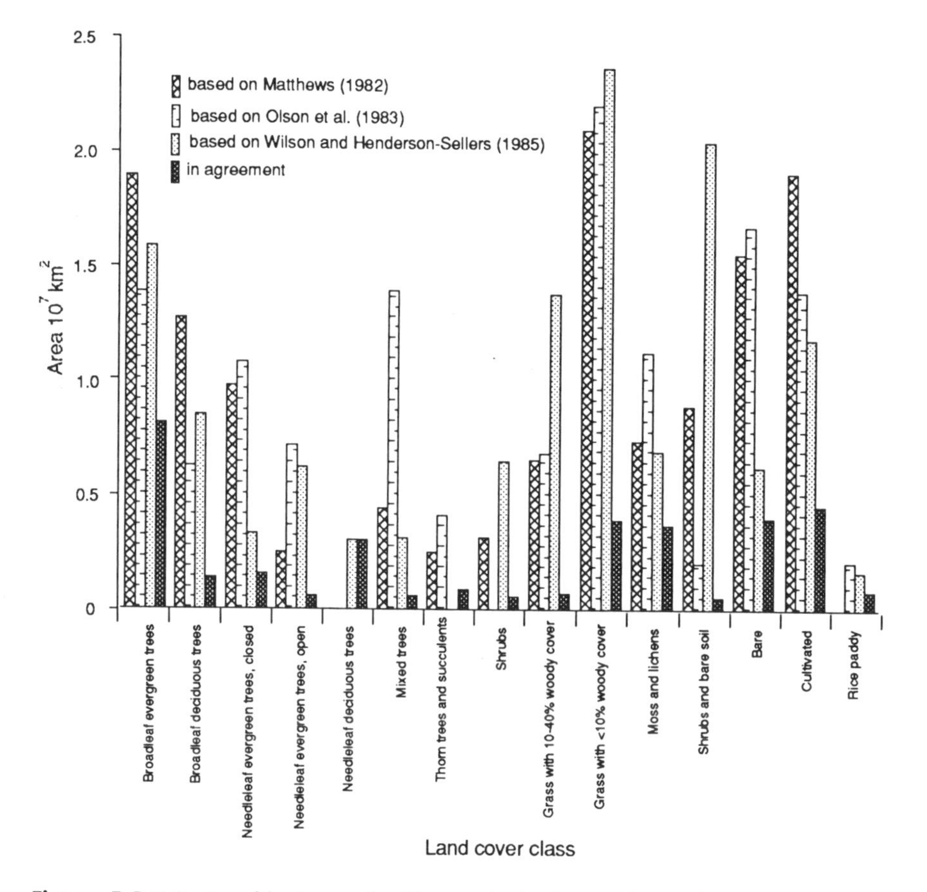

Figure 5.1 Estimates of land areas for fifteen major land cover classes derived from Matthews (1982), Olson *et al.* (1983) and Wilson and Henderson-Sellers (1985) according to groupings in Appendix A. Land area where the data sets agree that the same land cover class is present is shown by the dark bar. For some of the classes, only one or two of the data sets were used to indicate whether a certain class is present because the other data sets do not include comparable categories (Appendix A). These classes are: needle leaf deciduous trees, based on Wilson and Henderson-Sellers (1985) only; succulents and thorn trees, based on Matthews (1982) and Olson *et al.* (1983); shrubs, based on Matthews (1982) and Wilson and Henderson-Sellers (1985); and paddy rice, based on Olson *et al.* (1983) and Wilson and Henderson-Sellers (1985).

.... VERIFICATION OF GROUND-BASED SOURCES OF LAND COVER INFORMATION WITH AVHRR DATA

Given the difficulties in achieving consistent global descriptions of land cover, the differences between the three data sets and the lack of clear definitions of land cover types as reported in the previous section are not perhaps surprising. In terms of their use in global scale research, however, they are a matter of considerable concern.

To explore the internal consistency of these global land cover classifications

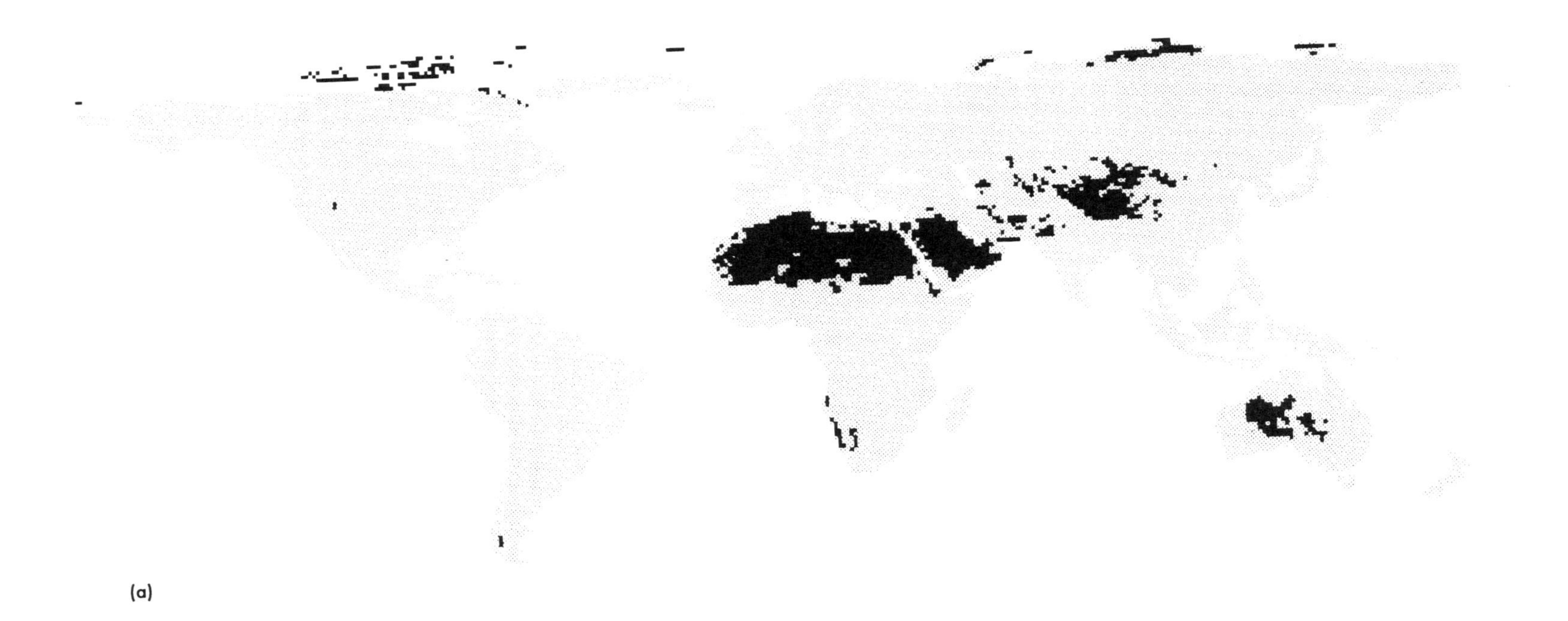

(a)

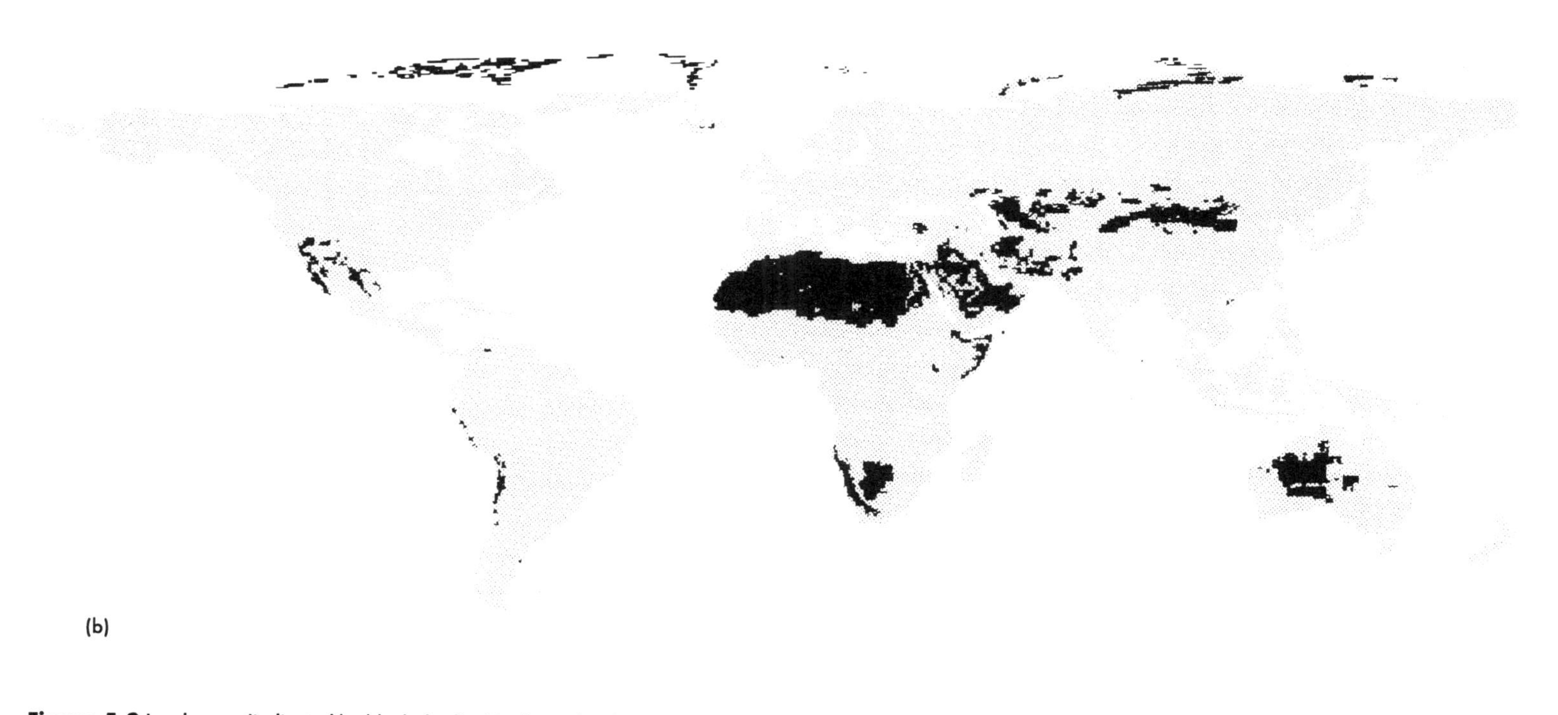

Figure 5.2 Land areas (indicated by black shading) in 'bare' land cover class derived from existing ground-based data bases of global land cover: (a) is derived from Matthews (1982), (b) from Olson *et al.* (1983), and (c) (*overleaf*) from Wilson and Henderson-Sellers (1985) according to groupings indicated in Appendix A.

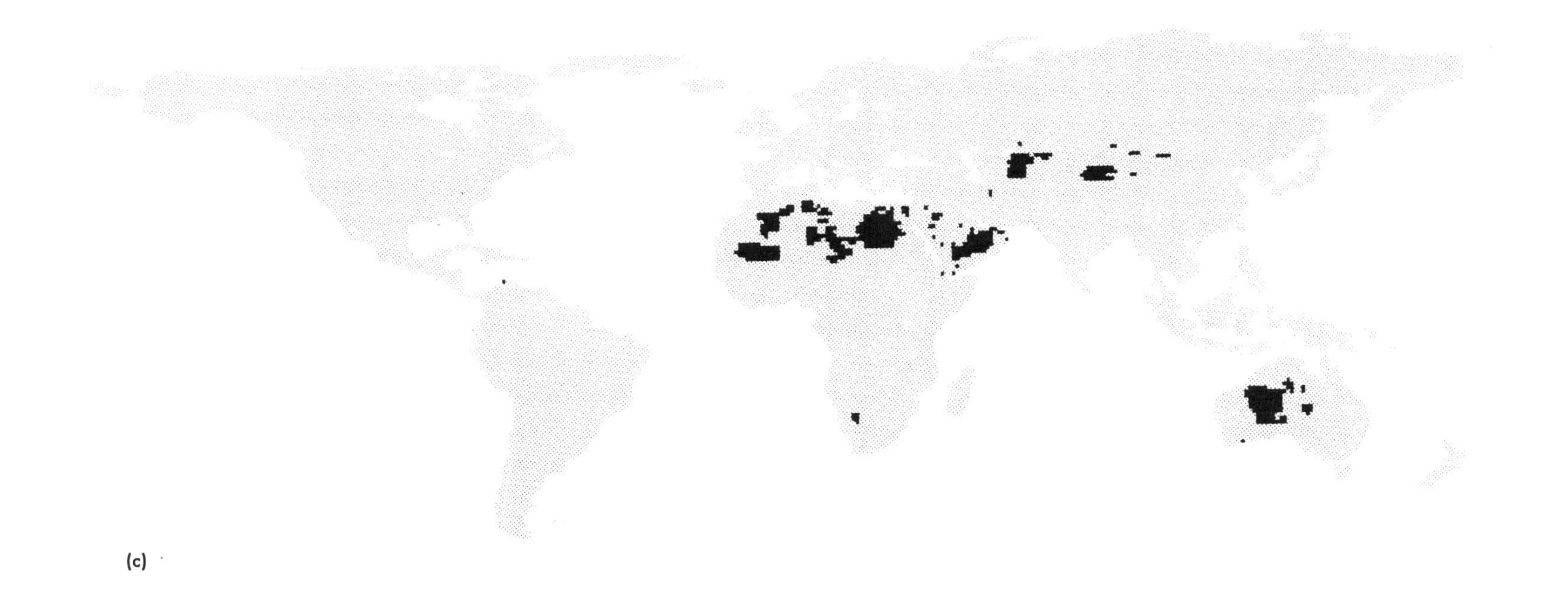

Figure 5.2 *Continued*

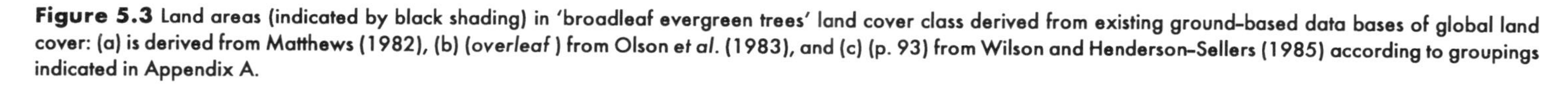
Figure 5.3 Land areas (indicated by black shading) in 'broadleaf evergreen trees' land cover class derived from existing ground-based data bases of global land cover: (a) is derived from Matthews (1982), (b) (*overleaf*) from Olson *et al.* (1983), and (c) (p. 93) from Wilson and Henderson-Sellers (1985) according to groupings indicated in Appendix A.

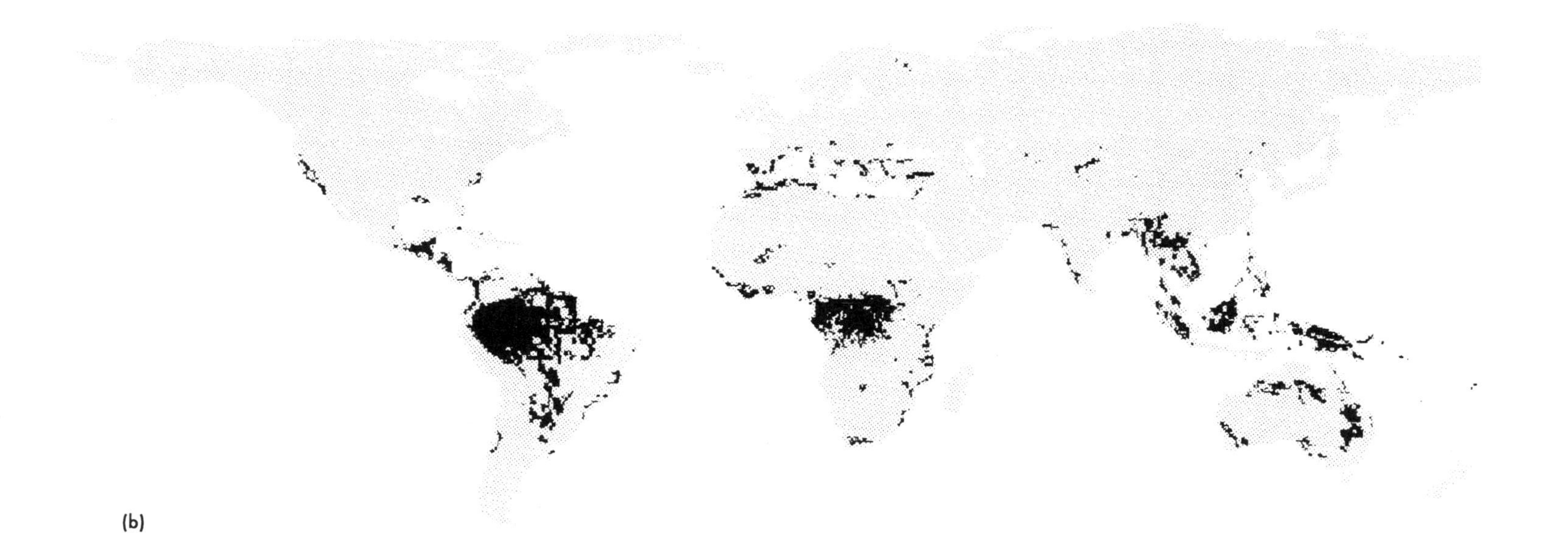

Figure 5.3 *Continued*

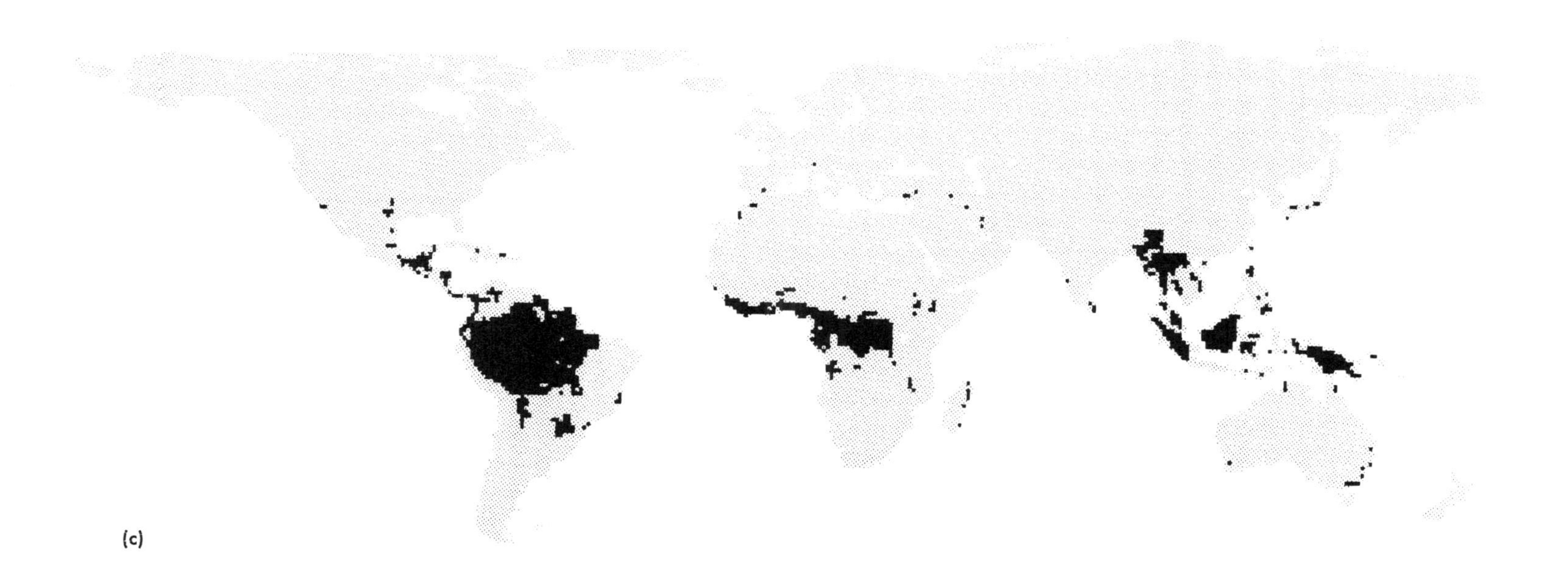

Figure 5.3 *Continued*

and the potential rôle of remotely sensed data, an automated classificatory procedure was carried out using AVHRR data. In outline, the procedure reclassifies each pixel to the class which it most resembles in terms of its remotely sensed properties. The remotely sensed properties are used to define a multi-dimensional space within which pixels of each cover type can be located. The mean vector and variance–covariance matrix for each cover type are then estimated using its worldwide population of pixels from the original data set. This was performed for all the cover types derived from each of the three ground-based data sets. Then, using the maximum likelihood rule (Swain and Davis, 1978), the multi-dimensional space was partitioned into sub-spaces, each uniquely associated with one land cover class. The whole of the global land mass was then reclassified according to the remotely sensed properties of each pixel. Thus, if a pixel of land cover class C_i, fell within the sub-space associated with that cover type, it retained the same label, but if it fell within the sub-space of another category, it was reclassified to another land cover class, C_j.

The underlying assumption of this approach was that training and testing sites for classification can be selected from locations where spectral information confirms that the land cover class indicated in the ground-based data sets was present. Because this approach treats each ground-based data set independently, it reduces errors introduced solely because different data sets were based on different, perhaps incomparable, definitions of cover types.

The satellite sensor data used for this analysis contained monthly averages over a one degree by one degree pixel, composited from maximum values for the month, of normalised difference vegetation index (NDVI) values derived from AVHRR data. AVHRR-NDVI, derived from a ratio of the remotely sensed response in the red and near infrared wave bands, may be linearly related to photosynthetic activity of vegetation. The data set, based on AVHRR Global Area Coverage (GAC) data compiled at a spatial resolution of approximately 8 km, was generated by NASA's Goddard Space Flight Center (Los *et al.*, 1993). A Fourier transform was applied to eliminate aberrant low values (Sellers *et al.*, 1993).

Because seasonal variations in photosynthetic activity depend on the type of vegetation present, land cover classes can be characterised by seasonal variations in AVHRR-NDVI. Land cover classifications have been successfully carried out using multi-temporal NDVI values derived from AVHRR data by Townshend *et al.* (1987), Tucker *et al.* (1985) and others, though these have been based mostly on NOAA's coarser 20 km spatial resolution Global Vegetation Index product, which is known to have significant limitations (Goward *et al.*, 1993).

Three maximum likelihood classifications based on six months of AVHRR-NDVI values (January, March, May, July, September, and November) were performed, with training sites for each class taken as all pixels in the respective ground-based data sets falling in that class. Statistics were generated for each land cover class, and the maximum likelihood classification indicated areas where the NDVI values for the six months were statistically similar to the training site. The classifications were run separately for each of the three data sets used in this analysis, with the land cover classes in Appendix A used as the training sites. Northern and southern hemispheres were run individually, and then combined, to account for differences in phase of the seasons in the two hemispheres.

By using the entire land cover class in each data set as the training site, it might be expected that the pixels would by definition be assigned to their original land cover class. This is not necessarily the case if some pixels of the training set are statistically dissimilar to the rest of the set. For example, Figure 5.4 shows the training site for the class 'needle leaf evergreen trees, open canopy' based on the Wilson and Henderson-Sellers (1985) data set, and the results of the classification for that class. While the major portion of the training site was indeed classified as 'needle leaf evergreen trees, open canopy', a significant portion was not statistically similar and was not classified as 'needle leaf evergreen trees, open canopy'. While the procedure is biased toward classifying the training site as the assigned land cover class, the results do indicate where the remotely sensed information confirms that the land cover class is or is not present. Of course, some of the lack of agreement between the remotely sensed information and the ground-based data sets may be due to deficiencies in the AVHRR data set.

Figure 5.5 shows results of the maximum likelihood classifications using the three data sets as training sites. For the data set derived from Matthews (1982), Olson *et al.* (1983) and Wilson and Henderson-Sellers (1985) there is agreement between the classification based on spectral information and the ground-based data set for approximately 42, 41, and 44 per cent of the land surface respectively. This indicates that the satellite sensor data fail to confirm the presence of the land cover class in question for some 60 per cent of the land surface.

To discern whether the classifications based on the AVHRR data and trained on the individual ground-based data sets could increase the accuracy of the land cover information, we compared the results of the three classifications to see if the areas classified as a given land cover class overlap. Figure 5.6(a) shows the areas of agreement when the comparison is based solely on the ground-based data sets, and Figure 5.6(b) shows the areas of agreement when the comparison is based on classifications using the ground-based data sets to identify training sites. Figure 5.7 shows these areas of agreement for the 15 land cover classes. Overall, there is agreement on approximately 58 per cent of the land surface when the comparison is based on the classification results, compared to approximately 35 per cent agreement when the analysis is based solely on the ground-based data sets without benefit of the spectral information. The latter percentage differs from the 26 per cent area of agreement indicated in the previous section because the former is based on comparison using the same groupings as the classification results. The classifications could not statistically distinguish between 'bare' and 'shrubs and bare' in the Matthews (1982) data set; 'bare' and 'succulents and thorns' in the Olson *et al.* (1983) data set; and between 'moss/lichens' and 'shrubs' in the Wilson and Henderson-Sellers (1985) data set. These classes were thus combined in the comparison of ground-based data sets to allow direct comparison with the classification results.

(a)

Figure 5.4 Comparison between (a) training set (indicated by black shading in Fig. 5.4(a)) and (b) classification result (indicated by black shading in Fig. 5.4(b)) for 'needle leaf evergreen trees, open canopy' derived from the Wilson and Henderson-Sellers (1985) data set. Note that all areas in the training set were not necessarily classified as 'needle leaf evergreen trees, open canopy'.

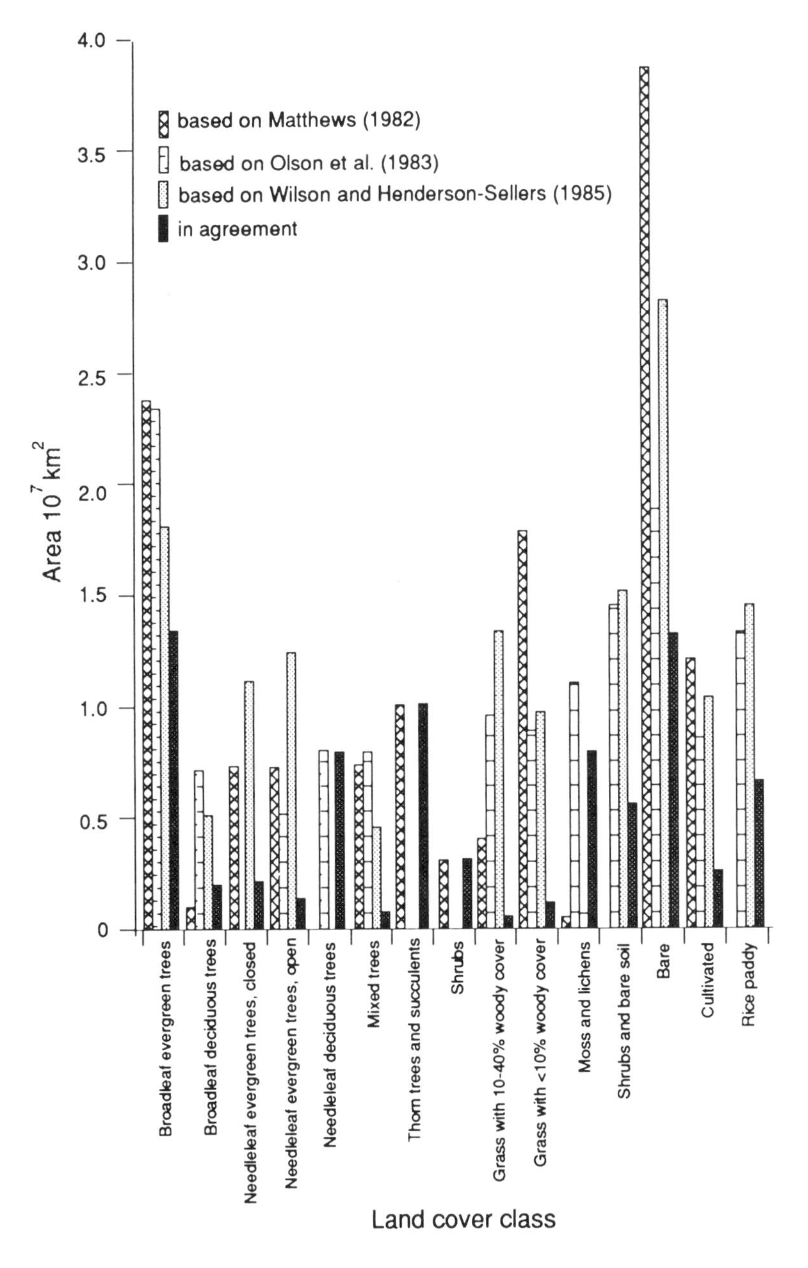

Figure 5.5 Land area for fifteen major land cover in classification results with training sites derived from Matthews (1982), Olson *et al.* (1983), and Wilson and Henderson-Sellers (1985). Land area where the classification results agree that the same land cover class is present is shown by the dark bar. Some of the classes were combined where it was not possible to statistically distinguish between land cover classes. 'Succulents and thorn trees' and 'bare' were combined and labelled as 'bare' in the result based on Olson *et al.* (1983); 'shrubs' and 'moss/lichens' were combined and labelled as 'moss/lichens' in the result based on Wilson and Henderson-Sellers (1985); and 'shrubs and bare' and 'bare' in the result based on Matthews (1982).

Figure 5.6 (a) Land areas (indicated by black shading in Fig. 5.6(a)) where existing ground-based data sets agree that the same land cover class is present, for fifteen major land cover classes derived from Matthews (1982), Olson *et al*. (1983), and Wilson and Henderson-Sellers (1985) as indicated in Appendix A. The classes were further grouped for the classes that could not be statistically distinguished (see caption to Fig. 5.5) to allow direct comparison with (b) (*overleaf*) land areas (indicated by black shading in Fig. 5.6(b)) where three classification results, each with training sites derived from the three data sets, agree that the same land cover is present.

Figure 5.6 *Continued*

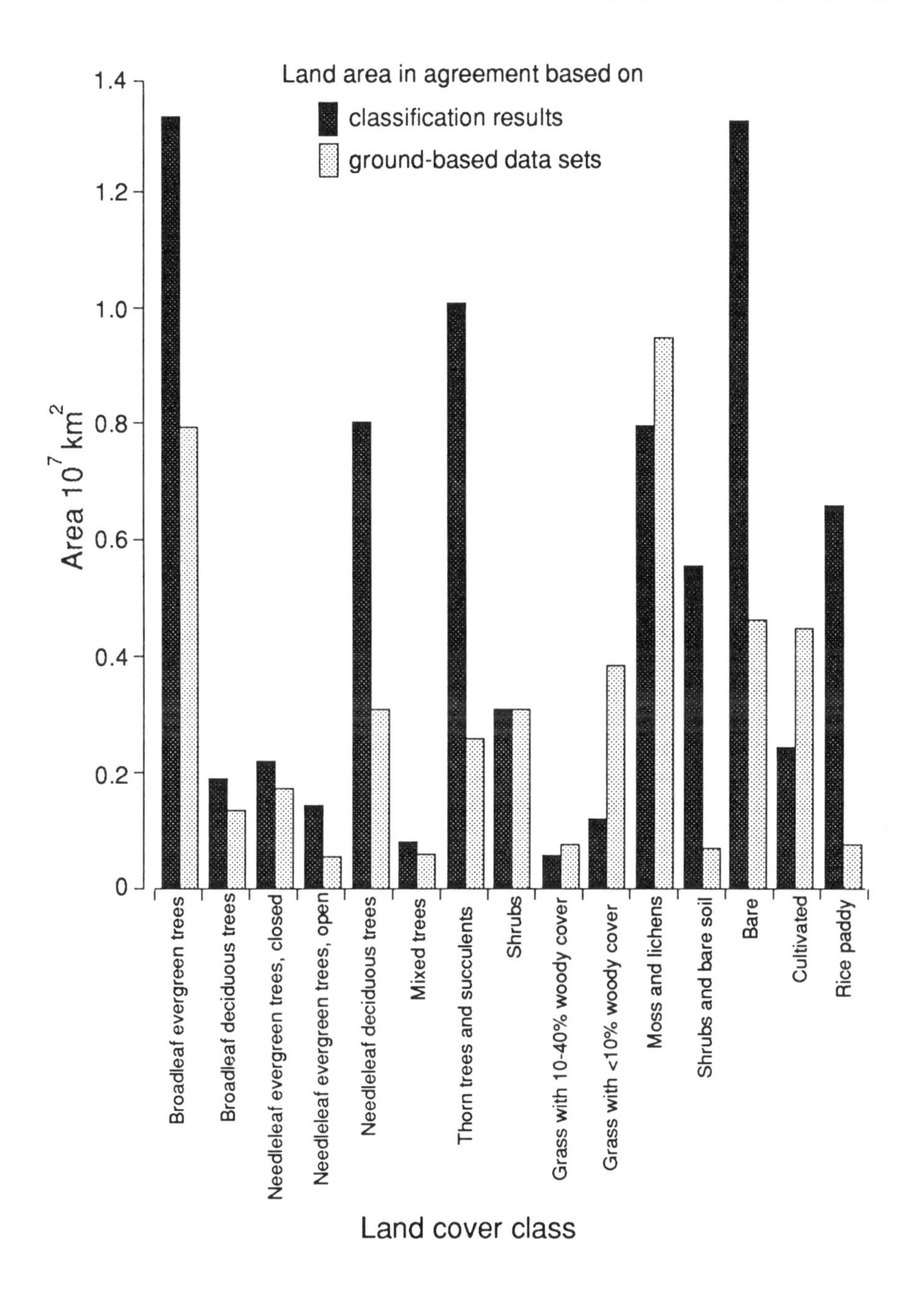

Figure 5.7 Land areas where existing ground-based data sets agree that the same land cover class is present for fifteen major land cover classes derived from Matthews (1982), Olson *et al.* (1983) and Wilson and Henderson-Sellers (1985), grouped as indicated in Appendix A and caption to Fig. 5.5, compared to land areas where three classification results, each with training sites derived from the three data sets, agree that the same land cover class is present.

.... DISCUSSION

The previous analysis shows that there is considerable uncertainty in characterising and quantifying global land cover, both when the information is derived from ground-based and from satellite sensor data. Some of the uncertainty results from the lack of a standard scheme for describing land cover that can be applied globally. But a major portion of the uncertainty results simply from lack of reliable, up-to-date sources of ground-based information. While ground-based sources might provide reliable information for distinct locales, the process of compiling information from local and regional to global scales strongly suggests that discrepancies will result. Using ground-based information as training and testing sites for global land cover characterisations based on satellite sensor data appears to be a more reasonable approach. Satellite sensor data have major advantages because the entire Earth is covered consistently and repeatedly over time.

The analysis in this chapter suggests that an AVHRR global data set can be used to verify ground-based sources of land cover data and increase the reliability of land cover information. However, even with the benefit of the AVHRR data set used in this analysis, quantified information on global land cover is not adequate for global change research.

Several methodological issues need to be tackled before satellite sensor data can be used to reliably quantify global land cover. First, methodologies to incorporate multi-year data sets would improve capabilities to characterise portions of the Earth such as semi-arid areas with high inter-annual variability in vegetative activity and to distinguish long-term changes in land cover from shorter-term fluctuations. Second, alternative approaches towards characterising land cover types need to be developed to capture gradients from one cover type to another and mixtures of different cover types. Approaches such as those suggested by Lloyd (1990), whereby land cover classifications are defined in terms of timing, duration, and intensity of photosynthetic activity, appear promising. Third, methodologies to incorporate portions of the spectrum in addition to the red and near infrared could be helpful to distinguish some cover types. For example, shortwave visible wave bands can distinguish between high latitude cover types (Petzold and Goward, 1988) and thermal infrared data can locate the boundary between tropical rain forest and savanna vegetation (Justice and Kendall, 1993). Finally, verification techniques based on independent ground or remotely sensed information will be crucial.

In addition to these methodological challenges, the ability to produce reliable land cover information from satellite sensor data will hinge on the quality of global data sets. The NOAA series of satellites provides potentially valuable data sets, though a substantial amount of work is needed to create internally consistent global data bases at even coarse spatial resolutions (Townshend, 1992; Goward *et al.*, 1993). Efforts are underway to assemble a global data set at 1 km spatial resolution from existing data sources (Townshend, 1992). Synthetic aperture radar (SAR) systems potentially provide data that could improve land cover characterisation (ESA, 1985; Krul, 1988). During the coming decade, new sensors such as the Moderate Resolution Imaging Spectrometer (MODIS) will provide valuable data sets for characterising global land cover.

.... SUMMARY AND CONCLUSIONS

The analysis presented in this chapter indicates that three published, digitised, ground-based data sets for global land cover agree that a common land cover type is present on only 26 per cent of the total land surface. Part of the discrepancy arises from varying definitions and classification schemes for land cover types used in the three data sets. However, the nature of compiling unverified and dated sources of local and regional land cover information to global scales suggests that discrepancies will inevitably result. The data sources compiled from ground-based sources, moreover, cannot be reproduced over time to monitor changes in global land cover.

To discern whether satellite sensor data can be used to verify whether the cover classes indicated in the ground-based data sources are present, maximum likelihood classifications were carried out using multi-temporal Normalised Difference Vegetation Index (NDVI) values from a monthly-averaged data set compiled from AVHRR data to a spatial resolution of one degree for a one year period. Three separate classifications were run, each using training sites for 15 land cover classes derived separately from the three ground-based data sets. Areas where the three classifications results overlap – suggesting that the remotely sensed data confirms that a common land cover type is present – cover approximately 58 per cent of the total land surface.

The remotely sensed spectral information increased the total area where all the three data sets agree from 35 to 58 per cent. This result indicates that such information can potentially improve our knowledge of global land cover. Several methodological issues need to be resolved before satellite sensor data can be fully utilised to provide reliable information on global land cover, including methods to incorporate multi-year data sets and data from other parts of the spectrum, alternative approaches to land cover classification schemes, and improved verification techniques.

.... ACKNOWLEDGEMENTS

The authors thank Chris Justice, Sietse Los and Piers Sellers for making the AVHRR data set available. Derek Pross and Vivre Koomanoff provided technical assistance. The work was carried out under the Global Mapping of Vegetative Land Cover project funded by the National Aeronautics and Space Administration (NASA).

.... REFERENCES

ESA, 1985, *Microwave remote sensing applied to vegetation*, European Space Agency SP-227, Paris

Goward, S., Dye, D., Turner, S., Yang, J., 1993, 'Objective assessment of the NOAA global vegetation index data product', *International Journal of Remote Sensing*, in press

Henderson-Sellers, A., Wilson, M.F., Thomas, G., Dickinson, R.E., 1986,

Current global land-surface data sets for use in climate-related studies, NCAR Technical Note 272: National Center for Atmospheric Research, Boulder, Colorado

Justice, C.O., Kendall, 1993, 'Assessment of the forests of Central Africa using coarse resolution satellite data', forthcoming

Koomanoff, V.A., 1989, *Analysis of global vegetation patterns: a comparison between remotely sensed data and a conventional map*, Biogeography Research Series, Report No. 890201, Department of Geography, University of Maryland, College Park

Krul, L., 1988, 'Some results of microwave remote sensing research in The Netherlands with a view to land applications in the 1990s', *International Journal of Remote Sensing*, **9**: 1553–64

Lloyd, D., 1990, 'A phenological classification of terrestrial vegetation cover using shortwave vegetation index imagery', *International Journal of Remote Sensing*, **11**: 2269–79

Los, S., Justice, C., Tucker, C.J., 1993, 'A global 1 × 1 degree NDVI data set for climate studies, part I', forthcoming

Matthews, E., 1982, 'Global vegetation and land use: new high resolution data bases for climate studies', *Journal of Climate and Applied Meteorology*, **22**: 474–87

Mueller-Dombois, D., 1984, 'Classification and mapping of plant communities: a review with emphasis on tropical vegetation', in Woodwell, G.M. (editor), *SCOPE 23: The role of terrestrial vegetation in the global carbon cycle: measurement by remote sensing*, Wiley, Chichester: 19–88

Olson, J.S., Watts, J.A., Allison, L.J., 1983, *Carbon in live vegetation of major world ecosystems*, Oak Ridge National Laboratory-5862, Oak Ridge, Tennessee, Oak Ridge National Laboratory

Petzold, D.E., Goward, S.N., 1988, 'Reflective spectra of sub-Arctic lichens', *Remote Sensing of Environment*, **24**: 481–92

Richards, J.F., 1990, 'Land transformation', in Turner II, B.L., Clark, W.C., Kates, R.W., Richards, J.F., Matthews, J.T., Meyer, W.B. (editors), *The Earth as transformed by human action: global and regional changes in the biosphere over the past 300 years*, Cambridge University Press and Clark University, Cambridge: 163–78

Sellers, P.J., Mintz, Y., Sud, Y.C., Dalcher, A., 1986, 'A simple biosphere model (SiB) for use within general circulation models', *Journal of Atmospheric Sciences*, **43**: 505–31

Sellers, P.J., Los, S., Justice, C., Tucker, C.J., 1993, 'A global 1 × 1 degree NDVI data set for climate studies, part II, corrections and products', forthcoming

Swain, P.H., Davis, S.M. (editors), 1978, *Remote sensing: the quantitative approach*, McGraw-Hill, New York

Townshend, J.R.G. (editor), 1992, *Improved global data for land applications*, International Geosphere-Biosphere Programme Report No. 20, Stockholm, IGBP/ICSU

Townshend, J.R.G., Justice, C.O., Kalb, V.T., 1987, 'Characterization and classification of South American land cover types using satellite data', *International Journal of Remote Sensing*, **8**: 1189–207

Townshend, J.R.G., Justice, C.O., Choudhury, B.J., Tucker, C.J., Kalb, V.T., Goff, T.E., 1989, 'A comparison of SMMR and AVHRR data for continental land cover characterization', *International Journal of Remote Sensing*, **10**: 1633–42

Townshend, J.R.G., Justice, C.O., Wei, L., Gurney, C., McManus, J., 1991, 'Global land cover classification by remote sensing: present capabilities and future possibilities', *Remote Sensing of Environment*, **35**: 243–55

Tucker, C.J., Townshend, J.G.R., Goff, T.E., 1985, 'African land cover classification using satellite data', *Science*, 227: 233–50
UNESCO, 1973, *International classification and mapping of vegetation*, United Nations, UNESCO, Paris
Watson, R.T., Rodhe, H., Oeschger, H., Siegenthaler, U., 1990, 'Greenhouse gases and aerosols', in Houghton, J.T., Jenkins, C.J., Ephraums, J.J. (editors), *Climate Change: The IPCC Assessment*, Cambridge University Press, Cambridge: 1–40
Wilson, M.F., Henderson-Sellers, A., 1985, 'A global archive of land cover and soils data for use in general circulation models', *Journal of Climatology*, 5: 119–43

.... APPENDIX A

Classes from existing global land cover data bases assigned to major land cover classes. In all three data sets, categories for ice were not included for this analysis. Categories 44 'coastal-major bog/mire, cool/cold climates' and 65 through 68 'coastal-shore and hinterland complexes' in the Olson *et al.* (1983) data set and category 2 'bog or marsh' in the Wilson and Henderson-Sellers (1985) data set were not included because this analysis did not contain a class for wetland vegetation. Category 64 'heath and moorland' in the Olson *et al.* (1983) data set was not included because of the extremely small and local areal extent. Category 80 'urban' in the Wilson and Henderson-Sellers (1985) data set was not included because it contained no entries.

Broadleaf evergreen trees

Matthews (1982)	Olson *et al.* (1983)	Wilson and Henderson-Sellers (1985)
1 tropical evergreen rainforest, mangrove forest	29,33,73 tropical/ subtropical broadleaved humid forest	5 mangrove tree swamp
2 tropical/ subtropical evergreen seasonal broadleaved forest	45,72 mangroves along some coasts	14 evergreen broadleaf woodland
3 subtropical evergreen rainforest	46 Mediterranean types	19 dense evergreen broadleaf forest
4 temperate/ subpolar evergreen rainforest	48 semiarid woodland or low forest	23 open tropical woodland
5 temperate evergreen seasonal broadleaved forest, summer rain		50 equatorial rainforest
6 evergreen broadleaved sclerophyllous forest, winter rain		52 tropical broadleaf forest
13 evergreen broadleaved sclerophyllous woodland		

Broadleaf deciduous trees

Matthews (1982)	Olson *et al.* (1983)	Wilson and Henderson-Sellers (1985)
9 tropical/subtropical drought-deciduous forest 11 cold-deciduous forest, without evergreens 15 tropical/ subtropical drought deciduous woodland 16 cold-deciduous woodland (includes needleleaf deciduous and mixed woodlands)	25,26 temperate broadleaf forest 32 tropical dry forest and woodland	20 dense deciduous broadleaf forest 21 open deciduous broadleaf woodland 24 woodland plus shrub 26 open drought-deciduous woodland

Needleleaf evergreen trees – closed canopy

Matthews (1982)	Olson *et al.* (1983)	Wilson and Henderson-Sellers (1985)
7 tropical/subtropical evergreen needleleaved forest 8 temperate/subpolar evergreen needleleaved forest	20,21 main taiga 22,27 other conifer 60,61 southern continental taiga	10 dense needleleaf evergreen forest

Needleleaf evergreen trees – open canopy

Matthews (1982)	Olson *et al.* (1983)	Wilson and Henderson-Sellers (1985)
14 evergreen needleleaved woodland	47 other dry or highland tree or shrub types 62 northern continental taiga	11 open needleleaf evergreen woodland

Needleleaf deciduous trees

Matthews (1982)	Olson *et al.*(1983)	Wilson and Henderson-Sellers (1985)
no separate category	no separate category	17 open deciduous needleleaf woodland 18 dense deciduous needleleaf forest

Mixed broadleaf deciduous and needleleaf evergreen trees

Matthews (1982)	Olson *et al.* (1983)	Wilson and Henderson-Sellers (1985)
10 cold deciduous forest, with evergreens	23,24 mixed: deciduous and evergreen broadleaf with conifer 28 tropical mountain: forest, grass, scrub, paramo, rock 55,58 trop/temp woods, fields, grass, scrub 56,57 2nd growth trop/subtrop, humid/temp/boreal forest	12 dense mixed forest 13 open mixed woodland

Succulents and thorn trees and shrubs

Matthews (1982)	Olson *et al.* (1983)	Wilson and Henderson-Sellers (1985)
12 xeromorphic forest/ woodland	59 succulent and thorn woods and scrub	no separate category

Shrubs dominant

Matthews (1982)	Olson *et al.* (1983)	Wilson and Henderson-Sellers (1985)
17 evergreen broadleaved shrubland/ thicket, evergreen dwarf shrubland 18 evergreen needleleaved or microphyllour shrubland/thicket 19 drought deciduous shrubland/ thicket 20 cold deciduous subalpine/ subpolar shrubland, cold deciduous dwarf shrubland	no separate category, combined with grassland category, only category is 64 heath and moorland	16 evergreen broadleaf shrub 27 deciduous shrub 28 thorn shrub 62 dwarf shrub

Grass with 10 to 40 per cent woody cover

Matthews (1982)	Olson *et al.* (1983)	Wilson and Henderson-Sellers (1985)
23 tall/medium/short grassland with 10 to 40% woody tree cover	43 tropical savanna and woodland	35 pasture plus tree 37 tropical savanna

Grass dominant with less than 10 per cent woody cover

Matthews (1982)	Olson *et al.* (1983)	Wilson and Henderson-Sellers (1985)
24 tall/medium/short grassland with less than 10 per cent woody tree cover or tuft plant cover 25 tall/medium/short grassland with shrub cover 26 tall grassland, no woody cover 27 medium grassland, no woody cover 29 forb formations	40 cool grassland, scrub 41 warm or hot scrub and grassland 42 siberian parklands, Tibetan meadows	30 temperate meadow and permanent pasture 31 temperate rough grazing 32 tropical grassland plus shrub 33 tropical pasture 34 rough grazing plus shrub 39 pasture plus shrub

Moss/lichens with tree/shrubs

Matthews (1982)	Olson *et al.* (1983)	Wilson and Henderson-Sellers (1985)
22 arctic/alpine tundra, mossy bog	53,54 tundra 63 wooded tundra	61 tundra

Shrubs and bare soil

Matthews (1982)	Olson *et al.* (1983)	Wilson and Henderson-Sellers (1985)
21 xeromorphic shrubland/ dwarf shrubland	52 semidesert scrub, cool	36 semi-arid rough grazing 71 scrub desert and semidesert 73 semidesert and scattered trees

Bare

Matthews (1982)	Olson *et al.* (1983)	Wilson and Henderson-Sellers (1985)
30 desert	49 sparse vegetation, rocky 50 sand desert 51,71 desert or semidesert 69 polar or rock desert	70 sand desert and barren land

Cultivated crops (except paddy rice)

Matthews (1982)	Olson *et al.* (1983)	Wilson and Henderson-Sellers (1985)
32 cultivation (categories 3,4,5 of cultivation intensity)	30 cool/cold farms/towns 31 warm/hot farms/towns 37,38,39 other irrigated dryland	15 evergreen broadleaf cropland 22 deciduous tree crop 40 arable cropland 41 dry farm arable 42 nursery and market gardening 43 cane sugar 44 maize 45 cotton 46 coffee 47 vineyard 48 irrigated cropland 49 tea 51 equatorial tree crops

Paddy rice

Matthews (1982)	Olson *et al.* (1983)	Wilson and Henderson-Sellers (1985)
no separate category	36 irrigated paddyland	4 paddy rice

6. Snow Monitoring in the United Kingdom Using NOAA AVHRR Imagery

Richard M. Lucas and Andrew R. Harrison

.... INTRODUCTION

Data on snow cover in the United Kingdom (UK) is currently collected by the National Snow Survey which relies on a voluntary network of approximately 160 stations across the country. Observers at each station record, on a daily basis, snow or sleet falling at the station at any time during the day together with the depth of old and fresh snow. The snowline altitude (to the nearest 150 m) on the surrounding hills is also approximated and rain-water equivalent measurements are taken. Records are forwarded to the Meteorological Office at the end of each month (for October to May) and are published annually in the *Snow Survey of Great Britain*.

There are, however, a number of limitations to the National Snow Survey. In particular, the spatial variation in snow area and volume cannot be assessed adequately as only point measurements of snow depth or water equivalent are recorded. Estimates of the snowline altitude are also relatively crude and observations are not immediately available to users. With an increased awareness of the contribution of snow area and snowmelt to major flooding, more timely and useful information is now required by organisations such as the National Rivers Authority in order to predict such events. Hydroelectric power companies, ski resorts and local councils undertaking snow clearance operations also require up-to-date and accurate snow cover data.

Satellite remote sensing of snow cover shows considerable potential for improving or even replacing the present method of the National Snow Survey. The major advantage of data derived from satellite sensor data is that near real time observations of snow distribution are available on a regular basis, particularly when using the NOAA AVHRR and sensors onboard geostationary satellites (e.g., GOES, METEOSAT). Furthermore, actual estimates of the areal extent of snow cover over large areas are feasible and snow data can be obtained from remote regions where ground observations are sparse or non-existent. Even so, reliable measurements of snow depth or water equivalent cannot be derived directly from optical satellite sensor imagery, although passive microwave sensors may allow estimation of these variables (Chang, 1986; Chang *et al.*, 1987). A further criticism of satellite sensor data is the difficulty of accurate geographical location within the imagery, particularly with coarse spatial resolution data. In addition, satellite sensor observation of snow cover in the UK is often inhibited by excessive cloud cover and the snowpack may be obscured by dense forest.

This chapter describes a procedure for the routine production of snow area estimates within the UK using NOAA AVHRR imagery. The research focused on the development of a multispectral rather than a single channel approach to the separation of snow and cloud and the delineation of the snow margin, and furthermore outlined techniques for the routine description of snow condition and surface characteristics.

.... GEOMETRIC PREPROCESSING OF AVHRR IMAGERY

A number of essential prerequisites to the routine production of snow cover maps were identified. These included the correction of panoramic distortion within AVHRR imagery, the development of techniques for registration and subsequent comparison of imagery between dates, and the integration of ancillary geographical information with the satellite sensor data.

CORRECTION OF PANORAMIC DISTORTION

The data employed for the snow mapping was the AVHRR High Resolution Picture Transmission (HRPT) obtained from the Dundee University receiving station for satellite sensor data. Considerable distortion in the across-track dimension was characteristic of the AVHRR imagery and removal of the panoramic distortion was achieved by linearising the AVHRR data. A linearisation algorithm was employed which took the maximum scan angle from nadir ($\pm$ 56°), calculated the maximum geocentric angle (Gm) from nadir, and divided the scan line, which ranged from –Gm to +Gm, into 6144 units (allowing each pixel at the centre of the scan to be incorporated in subsequent sampling). The scan angle was then calculated and the number of pixels assigned to each digital value within the scan line was determined by effectively resampling the data (to 2048 units) so that pixels were duplicated more frequently near the edge of the scan than in the centre. Every other pixel was then extracted from the linearised scan line resulting in an image that was virtually linear in both across track and along track dimensions with a nominal pixel size of 0.97 $\times$ 1.1 km (Brush, 1988). This linearisation procedure was performed on the data in all AVHRR channels.

NAVIGATION OF IMAGERY

Navigation, as applied to satellite sensor images, is the geometric rectification of digital image data to fit a selected map projection, thereby assigning a geographic location (e.g., a latitude/longitude coordinate) to each pixel across the scan line (Emery and Ikeda, 1984). The navigation procedure developed by Brush (1985; 1988) was employed in this study. Navigation was accomplished by merging a computer produced latitude/longitude grid and land outlines with the linearised scan lines of data. The location of the scan line centre, in terms of latitude and longitude coordinates, was determined both from information on the equatorial crossing time and location (°West) and the geocentric angle traversed by the satellite from the equatorial crossing point. Latitude and longitude coordinates were then identified for pixels in the across track dimension using the known altitude of the satellite's orbit and the Earth's radius.

BETWEEN-DATE REGISTRATION

Following navigation, coastal outlines were overlain onto each linearised AVHRR image. Registration of two scenes from different dates was achieved by selecting corresponding control points from the coastline overlays. A nearest neighbour interpolation then facilitated the co-registration of AVHRR data regardless of cloud cover within the scenes. A major criterion, however, was that at least part of the UK coast must be visible within the imagery to allow location of the coastal overlay. Where cloud cover completely obscured the terrain from the view of the satellite sensor, snow mapping was unlikely to be successful and the affected images were then omitted from the analysis. The use of coastal overlays allowed a full set of images for selected periods to be registered, even in conditions of excessive cloud cover.

INTEGRATION OF ANCILLARY DATA

The development of the linearisation and navigation algorithms for the registration of multi-temporal AVHRR images was vital for the construction of composite snow cover maps. In addition, these procedures allowed geographically referenced ancillary information to be integrated into the multitemporal image data set. Four main data sets were used: forest cover maps, elevation data, snow survey station locations and UK Water Authority Divisions. This information was essential for the assessment of the snow mapping results and the later description of snow cover distribution.

.... RADIOMETRIC PREPROCESSING OF AVHRR IMAGERY

Radiometric correction of the AVHRR imagery was undertaken primarily to increase image handling capabilities and also to optimise the description of the snowpack using single date imagery or through multi-temporal comparison.

DATA CALIBRATION

AVHRR imagery were calibrated to percentage albedo (channels 1 and 2) and temperature, °C, (channels 3, 4 and 5). The scaling of the images was adjusted so that the full range of albedo and temperature values for snow could be examined. Physical values were also produced for each AVHRR channel leading to a better understanding of the melt/accumulation dynamics of the snowpack. Of particular significance was the derivation of surface temperatures allowing a description of the metamorphic state of the snow. Likewise, the percentage albedo values for channels 1 and 2 were directly comparable, permitting an improved definition of snow surface characteristics using combinations of these channels.

ATMOSPHERIC CORRECTION OF THERMAL INFRARED DATA

An atmospheric correction procedure (McClain *et al.*, 1985) requiring no restrictions on solar angle, even in winter imagery and based on calibrated AVHRR data, was applied. A sixth atmospherically-corrected channel was produced from a combination of temperature calibrated AVHRR channel 4 and 5 imagery. Atmospheric absorption in these two channels is primarily due to water vapour and the temperature deficit in one AVHRR channel, resulting from atmospheric attenuation through absorption by water vapour, is a linear function of the brightness temperature difference of these two window channels such that:

$$T6 = 1.0346 \times T4 + 2.5779\ (T4 - T5) - 283.21 \qquad [1]$$

where T6 is the atmospherically corrected pixel digital number and T4 and T5 are calibrated AVHRR channel 4 and 5 digital numbers. Atmospheric correction of thermal infrared data was important as it enabled the accurate measurement of snowpack temperature and consistent monitoring of snow surface temperature between image dates.

Although the correction procedure was developed initially to estimate sea surface temperature, the accuracy of snow surface measurement was assumed to be similar, not least because both are water surfaces. The stated accuracy of the derived thermal measurements is +/– 1°C which, to some extent, limits the ability to locate the transition zone between melting and non-melting snow surfaces within AVHRR imagery.

.... DETECTION OF SNOW AREA USING AVHRR DATA

The geometric and radiometric correction procedures described provided adequate preparation of AVHRR imagery for operational snow cover monitoring on a daily or weekly basis. Reliable and reproducible techniques for the discrimination of snow and cloud, detection of snow obscured by dense vegetation canopies (e.g., conifer forest) and definition of the snow margin were then developed.

THE IMPACT OF CLOUD COVER ON SNOW OBSERVATIONS FROM SATELLITE SENSORS

The discrimination between snow and cloud using AVHRR imagery has proved a major barrier to the development of operational snow mapping procedures. The major difficulty with extensive cloud occurrence is that the underlying snow surface is often completely obscured from the view of the satellite sensor. With persistent cloud cover, observation of snow may be prevented for considerable periods of time during which significant changes in snow extent and condition (due to melting and/or accumulation) may take place. Direct observation of snow from the AVHRR is therefore only possible in cloud-free conditions or where snow can be observed beneath the cloud layer (e.g., through thin stratus). Extraction of the snow area then becomes the main obstacle to snow mapping as the overlap in the response of snow and cloud in the spectral wavebands sensed prevents reliable separation of snow covered ground resolution elements.

GENERAL CHARACTERISTICS OF SNOW, CLOUD AND LAND SURFACES

In the visible and near infrared wavebands, snow cover and cloud have a similar spectral response and may be confused, although the contrast between snow covered and snow free surfaces is greatest in these channels, particularly in the visible wavebands. AVHRR channel 3 (3.55μm–3.93μm) is sensitive to both reflected solar radiation and emitted terrestrial radiation (Knottenberg and Raschke, 1982; Holroyd *et al.*, 1989). In channel 3 imagery, snow has near zero reflectance (Liljas, 1987), and is dark on calibrated imagery. The response of snow shows some similarity, but also a certain degree of separation, with clouds composed of larger ice particles (e.g., cumulonimbus) and large liquid water droplets (e.g., cumulus) which exhibit both high spectral absorption and low thermal emission and, like snow, are dark in appearance (Scorer, 1987). Clouds exhibit a bright appearance in channel 3 where the particles of which they are composed are of a diameter less than the wavelength. Snow therefore contrasts dramatically with clouds composed of very small ice particles (e.g., cirrus) and small water droplets (e.g., altostratus) which have a greater response in channel 3 (Scorer, 1987; Kidder and Wu, 1984; Ebert, 1987).

In AVHRR channels 4 and 5, higher elevation clouds have a dark appearance in calibrated imagery, due to low thermal emission, and contrast with the brighter snow surfaces while low altitude, warmer clouds exhibit some spectral overlap with snow surfaces. Channels 4 and 5 are useful for snow mapping as shadows are minimal (Scorer, 1987) and snow obscured by forest canopies can also be detected.

PROCEDURE FOR THE DISCRIMINATION OF SNOW COVERED SURFACES

Various methods were tested for the discrimination of snow and cloud, including interactive thresholding and supervised classifications. However, the

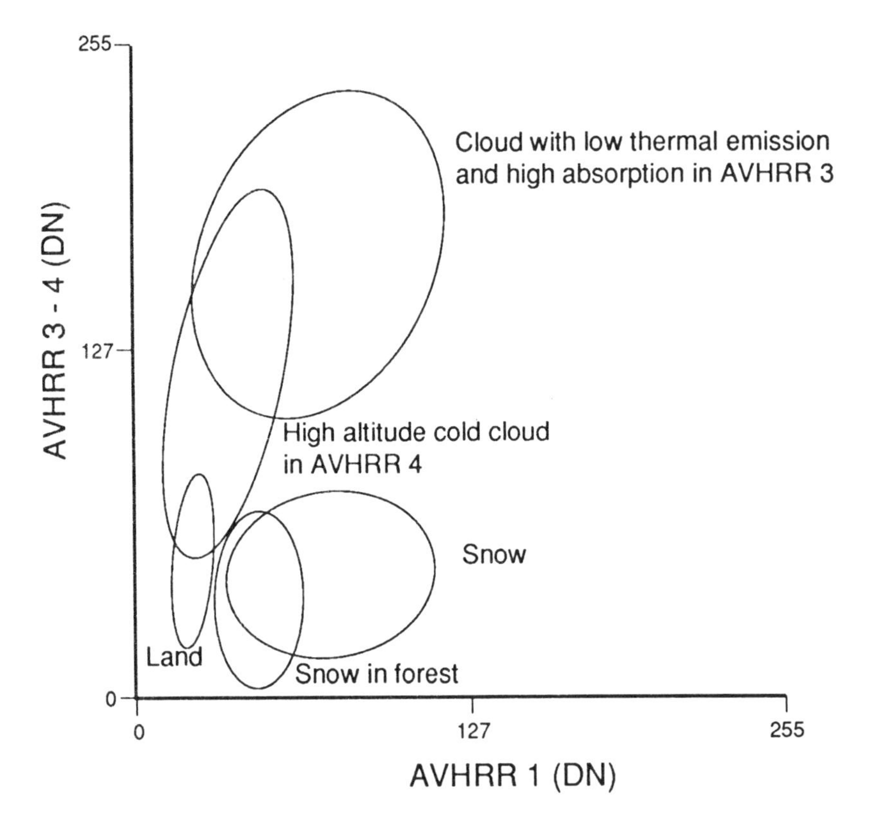

Figure 6.1 Modelled ellipses of class distribution within the feature space of AVHRR 3–AVHRR 4 and AVHRR 1 for northern Britain (10 February, 1986).

most successful approach involved an unsupervised cluster analysis of AVHRR channels 1, 3 and 4.

The potential of the unsupervised procedure for snow/cloud discrimination is illustrated in Plate 3 which shows a clear definition of snow area observed in conditions of radiation scattered though extensive cloud cover and also demonstrates the potential for separating snow covered from snow free surfaces. A 'snow in forest' class was also defined.

The significance of combining AVHRR channels 1, 3 and 4 in an unsupervised classification is illustrated in Figure 6.1. This feature space plot confirms both the separation of snow from cloud surfaces and snow in forest from other non-snow covered vegetated surfaces. The detection of snow in forest was allowed primarily through increases in spectral reflectance in AVHRR channels 1 and 2 caused by snow accumulating on the canopy or within forest rides or cleared areas. Thermal emissions observed in channels 3 and 4 were not affected by tree cover to the same extent as spectral reflectances and the thermal infrared channels therefore played an important rôle in defining the actual area of snow lie in forested areas. Channels 3 and 4 were also primarily the snow/cloud discrimination channels and, as the contrast between snow covered and snow free surfaces was greatest in the visible wavebands, AVHRR channel 1 was most suited to delineating the snow boundary.

The main advantages of the multispectral unsupervised classification over other techniques (e.g., brightness intensity thresholding) was that automatic delineation of the snowpack boundary was achieved, a 'snow in forest' class was often obtained and the technique was independent of normalisation or atmospheric correction procedures (i.e., the technique relied entirely on information contained within the imagery).

A criterion for obtaining optimal snow maps was that the redistribution of AVHRR digital values within the natural clusters, generated during the unsupervised clustering procedure, was minimised such that the location of classes was unlikely to change with further iterative classification. Where the clustering procedure was prematurely terminated, considerable inaccuracies in snow area definition were evident. Stabilisation of the clusters was, therefore, essential and was assessed by interactively monitoring both the variation in snow area and the reduction in mean vector migration between iterations. Where no further change was observed, the classification result was accepted.

To increase the performance of the snow mapping algorithm, a semi-supervised approach was adopted whereby training statistics, derived from a previous unsupervised classification of AVHRR data, were introduced into the classification. A decrease in time for cluster stabilisation was often observed as starting class seed locations were positioned close to the means of the naturally occurring data clusters. However, this technique demands the application of consistent scalings and offsets for image display and an informed selection of training data. A further advantage of the semi-supervised approach was that often classes were similarly labelled to those providing the training statistics in subsequent classifications of AVHRR data. The semi-supervised approach therefore shows considerable potential for the automation of the classification procedure.

.... DESCRIPTION OF THE SNOWPACK

With the confident detection of image pixels for which the ground resolution element was snow covered within NOAA AVHRR data, a number of procedures were developed to describe the surface conditions of the snowpack. Each technique relies upon physical measurements, such as the AVHRR-Normalised Difference Vegetation Index (NDVI) and surface temperature, rather than inconsistent and less reproducible approaches such as classification.

APPLICATION OF THE NDVI

To study the effect of snow cover on vegetation response, the NDVI was calculated using calibrated AVHRR channels 1 and 2 (equation 1):

$$\text{NDVI} = (\text{Channel } 2 - \text{Channel } 1) / (\text{Channel } 1 + \text{Channel } 2) \qquad [2]$$

The AVHRR-NDVI is related, under certain conditions, to the photosynthetically active radiation absorbed by vegetation canopies and is positive for vegetation. However, where a snow cover exists the amount of vegetation sensed will be reduced, resulting in zero or negative AVHRR-NDVI values being produced. This was confirmed by comparing classified Landsat Thematic Mapper (TM) with NOAA AVHRR-NDVI data (27 February, 1987) of the Brecon Beacons National Park, South Wales. In this analysis, an unsupervised classification of Landsat TM bands 2, 4 and 5 enabled confident discrimination of snow free surfaces, dominantly snow free ground with patches of snow, dominantly snow covered ground with vegetation emerging through the snowpack, completely snow covered ground, conifer forest and lakes. The categorisation of the amount of snow mixed with vegetation was assessed by visual analysis of the proportion of 'red' vegetation and 'white' snow within the false colour composite.

The class map produced using Landsat TM data was co-registered to same-date NOAA AVHRR imagery and the percentage of different TM-derived cover types within each individual AVHRR pixel was computed. This cover type data was then compared with the AVHRR-NDVI. The results (Figure 6.2) indicated that as snow gradually concealed the vegetation cover, AVHRR-NDVI values declined and eventually became negative when snow blanketed the area.

The changeover from positive to negative AVHRR-NDVI values was therefore used to infer near-complete masking of vegetation by snow and vice versa. However, where snow cover completely concealed the vegetation, variations in the AVHRR-NDVI still occurred but were considered to be mainly a function of the condition of the snowpack (O'Brien and Munis, 1975a; 1975b). The AVHRR-NDVI was, however, affected considerably by deep shadows and waterbodies which cause the near infrared response to be reduced dramatically relative to the visible response.

The relationship of cover type and shadowing to the near infrared and visible response has major implications for snow mapping as, within a number of AVHRR images, the forested areas with snow cover were spectrally similar to land and cloud surfaces in shadow. Both categories were often represented as a 'snow in forest/shadowed surface' class. By isolating this class and assigning all pixels with positive AVHRR-NDVI to areas of 'snow in forest' and negative AVHRR-NDVI to 'shadowed surfaces', the algorithm could more accurately detect snow obscured by dense canopies and separate snow-covered surfaces from cloud. A similar separation of snow and cloud classes was obtained where the identified snow area within a particular class, generated in the clustering procedure, showed exclusively positive AVHRR-NDVI values. Further increases in the accuracy of snow/cloud discrimination were achieved by applying a majority filter, whereby isolated pixels (e.g., cloud pixels completely surrounded by a snow class) were reclassified to the class within which they were contained.

MELT/ACCUMULATION ESTIMATION USING AVHRR INFORMATION

Zones of snow melt or accumulation were defined within daily AVHRR imagery by examining the calibrated data values or were located through

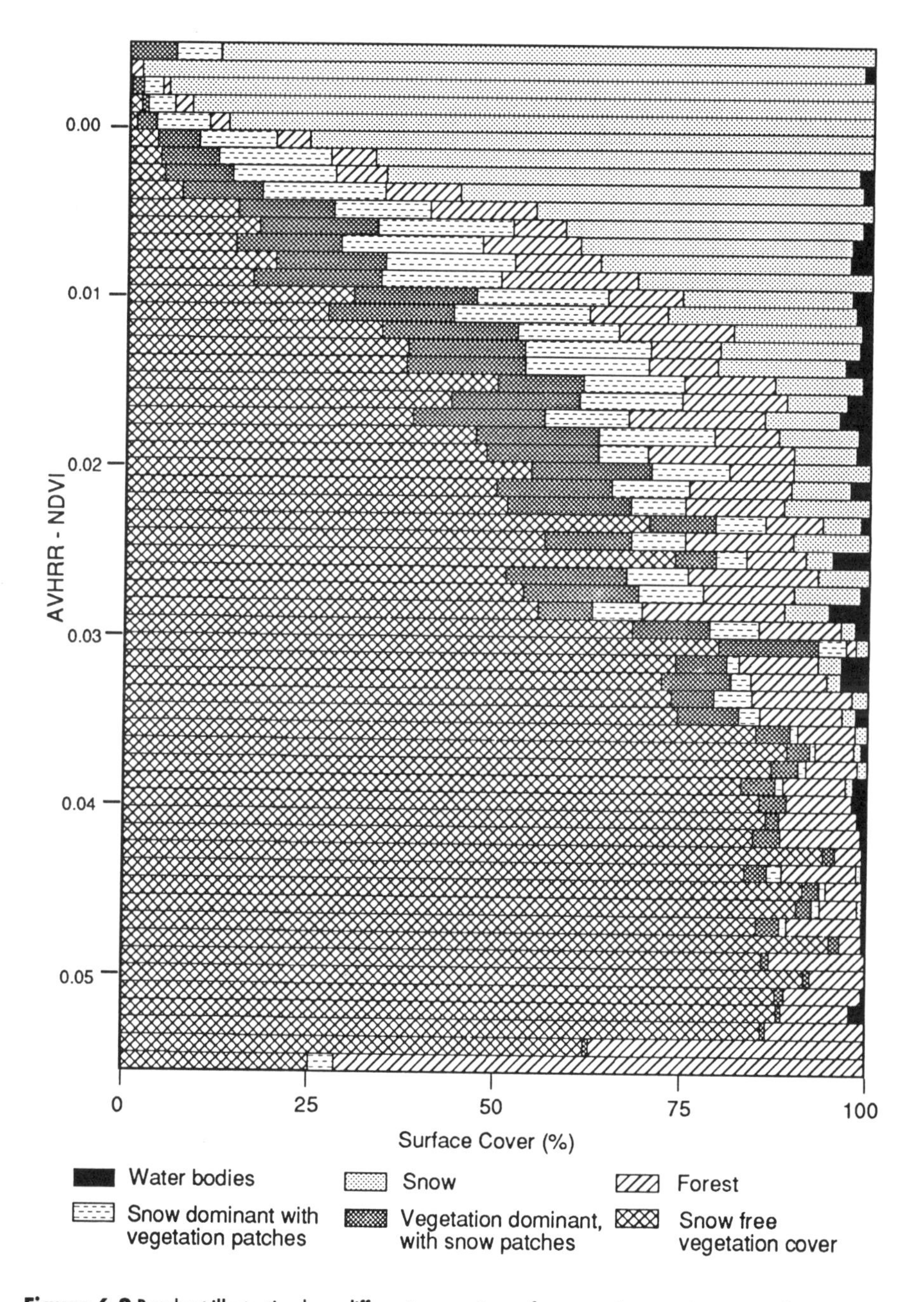

Figure 6.2 Barchart illustrating how different proportions of snow and vegetation cover observed by the AVHRR sensor in the Brecon Beacons, South Wales (27 February, 1987) influence NDVI values.

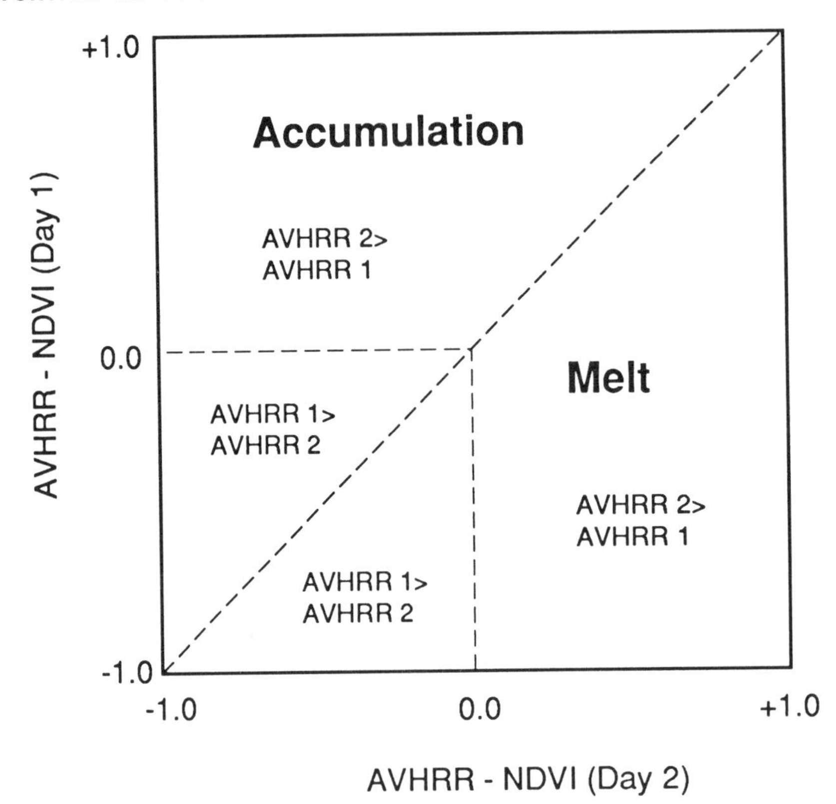

Figure 6.3 Generalised diagram illustrating the technique for detecting melt or accumulation zones based on differences in AVHRR-NDVI between dates.

multi-temporal comparison of calibrated AVHRR data. On the basis of measurements from atmospherically corrected AVHRR thermal infrared imagery, non-melting or accumulating snowpacks were assumed to exhibit temperatures below 0°C while areas of partial snow melt were likely to correspond with the 0–1°C isotherm. Surfaces warmer than 1°C were considered to represent either snowpacks undergoing extensive melt or snow-free areas.

Numerical multi-temporal comparisons of single channel spectral data were generally unreliable due to between-date variations arising from differences in scan angle, path length, atmospheric scattering and absorption and illumination conditions. Normalisation of calibrated spectral data, to a certain extent, caters for within and between-date variations in viewing and illumination geometry (Gutman, 1987). Consequently, the AVHRR-NDVI was employed to study snowpack dynamics as variations in the AVHRR-NDVI between consecutive dates was assumed to be related directly to changes in snow depth and condition due either to melt or to the general ageing of the snowpack (Figure 6.3).

By employing multi-temporal AVHRR imagery a melting snowpack was indicated by:

(i) an increase in the percentage of the snowpack above 0°C or a general rise in temperature;
(ii) an increase in AVHRR-NDVI values on the second day, suggesting a reduction in snow depth or optical thickness;
(iii) a reduction in the extent of the total observed snow area; and
(iv) a decrease in the proportion of complete snow cover (i.e., AVHRR-NDVI < = 1.0) relative to partial cover within the snow field.

Snow accumulation was indicated by inverse relationships to those described above. These relationships were, therefore, used to describe the daily and weekly patterns of snow melt and accumulation across the UK.

.... CLASSIFICATION ACCURACY

Assessing the accuracy of snow cover maps generated from AVHRR data was inhibited by the lack of comparative ground measurements. Two methods were therefore developed to estimate errors of snow detection.

SNOW CLOUD DISCRIMINATION

Absolute error ratings for the discrimination of snow and cloud were estimated by calculating the proportion of cloud misclassified as snow over sea areas. The percentage confusion between snow and cloud classes was generally less than 2 per cent (Table 6.1). An increase in classification accuracy was observed when using the AVHRR-NDVI to separate cloud shadows from snow obscured by forest and by applying a majority filter.

SNOW BOUNDARY LOCATION

The accuracy of snow boundary delineation was tested by comparing snow extent estimated from Landsat TM imagery of both South Wales (27 February, 1987) and the Southern Uplands (24 January, 1986) with the snow area classifications obtained using same-date NOAA AVHRR imagery. Snow cover was located within Landsat TM band 2 data by applying a brightness intensity threshold, while the semi-supervised classification procedure was used to define snow distribution in the AVHRR scene. The AVHRR image of the Southern Uplands was partially cloud covered and comparison of snow extent was based upon the cloud free portions of both the Landsat TM and NOAA AVHRR images.

The accuracy of the AVHRR-based classification in delineating snow extent was assessed by constructing a contingency table (Table 6.2) of the proportion of snow covered and snow free areas correctly classified within AVHRR imagery, using the Landsat TM data to provide a 'true' indication of the spatial distribution of snow cover. In both cases, the classification of the AVHRR data overestimated the snow area compared to the Landsat TM thresholding technique, although some areas identified within the Landsat TM scenes were assessed as being snow free by the AVHRR snow mapping algorithm.

Table 6.1 Accuracy assessment of the discrimination of snow and cloud using the semi-supervised multispectral classification. Error rates were obtained by calculating the proportion of cloud misclassified as snow over sea areas.

Date	Area	Initial errors (%)	Single class segregation using AVHRR-NDVI (%)	Majority filter applied (%)
22/01/86	North UK	3.80	1.33	1.01
22/01/86	South UK	0.99	—	0.92
23/01/86	North UK	0.59	0.20	0.17
23/01/86	South UK	0.08	—	0.04
24/01/86	North UK	1.46	—	1.03
24/01/86	South UK	0.82	—	0.66
25/01/86	North UK	0.13	—	0.09
25/01/86	South UK	0.34	—	0.21
26/01/86	North UK	0.68	—	0.57
27/01/86	South UK	1.83	1.47	1.01
27/01/86	North UK	3.30	2.24	2.02
28/01/86	South UK	2.50	0.20	0.10
29/01/86	North UK	0.14	—	0.05
29/01/86	South UK	0.10	—	0.07

Table 6.2 Accuracy assessment of snow boundary delineation using the multispectral AVHRR classification.

Area		Landsat TM Snow	No Snow
Southern Uplands	NOAA AVHRR		
	Snow	85.8%	24.6%
	No Snow	14.2%	75.4%
South Wales	Snow	92.8%	27.3%
	No Snow	7.2%	72.8%

These variations in measured snow area were due mainly to differences in sensor spatial resolution. While small areas (< 1 km^2) of snow free ground (e.g., in deep valleys) were identified by the fine spatial resolution of the Landsat TM sensor, the spatial resolution of the NOAA AVHRR sensor was too coarse to allow segregation of snow free from snow covered surfaces. Furthermore, the thresholding procedure employed for the mapping of snow extent in Landsat TM imagery failed to incorporate areas of snow in conifer forest or deep shadow. The occurrence of small patches of snow within and adjacent to forest/shadow in the ground resolution element was at times sufficient to raise AVHRR pixel intensity to a level appropriate for inclusion into a snow class, thereby leading to an apparent overestimate of snow extent.

Perhaps the most convincing demonstration of the ability of the multispectral algorithm to detect snow cover was the visual comparison of the NOAA AVHRR-defined snow margin with the observed distribution of snow accumulation in colour composite Landsat TM imagery. Plate 4 compares the snow area observed by the Landsat TM in New Galloway, Scotland with the NOAA AVHRR estimate of snow extent. At the time of the Landsat satellite overpass (c. 0900 GMT) the area was cloud free. However, scattered cloud was observed at the time of the NOAA satellite overpass (c. 1325 GMT). Even so, the AVHRR classification procedure was able to identify cloud pixels and delineate the snow/snow free boundary.

.... SNOW DISTRIBUTION AND ALTITUDE RELATIONSHIPS

The snow maps generated using AVHRR data were refined by exploiting the relationship between snow cover distribution and altitude. As the distribution of snow cover is often controlled by elevation (Lichtenegger *et al.*, 1981; Meier, 1975; Martinec, 1987), terrain information needed to be included in the snow mapping procedure.

USE OF ELEVATION DATA

Most techniques relating snow area distribution to elevation are generally only applicable to high altitude regions where snow coverage exhibits a typical pattern of successive accumulation during late autumn and winter followed by subsequent melt in the spring. Consequently the location of snow cover on one day will remain similar to the snow distribution on previous or successive days, although a gradual change in total area coverage is often evident.

The temporary and unpredictable nature of snow accumulation within the UK identifies snowpacks more with the marginal areas of these high altitude snowfields and, consequently, consistent deployment of classic snow area/altitude relationships is often limited. Elevation data were, however, beneficial for the routine description of snow area in terms of snowline location, retreat and advance. In the event of complete masking of the snowpack by forest, altitude information was used to extend the snowline into forest regions, although the technique was inappropriate in level terrain as a number of other variables control the wide-area distribution of snow cover. The majority filter was also useful in incorporating areas of snow within small forest plantations or dense forest into the total snow cover map. Furthermore, the area of snow concealed by cloud was approximated by using snowline altitude estimates in upland areas. The observed history (e.g., melting or accumulation state) of the snowpack was also used to infer snow distribution where satellite sensor observations were not possible.

Although altitude was an important control on snow distribution, an analysis of AVHRR imagery from the winter of 1985/6 revealed that there was extreme variability in snow distribution across the UK and that synoptic weather systems were equally influential in determining snow occurrence.

.... THE OPERATIONAL PROCEDURE

The procedures described in previous sections were combined to generate a NOAA AVHRR-based snow mapping algorithm which has been employed within the UK and tested with good results in both Australia and Pakistan. The multispectral snow mapping algorithm is summarised in Figures 6.4–6.7 and consists of routines performing geometric and radiometric correction, integration of ancillary data, basic snow map production, and snow cover description. By following these procedures, which have been described in this chapter, detailed snow cover maps for the UK and for other cold regions, can be derived.

.... CONCLUSIONS

The research described in this chapter has led to the generation of a potentially operational daily and/or weekly snow area mapping procedure presently available for use in the UK. Furthermore, this technique includes subroutines relating to snow condition assessment and is also independent of ground information or alternative imagery.

However, in its present state, the algorithm is unable to replace completely the present National Snow Survey as measurements of snow depth or rain water equivalent cannot be provided. Furthermore, the technique is limited in application where cloud obscures the snowpack and there are also difficulties associated with the use of the AVHRR-NDVI and thermal measurements (Goward *et al.*, 1991; Gutman, 1991). The technique has potential for refinement especially with the inclusion of the 1.58–1.79μm channel on future AVHRR sensors, improved radiometric and geometric correction procedures, and integration of active and passive microwave information derived from, for example, the ERS-1 Synthetic Aperture Radar (SAR) and Scanning Multichannel Microwave Imaging (SMMI) data. Currently, the spatial information derived from multispectral AVHRR imagery exceeds, to some extent, the capabilities of the National Snow Survey and its use in an operational system is recommended.

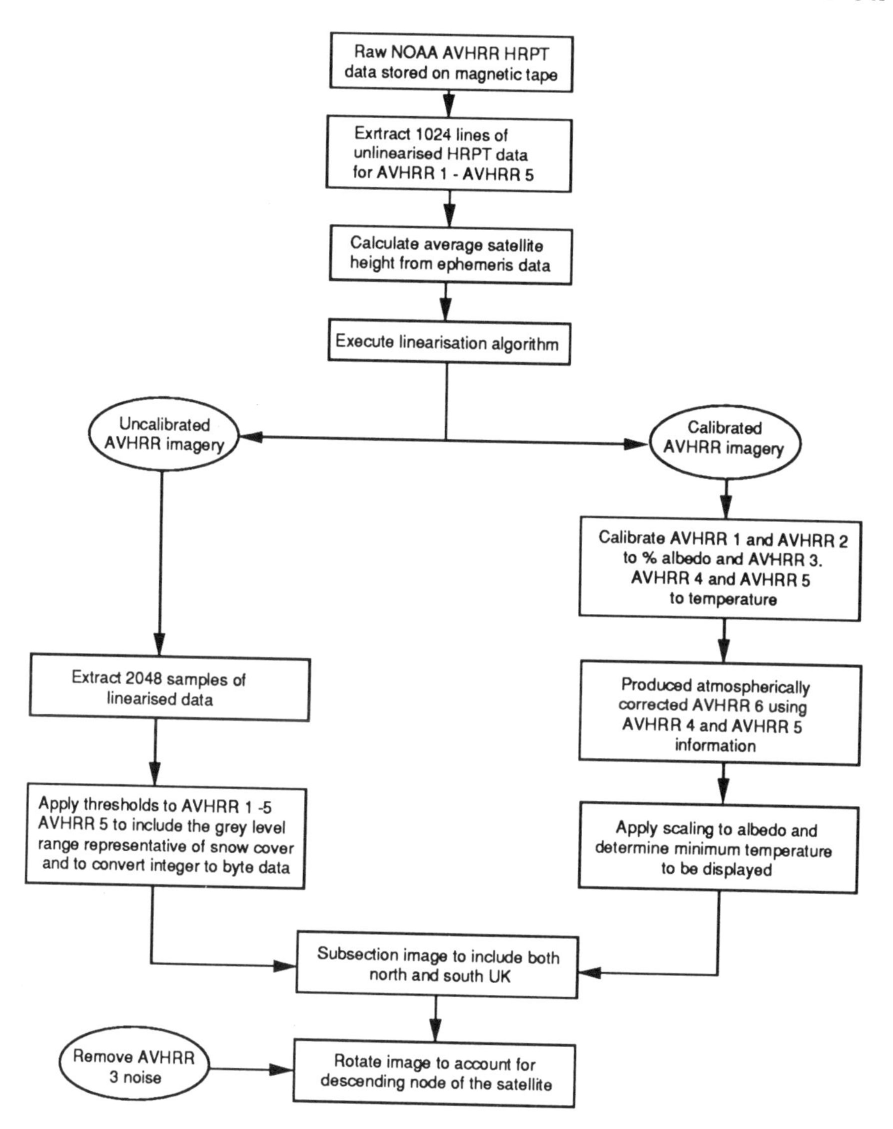

Figure 6.4 The geometric and radiometric correction procedure.

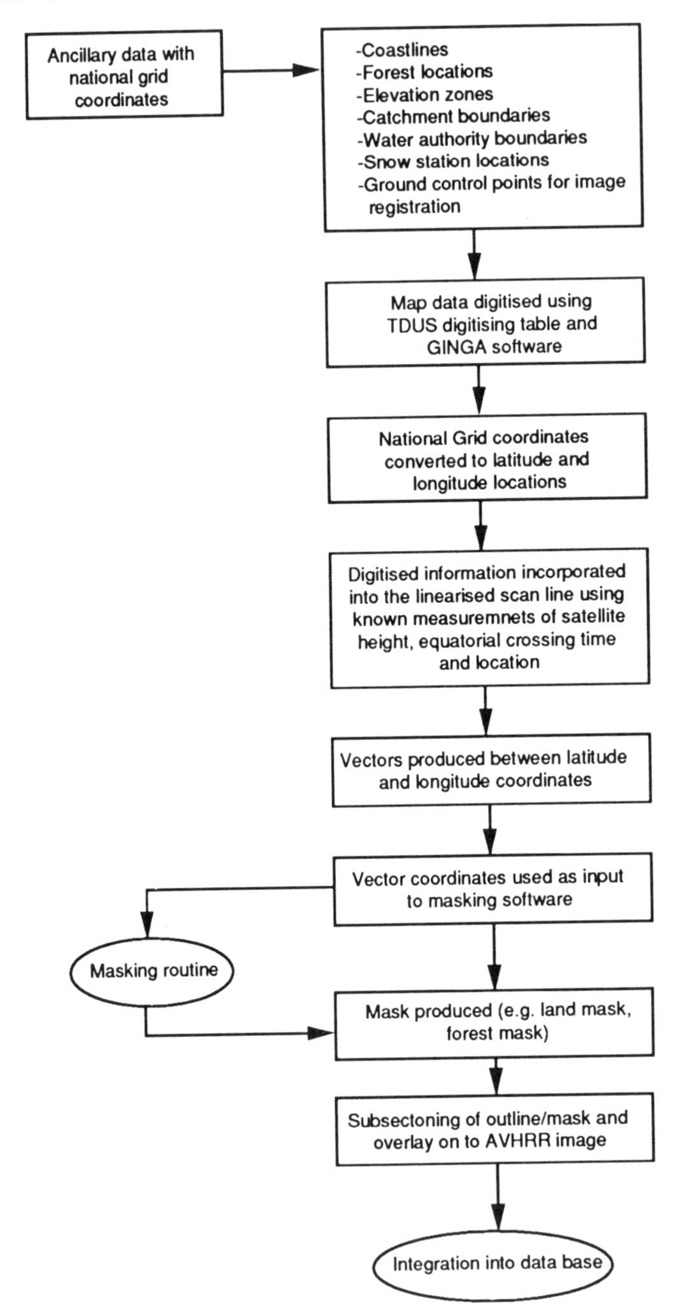

Figure 6.5 The procedure for integrating ancillary information with image data.

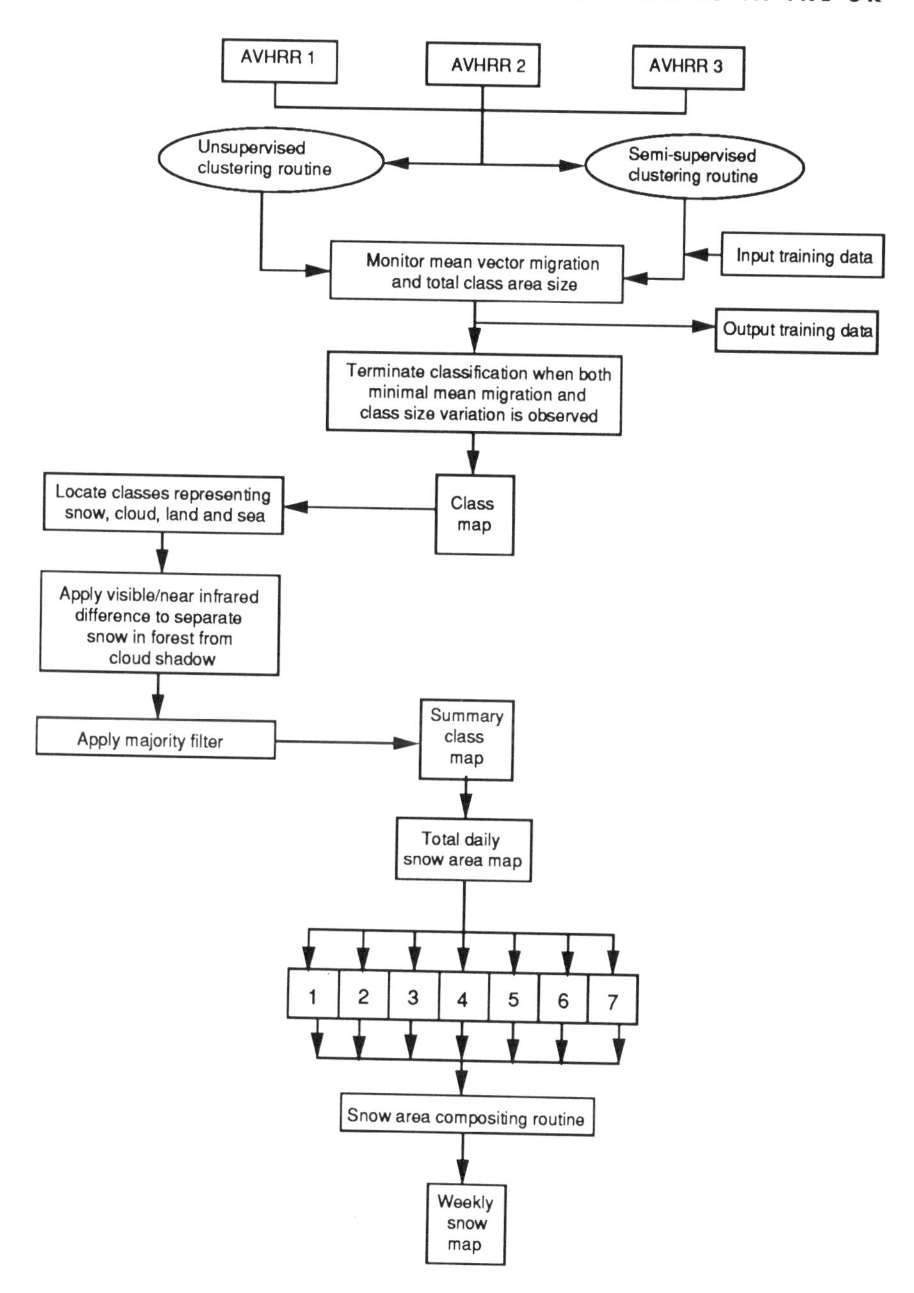

Figure 6.6 The operational procedure for the production of daily and weekly snow area maps of the UK using NOAA AVHRR imagery.

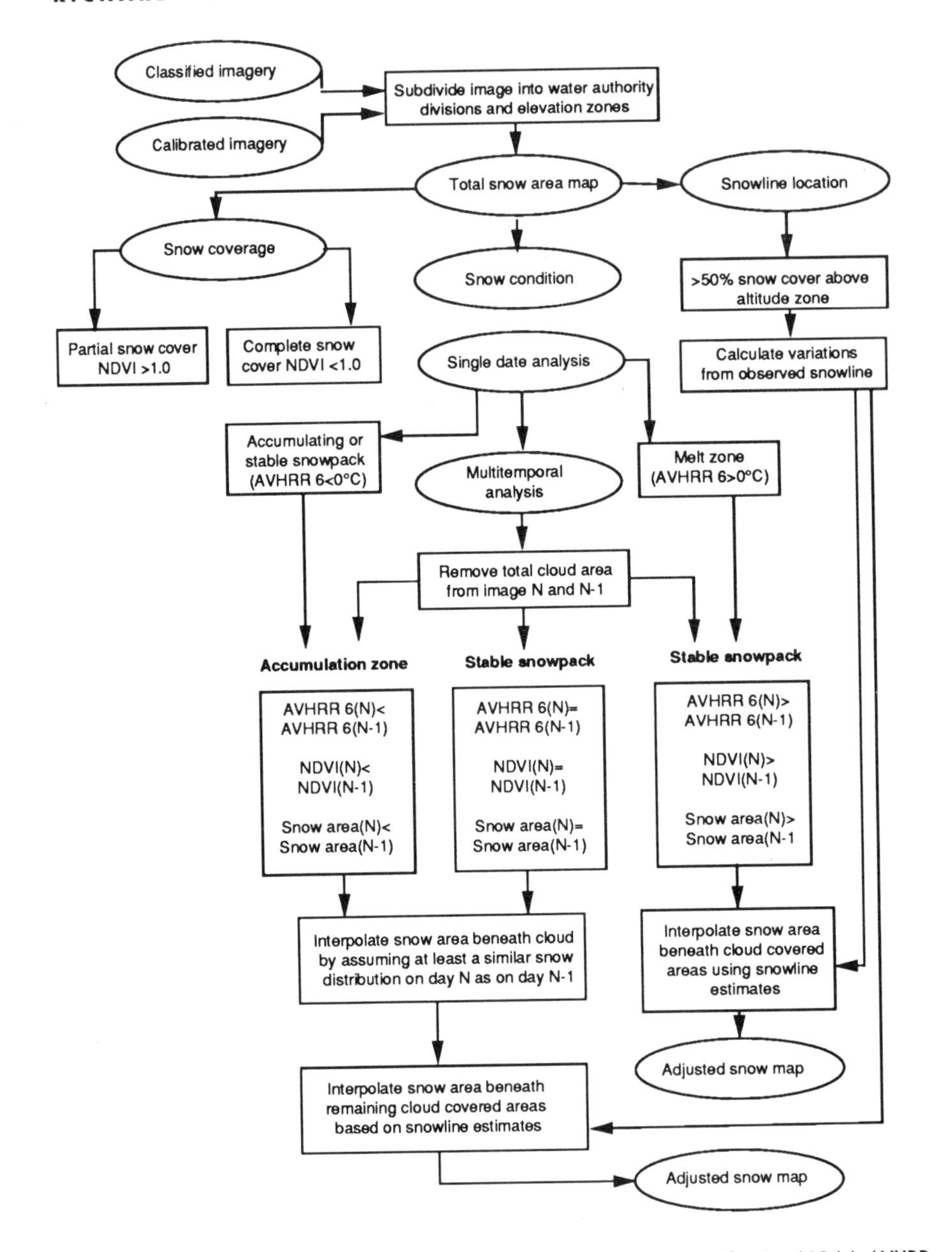

Figure 6.7 The operational procedure for the description of the snowpack using NOAA AVHRR imagery.

.... ACKNOWLEDGEMENTS

This work was carried out as part of a commissioned research project entitled 'Satellite Remote Sensing in Hydrology and Water Management' funded by the UK Department of the Environment managed in the University of Bristol Remote Sensing Unit by Dr E.C. Barrett and Dr R.W. Herschy. The views expressed are those of the authors and do not necessarily reflect those of the Department of the Environment

.... REFERENCES

Brush, R.J.H., 1985, 'A method for real time navigation of AVHRR imagery', *IEEE Transactions on Geoscience and Remote Sensing*, **23**: 876–87

Brush, R.J.H., 1988, 'The navigation of AVHRR imagery', *International Journal of Remote Sensing*, **9**: 1491–1502

Chang, A.T.C., 1986, 'Remote sensing of snowpack parameters by microwave radiometry', *Hydrological applications of space technology*, International Association of Hydrological Sciences Publication 160, Wallingford: 215–24

Chang, A.T.C., Foster, J.L., Hall, D.K., 1987, 'Nimbus 7 SMMR derived global snow cover parameters', *Annals of Glaciology*, **9**: 39–44

Ebert, E., 1987, 'A pattern recognition technique for distinguishing surface and cloud types in the polar regions', *Journal of Climate and Applied Meteorology*, **26**: 1412–27

Emery, W.J., Ikeda, M., 1984, 'A comparison of geometric correction methods for AVHRR imagery', *Canadian Journal of Remote Sensing*, **10**: 46–56

Goward, S.N., Markham, B., Dye, D.G., Dulaney, W., Yang, J., 1991, 'Normalized Difference Vegetation Index measurements from the Advanced Very High Resolution Radiometer', *Remote Sensing of Environment*, **35**: 257–77

Gutman, G., 1987, 'The derivation of vegetation indices from AVHRR data', *International Journal of Remote Sensing*, **8**: 1235–43

Gutman, G., 1991, 'Vegetation Indices from AVHRR: An update and future prospects', *Remote Sensing of Environment*, **35**: 121–36

Holroyd, E.W., Verdin, J.P., Carrol, T.R., 1989, 'Mapping snow cover with satellite imagery: Comparison of results from three sensor systems', *Western Snow Conference*, Fort Collins, Colorado

Kidder, S.Q., Wu, H., 1984, 'Dramatic contrast between low clouds and snow cover in daytime 3.7μm imagery', *Monthly Weather Review*, **106**: 2345–6

Knottenberg, H., Raschke, E., 1982, 'On the discrimination of water and ice clouds in multispectral AVHRR data', *Annals of Meteorology*, **18**: 145–7

Lichtenegger, J., Seidel, K., Keller, M., Haefner, H., 1981, 'Snow surface measurements from digital Landsat MSS data', *Nordic Hydrology*, **12**: 275–88

Liljas, E., 1987, 'Multi-spectral classification of cloud, fog and haze', in Vaughan, R.A. (editor), *Remote Sensing Applications in Meteorology and Climatology*, Reidel, Dordrecht: 301–20

Martinec, J., 1987, 'Importance and effects of seasonal snow cover', *Large scale effects of seasonal snow cover; Proceedings of the Vancouver symposium*, International Association of Hydrological Sciences Publication 166, Wallingford: 107–20

McClain, P.E., Pichel, W.G., Walton, C.C., 1985, 'Comparative performance of AVHRR-based multi-channel sea surface temperatures', *Journal of Geophysical Research*, **90**: 11587–601

Meier, M.F., 1975, 'Applications of remote sensing techniques to the study of seasonal snowcover', *Journal of Glaciology*, 15: 251–65

O'Brien, H.W., Munis, R.H., 1975a, 'Red and near infrared spectral reflectance of snow', *USA Cold Regions Research and Engineering Laboratory*, Research Report 332, 18pp.

O'Brien, H.W., Munis, R.H., 1975b, 'Red and near infrared spectral reflectance of snow', in Rango, A. (editor), *Operational applications of satellite snowcover observations*, National Aeronautics and Space Administration SP-391, Washington D.C.: 361–73

Scorer, R.S., 1987, 'Cloud formations seen by satellite', in Vaughan, R.A. (editor), *Remote sensing applications in meteorology and climatology*, Reidel, Dordrecht: 1–18

7. A Near-real-time Heat Source Monitoring System Using NOAA Polar Orbiting Meteorological Satellites

Geoffrey M. Smith and Robin A. Vaughan

.... Introduction

A system was developed to monitor heat sources with Advanced Very High Resolution Radiometer (AVHRR) data from NOAA meteorological satellites. When in operation the system used as little human input as possible. Remotely sensed data from satellite sensors have already been used to monitor heat sources in various environments. Muirhead and Cracknell (1985) monitored strawburning using NOAA AVHRR data and the National Remote Sensing Centre produced a feasibility study (Wooding, 1985) that confirmed the possibilities of such a system. The burning of tropical rainforest in South America has been monitored using NOAA AVHRR data (Matson *et al.*, 1987; Matson and Holben, 1987). Industrial heat sources caused by gas flares (Muirhead and Cracknell, 1984), factories (Matson and Dozier, 1981) and power stations (Cracknell and Dobson, 1986) have also been detected using NOAA AVHRR data. Fires and heat sources are now to be evaluated on a global scale for input to environmental models (Robinson, 1991).

The research reported in this chapter describes the integration of various pieces of work associated with the detection of heat sources. The overall aim of the research was to monitor various types of heat sources from short duration agricultural burning and forest fires to less transient heat sources such as steelworks, gas flaring and power stations. The system had to provide synoptic information on the number and distribution of heat sources in the United Kingdom (UK). This had to be done on a regular basis as the more transient sources of heat were only in existence for a few hours. Finally, the processing had to be performed in near-real-time, faster than data acquisition, to prevent a backlog of unprocessed data and to provide information for time-critical applications such as fire control.

.... HEAT SOURCES

Any object which has a temperature greater than absolute zero (–273°C) will emit electromagnetic radiation due to the kinetic energy associated with the random motions of atoms and molecules within the object. As the temperature of the object increases, then so does the random motion of the atoms. This increases the energy and reduces the wavelength of the radiation emitted. Radiation is not emitted at a single discrete wavelength but over a range of wavelengths with the wavelength of peak emission related to the temperature of the object. The heat sources monitored by this system emit radiation in middle infrared wavelengths (Table 7.1).

Table 7.1 Characteristics of some heat sources.

Source	Temperature (K)	Peak emission wavelength (μm)
Earth	~300	~10.0
Top of blast furnace	475–525	~5.8
Gas flares	~600	~4.8
Red hot material	870–970	~3.1
White flame	1270–1470	~2.1
Blue flame	>1470	~1.9
Sun	~6000	~0.5

.... WHY NOAA AVHRR?

The only sensor that comes close to fulfilling the requirements of the system outlined above is the AVHRR flown on the NOAA meteorological satellites (Schwalb, 1982). The AVHRR channel 3 is in the middle infrared with its response centred on 3.7 μm. Most of the heat sources in Table 7.1 can be detected in data acquired in this channel due to the distribution of emitted radiation around the peak wavelength. The AVHRR sensor images the UK five times each day with a 3,000 km swath width. Although the spatial resolution is relatively coarse, pixels represent a ground resolution element of 1.1 km × 1.1 km in size, the response function of the AVHRR detector in the middle infrared allows pixels containing relatively small heat sources to be identified. The background intensity (that emitted by objects at the ambient temperature) is typically so low in middle infrared wavelengths in comparison to that of heat sources that even heat sources that cover only a small fraction of the pixel area can be detected.

.... HARDWARE SYSTEM

To monitor heat sources in near-real-time, AVHRR data have to be readily available as soon after it had been received by a ground station as possible.

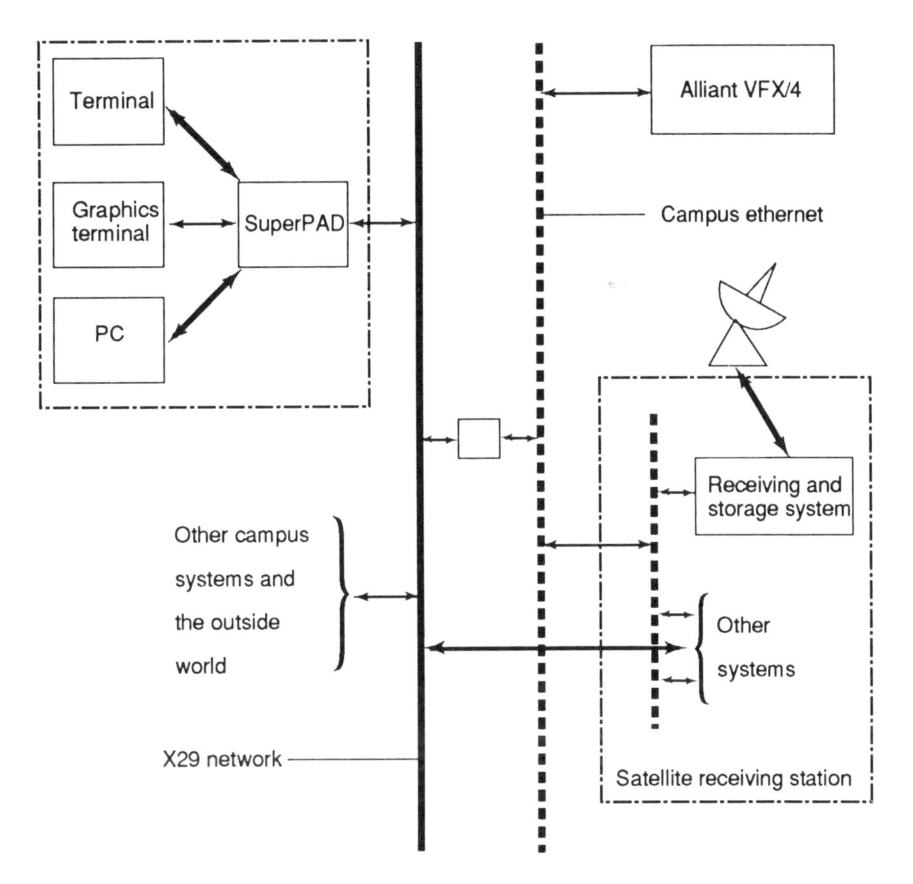

Figure 7.1 Hardware system at the University of Dundee used to process AVHRR data in near-real-time.

High speed processing was made possible by hardware at the University of Dundee (Figure 7.1) where the Satellite Receiving Station (SRS) collects AVHRR data on a routine basis. The High Resolution Picture Transmission (HRPT) stream sent down by the satellite, which contains the AVHRR images, is received and then stored on hard disks for a short period of time after each pass before being archived to magnetic tape. To process the data quickly an Alliant VFX/4 computer, which employs parallel processing techniques, was used to allow fast execution of the computationally-intensive software within the monitoring system. The SRS computers and the Alliant VFX/4 are connected by an Ethernet high speed network which transfers data at 10 million bits per second. Protocol running on the Ethernet allowed the SRS hard disks to be linked to the Alliant VFX/4 allowing it transparent access to the HRPT data, once a complete pass had been received.

.... SOFTWARE SYSTEM

The software system contained four main components: the automatic extraction of the UK from each AVHRR image that contains all or part of it; the production of data concerned with cloud cover; the automatic detection of heat sources; and the reporting of results. All the software was written in FORTRAN and ran on the Alliant VFX/4. The inputs to the software consisted of binary files of HRPT data and text files containing orbital parameters required to navigate the imagery. The output was in the form of a text report file which contained information generated by the three main processing steps.

EXTRACTING AVHRR COVERAGE OF THE UK

As the UK covers only a small fraction of a complete AVHRR image it could be extracted to reduce the amount of data and the amount of storage required, and therefore increase the processing speed of the system. To locate the UK within the AVHRR imagery a conversion algorithm, developed by Mr Khan at the University of Dundee, was used which related latitude and longitude to pixel and scan line number. The algorithm required satellite ephemeris data, current orbital data and timing information from the HRPT stream. Four points that form a rectangle around the UK were located on the image using the algorithm, and the minimum and maximum pixel and scan line values of these points selected. The window defined by the selected pixel and scan line values was then extracted from the whole AVHRR image, within the HRPT stream, for channels 1 to 4.

CALCULATING CLOUD INFORMATION

Cloud cover is a major problem for remote sensing in optical wavelengths (Saull, 1986). The amount of cloud cover will affect the number of heat sources that are visible at any one time. The cloud detection algorithm is based on two tests put forward by Saunders and Kriebel (1988), both of which must be passed by a pixel for it to be classified cloud-free. The first test uses a brightness temperature (the apparent temperature of an object recorded at the satellite) threshold to determine whether the ground resolution element has a low temperature and was therefore contaminated by cloud. Software developed for the estimation of sea surface temperature analysis by Ramsey (1990), was used to convert AVHRR thermal infrared, channel 4, data to surface brightness temperatures. If the brightness temperature was below the threshold, the pixel was classed as cloud. This threshold will vary diurnally and annually and must be set to suit the conditions during the monitoring period. The second test calculated the ratio of the near-infrared bi-directional radiation to visible bi-directional radiation for the pixel. Cloudy pixels have a value close to 1.0, while pixels over sea are around 0.5 and pixels over land are greater than 1.6. The second test was therefore used to classify the pixel into either cloud, sea or land. The algorithm then gives percentages of cloud, sea and land for the extracted UK window which are useful when assessing the number of heat sources detected.

DETECTING HEAT SOURCES

The heat source detection algorithm uses the AVHRR middle infrared, channel 3, data. These data are recorded in such a way that larger emissions from the surface are represented by smaller digital numbers (DN). The heat source detection algorithm was based on the scheme put forward by Hayes and Cracknell (1988). It comprises a series of tests, all of which must be passed by a pixel for it to be classified as a heat source. The tests were rearranged so that they made the most efficient use of computer time and therefore increased the processing speed of the system. This was achieved by performing the simplest tests first and once a test had been failed the algorithm moved on to the next pixel. The first test was a simple threshold which was set to 100 DN, as it was found that most heat sources had values below 50 DN. The second test was a relative threshold and measured the difference between the background DN and the DN of the pixel containing the suspected heat source. The suspected heat source pixel DN was subtracted from the mean of the remaining pixels in a five by five pixel kernel centred on the suspected heat source pixel. The threshold was set at 100 DN and most heat sources were at least 200 DN below the background value. The final test determined if the heat source affected the pixels around the one it was located in by calculating the standard deviation of the remaining pixels in a five by five pixel kernel centred on the pixel containing the suspected heat source. If the standard deviation of these pixels was small then it was more likely that the pixel value may be considered to be noise, because heat sources tend to influence surrounding pixels and produce an inverted bell shaped distribution in the remaining pixels of a five by five pixel kernel. When a pixel had passed all of these tests, then it was classified as a heat source.

.... PERFORMANCE OF THE HEAT SOURCE MONITORING SYSTEM

EXTRACTION OF THE AVHRR COVERAGE OF UK

The latitude to pixel and longitude to scan line conversion algorithm used the predicted satellite orbit which may be inaccurate due to slow changes over time. There are also errors in the satellite clock which cause errors in the timing accuracy of the HRPT frames and therefore the AVHRR line locations relative to the surface (Brush, 1988). The satellite experiences rolling and pitching motions which move the centre of its view from nadir, and yaw motions which cause the lines to be non-perpendicular to the satellite ground track. Figure 7.2 shows the required window and four consecutive daily windows extracted by the system. The extracted windows drift each day due to the procession of the Sun synchronous satellite orbit with time. The mismatches between the required and extracted windows are due to the errors within the navigation of the imagery described above. To obtain an estimate of the error involved, the locations of two known steelworks were used. The

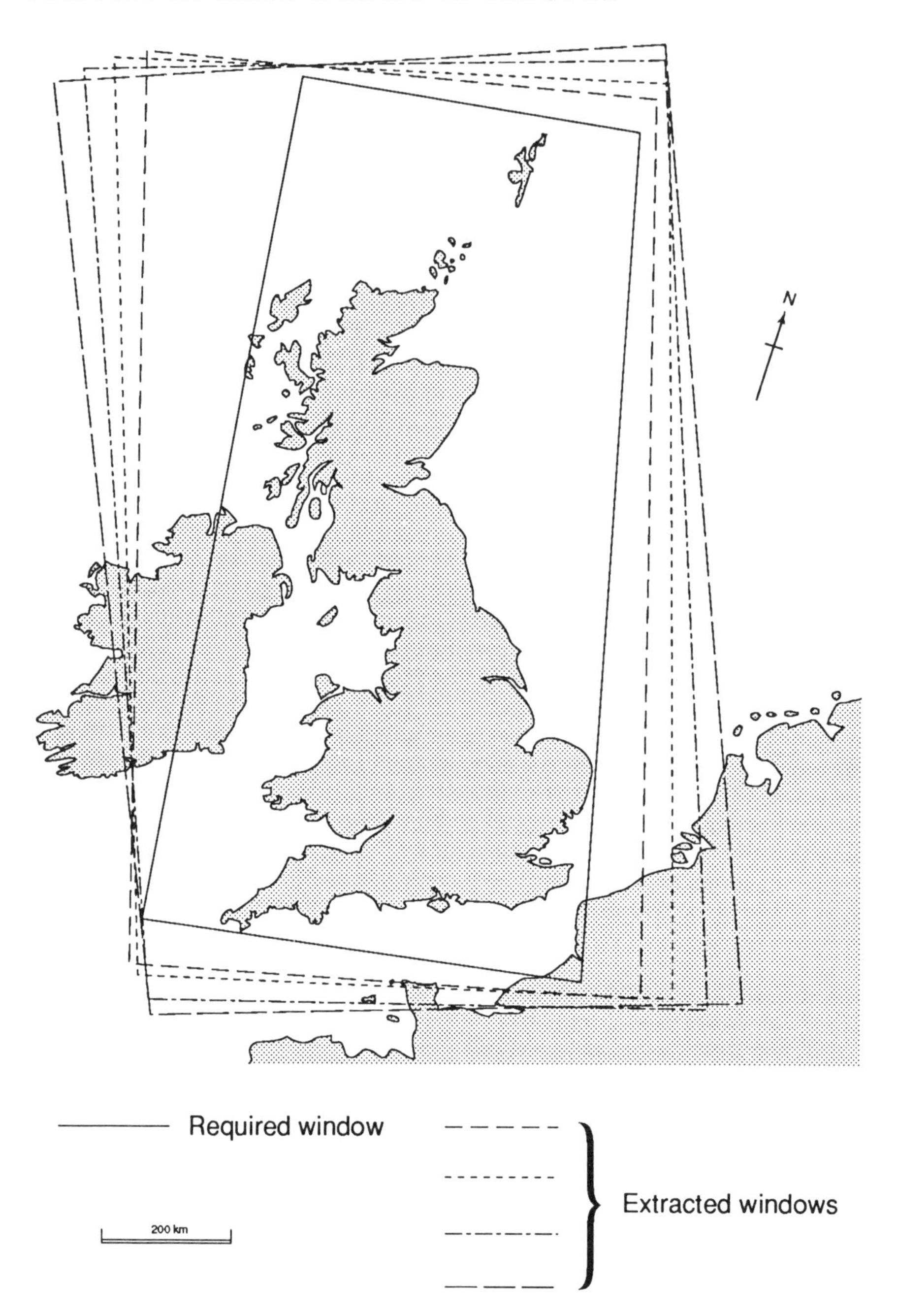

Figure 7.2 Comparison of the required UK window and the window selected by the image navigation algorithm.

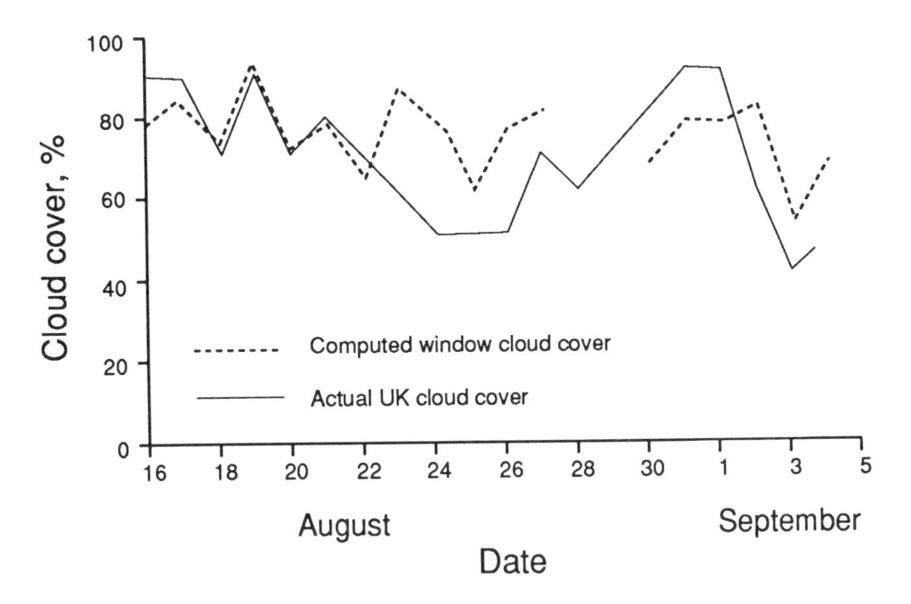

Figure 7.3 Comparison of computed window cloud cover and the actual UK cloud cover.

maximum errors found in the calculated locations of the steelworks were 0.085 degrees in longitude and 0.164 degrees in latitude. These correspond to distances of approximately 5.8 km and 18.2 km respectively. Larger errors in the latitude direction suggest the major cause of errors was timing problems associated with the AVHRR scan lines in the HRPT stream.

CLOUD INFORMATION

The cloud detection algorithm produced results with similar trends and values to those estimated from quick look images (Figure 7.3), but there were differences and these were caused by a number of factors. The algorithm detected cloud contaminated rather than cloud covered pixels, so those pixels containing thin high altitude cloud or small patches of cumulus were also counted as cloud covered. The second test can return a mixed pixel that contains both land and sea as cloud contaminated as their composite response was similar to that of cloud. The actual cloud cover was estimated only over the UK from quick look images whereas the window extracted from the AVHRR image also contained parts of the North Sea, Irish Sea, Ireland and Europe.

HEAT SOURCE MONITORING

To test the heat source detection algorithm, the system was used to monitor the 1990 strawburning season in the UK. Strawburning is a process for

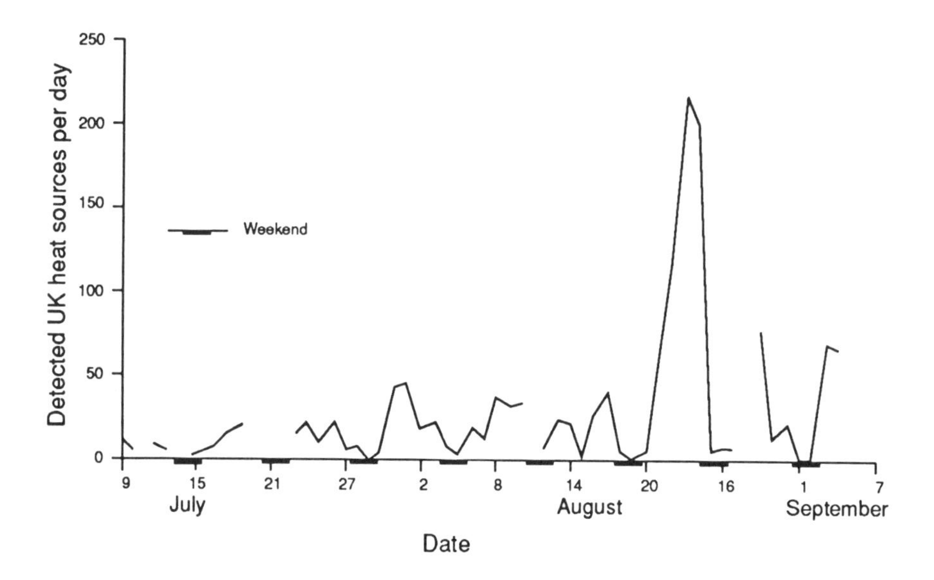

Figure 7.4 The number of detected UK heat sources per day during the monitored part of the 1990 strawburning season.

disposing of straw and stubble waste from cereal farming. The need to burn straw and stubble in large quantities had arisen due to a change in farming practice. A surplus of straw was often produced and the cheapest and simplest way to dispose of it was to burn it. It has been estimated that each year, during the July to September burning season in the UK, several million tonnes of straw are disposed of in this way (MAFF, 1984). Strawburning can cause a local loss of nutrients and the resulting fires can be a danger to lives and property. To reduce some of the effects of strawburning, the National Farmers Union (NFU) introduced a Voluntary Code of Practice in 1984. This code of practice recommends, among other things, that burning should be avoided at weekends and on statutory holidays. The government has now introduced legislation to ban strawburning as a farm management practice, although burning may still occur naturally or by accident.

The heat source detection algorithm classifies each pixel but the number of pixels classified as heat sources is not necessarily the number of heat sources. First, one heat source may influence more than one pixel if it overlaps the pixel boundaries or its hot combustion products, in the case of fires, are blown into an adjacent pixel. Second, more than one heat source may occur within the same 1.1 × 1.1 km pixel and just produce a stronger response. It was therefore decided to identify heat sources in terms of target areas that may be made up of one or more adjacent pixels.

Figure 7.4 shows the number of heat source target areas within the UK during the monitored part of the 1990 strawburning season. A slight increase occurred during the end of July and the first half of August, with occasional peaks which could probably be attributed to an increase in strawburning. A large peak occurred during the third week in August, which was caused by

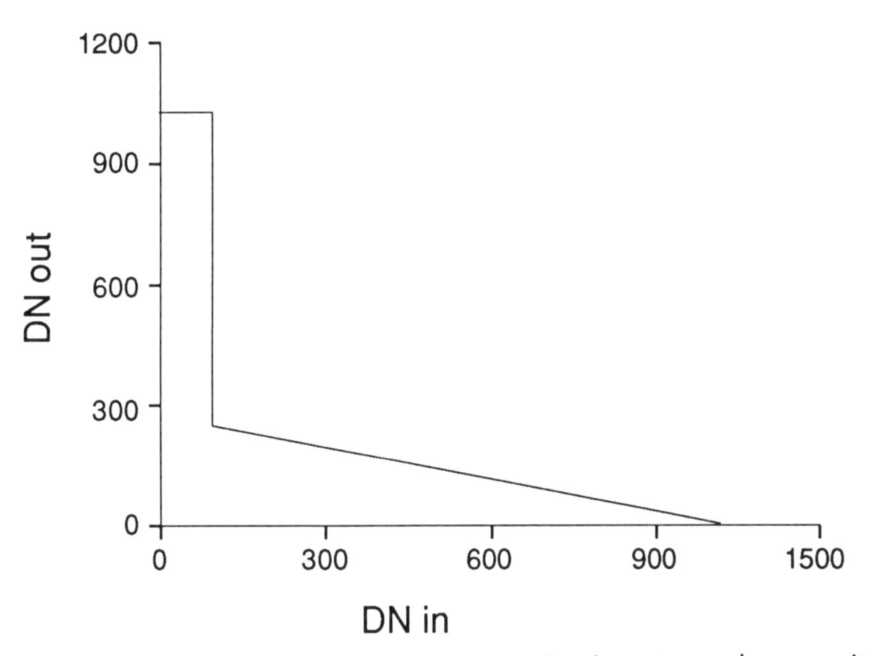

Figure 7.5 Transfer function to process AVHRR channel 3 data prior to colour composite generation.

strawburning. The increase in strawburning during the summer was accompanied by a cyclical pattern with its troughs centred on the weekends. This demonstrated the compliance of most farmers with the NFU guidelines.

.... COLOUR COMPOSITES

Colour composites of part of the extracted UK window were produced as part of the testing procedures for the heat source monitoring system. It was then possible to test if the UK was within the window, and whether the cloud cover estimate and the number of detected heat source pixels were correct. To produce the colour composites AVHRR channels 1, 2 and 3 were mapped to the blue, green and red planes respectively of an image processing system. Due to the inverted nature of channel 3, it was necessary to filter it prior to display. The transfer function to filter channel 3 is shown in Figure 7.5. It set the possible heat source pixels to red and removed a red cast that was present in the images.

Plates 5 and 6 are colour composites from Friday 24 August, 1990 and Saturday 25 August, 1990 respectively. The change in heat source density between the week day and weekend over England and Wales is clearly visible, and was due to farmers obeying the NFU guidelines. On Plate 5 the effects of cloud cover on the detection of heat sources can be seen. A large cereal growing area in southern England was obscured and no estimate of strawburning in this area could be made.

.... CONCLUSIONS

The results produced by the heat source monitoring system for the 1990 strawburning season demonstrated that it was possible to monitor heat sources in near-real-time if the appropriate hardware and software were available. The data did not give absolute numbers of heat sources because of issues such as cloud cover, timing of the satellite pass and the problem of multiple heat sources and pixels. The normal running time for the system was around twenty minutes per pass. Muirhead and Cracknell (1984) considered a near-real-time system to locate oil and gas platforms in the North Sea using ground control points (GCP) and estimated a running time of two and a half hours per pass. The system described here does not use GCPs to locate the heat sources but provides additional cloud cover information and, on the estimates of Muirhead and Cracknell (1984), still ran eight times faster. There was scope for the software to be optimised even further when running in the parallel processing environment of the Alliant VFX/4. A dedicated computer rather than a multi-user system would also improve performance.

As was appreciated by Saull (1986), and Djavadi and Cracknell (1986), cloud cover in the UK is a major handicap to any heat source monitoring system. The simple problem of cloud cover is complicated by the fact that heat sources appear through certain cloud types and that fires may generate conditions which appear to be cloudy due to the smoke they emit. When cloud contamination was added as a criterion to the heat source detection algorithm all but 2 per cent of the heat sources went undetected. Methods of cloud masking would therefore remove heat sources that are visible in imagery in middle infrared wavelengths. Cloud information given by this system was for the whole extracted window when cloud cover, for instance over the cereal growing areas when monitoring strawburning, is far more important.

One of the major problems in the system was the accuracy with which it located points on the images and the surface using the latitude to pixel and longitude to scan line conversion algorithm. These errors were acceptable for locating the UK window but, when the algorithm was used to locate individual heat sources, the error became more important. The conversion algorithm is currently under review and the SRS at the University of Dundee intend to include information on the differences between the orbital model predictions and the actual results for each pass. The changes and additions mentioned above will help to improve the system's location abilities, but an error of a number of pixels may still exist.

Although the system was developed for the monitoring of strawburning in the UK, there is no reason why it cannot be applied to other heat sources in other parts of the world. This system could be used to support any of the applications described earlier, such as the burning of tropical forest, fire monitoring of remote forests or gas flaring in the Persian Gulf, and may potentially be applied anywhere in the world.

.... ACKNOWLEDGEMENTS

The authors wish to thank Mr B. Khan for providing the software to navigate the NOAA imagery and Dr L. Hayes and Mr K.M. Lovell for their help with the hardware system.

.... REFERENCES

Brush, R.J.H., 1988, 'The navigation of AVHRR imagery', *International Journal of Remote Sensing*, 9: 1491–1502

Cracknell, A.P., Dobson, M.C., 1986, 'On the use of AVHRR channel 3 data for environmental studies', *ESA/EARSel Symposium on Europe from space*, European Space Agency SP-258, Noordwijk, 137–44

Djavadi, D., Cracknell, A.P., 1986, 'Cloud cover and the monitoring of strawburning using AVHRR data', *International Journal of Remote Sensing*, 7: 949–51

Hayes, L., Cracknell, A.P., 1988, 'A fast algorithm for the automatic recognition of heat sources in satellite images', *Lecture notes in Computer Science*, 38–47

MAFF, 1984, *Straw use and disposal, Booklet 2419*, Ministry of Agriculture, Fisheries and Food, Alnwick, Northumberland

Matson, M., Dozier, J., 1981, 'Identification of subresolution high temperature sources using a thermal IR sensor', *Photogrammetric Engineering and Remote Sensing*, 47: 1311–18

Matson, M., Holben, B., 1987, 'Satellite detection of tropical burning in Brazil', *International Journal of Remote Sensing*, 8: 509–16

Matson M., Stephens, G., Robinson, J., 1987, 'Fire detection using data from the NOAA-N satellites', *International Journal of Remote Sensing*, 8: 961–70

Muirhead, K., Cracknell, A.P., 1984, 'Identification of gas flares in the North Sea using satellite data', *International Journal of Remote Sensing*, 5: 199–212

Muirhead, K., Cracknell, A.P., 1985, 'Strawburning over Great Britain detected by AVHRR', *International Journal of Remote Sensing*, 6: 827–33

Ramsey, J.M.W., 1990, *Calculation of sea surface temperature using data transmitted from the NOAA series of satellites*, M.Sc. Thesis, Department of Applied Physics and Electronic and Manufacturing Engineering, University of Dundee, Dundee, UK

Robinson, J.M., 1991, 'Fires from space: Global fire evaluation using infrared remote sensing', *International Journal of Remote Sensing*, 12: 3–24

Saull, R.J., 1986, 'Strawburning over Great Britain detected by AVHRR, a comment', *International Journal of Remote Sensing*, 7: 169–72

Saunders, R.W., Kriebel, K.T., 1988, 'An improved method for detecting clear sky and cloudy radiances from AVHRR data', *International Journal of Remote Sensing*, 9: 123–50

Schwalb, A., 1982, *The TIROS-N/NOAA A-G satellite series*, NOAA Technical Memorandum NESS 95, Washington D.C.

Wooding, M., 1985, 'Strawburning detection using AVHRR data', *NRSC Newsletter Issue 4*, National Remote Sensing Centre, Farnborough

8. Per-zone Classification of Urban Land Cover for Urban Population Estimation

Christiane Weber

.... Introduction

The recent generation of satellite sensing systems has generated an interest in the remote sensing of urban areas. Satellite sensors with a relatively fine spatial resolution, such as those carried on the SPOT or Landsat satellites, have considerable potential in urban studies.

To comprehend the problems associated with the remote sensing of urban areas, the spatial, spectral and temporal properties of remotely sensed data must be considered with respect to the morphological characteristics of these areas (Jensen, 1981).

One of the most important issues is the heterogeneity of urban areas in relation to the spatial resolution of the sensor. Because of the variety of objects in an urban area (roofs, roads, trees etc.), there is a need to deal with a complex mixture of spectral responses (Forster, 1985) and as a result contextual information becomes critical for the accurate identification of these urban objects. The integrating effect of the sensor, can influence significantly the response from a single land cover class if the surrounding cover is dissimilar (Paulsson, 1991). In addition, the identification of land surfaces with a conventional image classification approach is difficult, as the continuum of land cover classes cannot be divided readily into discrete classes. Misclassifications can occur for a variety of reasons. For example, the spectral response of different types of urban land cover could be similar (e.g., harbour banks and housing roofs) and the spectral responses of some classes may be very broad (e.g., housing roofs made from different materials and of different ages).

Fine spatial resolution data should be appropriate for the study of urban areas (Forster, 1985) as each pixel is more likely to contain a single land cover type. However, the additional detail will increase the potential range of any one signal and thereby the spectral variability of each land cover class.

With fine spatial resolution SPOT HRV panchromatic imagery (10 m spatial resolution), textural differences in the spatial structure of different types of residential housing and particularly between urban and non-urban land-use are highlighted. The discrimination between urban and non-urban areas is more accurate when using SPOT HRV panchromatic data than when using the coarser spatial resolution multispectral channels (20 m spatial resolution). The characteristic variations of texture provide potential for identifying particular aspects of urban spatial structure (Haralick, 1979; NRSC, 1989).

A number of factors may degrade the ability of satellite sensors to classify accurately land cover (Carbiener, 1986). Shadows in urban areas are a major problem, especially when fine spatial resolution data, such as SPOT HRV panchromatic data, are used. Shadows provide some indication of urban network shape, this information is, however, difficult to extract.

The repetitivity of data acquisition by satellite sensors makes them attractive for up-dating information on urban areas. The major prerequisite for such use is the accurate geometric correction of the data (Verger, 1982). Monitoring urban growth could be considered as an important application of remotely sensed data (Martin and Howarth, 1989). Remote sensing may be used to evaluate the dynamism of the urban area through the rate of land cover variation (from agricultural land to industrial complexes, for instance) or on the effect of urban growth on natural resources (Gong and Howarth, 1990). By relating population density to building characteristics (e.g., building type and coverage) remotely sensed data may also be used to derive estimates of demographic variables (Lo and Welch, 1977).

Remotely sensed data have been used for many urban applications, ranging from mapping to modelling (Lo and Shipmann, 1990; Nicoloyanni, 1991; Paulsson, 1991). Topographic and thematic maps at scales of 1:10,000–1:50,000 are used typically by local authorities for planning purposes and these could be derived from satellite sensor imagery (ACSM-ASPRS, 1989).

Studies involved with topographic mapping using SPOT HRV data at 1:25,000 to 1:50,000 scale have found that when using simple visual interpretation some features can be difficult to identify and in particular linear objects and certain point objects cannot be mapped accurately (Manning and Evans, 1989). This conclusion points to the digital analysis of imagery.

A major application for satellite sensor data is land cover mapping (Bertaud, 1989). Different methods of land cover mapping have been tested: some include reflectance data alone while some include texture and context information (Gong and Howarth, 1990), sometimes within specific SPAtial Re-classification Kernel (SPARK) procedures (Barnsley *et al.*, 1991), or within 'optimal spatial zones' like census tracts (Michel, 1990).

Remotely sensed data may be used to estimate population density. This application has been developed previously from aerial photography. The first approach involved development of a generalised relationship between total built-up area and population size and is based on regression techniques (Holz *et al.*, 1969; Anderson and Anderson, 1973). One of the major problems identified was (and still remains) the differentiation of the rural/urban boundary (Lo and Welch, 1977; Jensen and Toll, 1982). The second approach involved the estimation of population size based on the identification and the measurement of areas of different land cover classes and the correlation of these measures with the average population densities for these land cover classes. The third approach involved a detailed count of individual dwelling units which is then multiplied by an average per-dwelling population density extracted from census data.

The approach discussed here draws upon all three of these approaches and is called the zonal approach, as it uses per-zone as opposed to per-pixel classifications.

RESEARCH AIM

The remaining part of this chapter will aim (i) to test this zonal approach to classification and to compare it against a conventional per-pixel classification, and (ii) to compare the results of population estimation derived using these two classification procedures.

The study area is Strasbourg (France) which is located at the German border on the Rhine Valley. The landscape is relatively flat, bordered on the east by the river and on the west by the Vosges mountains. Strasbourg and the surrounding towns cover some 123 km^2 and are bordered by two forests, one in the north (Roberstau) and one in the south (Illkirch). The urban areas characterised by the census units discussed here are those close to Strasbourg in a built-up continuum: Schiltigheim, Bischheim and Hoenheim in the north, Ostwald and Illkirch-Graffenstaden in the south-west.

Two SPOT HRV images, XS (multispectral) and P (Panchromatic), were used; the XS image was recorded on 8 June, 1986 and the P image was recorded on 10 September, 1986. The two images were atmospherically, radiometrically and geometrically corrected by the Remote Sensing Center, Toulouse.

The map of census units was digitised to a scale of 1:2,000 in the Sausheim coordinate system and 2812 census units were extracted from the resultant cartographic data-base for the six cities cited above.

.... METHODOLOGY

The study involved the combination and analysis of SPOT HRV XS and P data and the integration of cartographic data and census data. The SPOT HRV XS and P data were combined to preserve the spatial and spectral information of the two data sets. To achieve this the SPOT HRV XS data were resampled to a 10 m spatial resolution using the nearest-neighbour resampling routine. The images were integrated with the cartographic data and this allowed the automatic extraction of different digital values for the 2812 census units. The mean and standard deviation of the population density for each census unit were extracted and registered in a file along with the remotely sensed data.

Four new remotely sensed variables were formed from the original data set, red radiance, near infrared radiance, standard deviation of red radiance and standard deviation of near infrared radiance (Table 8.1). These variables were used for per-pixel and then per-zone classification. In total 90 training-zones were selected to train the supervised stepwise discriminant analysis, on the assumption that they possessed a minimal amount of intra-class variability.

Table 8.1 Summary of data (relative units).

Variable	Mean	Standard deviation	Coefficient of variation
Red radiance	48.1	6.2	0.1
Standard deviation of red radiance	6.4	3.1	0.5
Near infrared radiance	67.1	12.8	0.2
Standard deviation of near infrared radiance	8.0	4.3	0.5

It was difficult to select a representative sample of training data. The housing type spanned a wide range from dense housing (DH) in the inner city with a large population density, highrise buildings (HB) with a medium population density to suburban housing (SH) with semi-detached houses and smaller population density. The industrial group (IND) included harbour wharfs, industrial complexes and coal yards that were disseminated along the river banks or the railway network; new industrial areas in dedicated zoning near the highway network, and bare soil and parking inside the urban area. Vegetation (VEG) included gardens, forests, meadows, orchards and parks. Water (WATER) included rivers, ponds and ship-canals. The confusion which could arise with the land cover in the rural open space was avoided by working only on the urban space delimited by the map.

The classification results are summarised in Tables 8.2 and 8.3, along with classification accuracy. Comparison of the results for a per-pixel and a per-zone classification reveal that most of the values were similar for the two methods, but some differences were apparent, for instance the highrise buildings were classified with a greater accuracy when using the per-zone classification method but the industrial and commercial areas were classified with less accuracy when using the per-zone classification. The latter could be explained by the type of training-zones selected for the per-zone classification. One of the difficulties with the per-pixel classification of Strasbourg was the influence of the railway network on the radiance of the areas surrounding the railway (Plate 7(a)).

The method used for the population estimate relied on relationships between housing characteristics derived from classification results and population density counts. Regression equations were performed using both per-pixel classification and per-zone classification results (Chrisman, 1987).

For the per-zone classification results, only the pixels which were labelled as containing some form of inhabited housing were taken into account, and these were labelled according to the group of spatial units they belonged to (Plate 7(b)).

Regression functions were run without constant values, on the basis that the population had to be located in the housing areas and that negative terms would be inappropriate. The estimations were run in a two step way:

Table 8.2 Summary statistics for the training-area data used in the per-zone classification; a description of the urban land cover classes is given in the text.

Actual	Predicted							
	CH	HB	SH	IND	VEG	WATER	Ommission errors %	Classification accuracy %
CH	133			2			1.5	97.8
HB		24	6	6			33.3	66.7
SH	16	8	80	20			35.5	64.5
IND				249	1		0.4	99.6
VEG			4		384		1.0	99.0
WATER	1					71	1.4	98.6
Commission errors %	11.3	25.0	7.87	3.1	5.1	0.0		

Table 8.3 Summary statistics for the training-area data used in the per-pixel classification; a description of the urban land cover is given in the text.

Actual	Predicted							
	CH	HB	SH	IND	VEG	WATER	Ommission errors %	Classification accuracy %
CH	35	2	2				10.3	89.7
HB		18	2	2			18.1	81.8
SH		2	21				8.7	91.3
IND	2	2		15	1		25.0	75.0
VEG			1		7		22.2	77.8
WATER						2	0.0	100.0
Commission errors %	7.9	25.0	16.0	16.7	12.5	0.0		

(i) with a full identification of housing land use areas according to Tables 8.2 and 8.3 (CH (crowded housing), HB and SH),
(ii) with a narrowed option (DI = sum of the CH, HB and SH areas).

Comparing the results (Table 8.4) of the two classification methods, per-zone classification was observed to produce a more accurate population estimation. Thus, the regression coefficients were larger and the root mean square error values smaller. Residual values were also smaller in the per-zone estimation procedure. The difficulties of estimating the population density of small settlements surrounded by vegetation were reduced because these were isolated through the map.

Per-pixel classification – Step 1

$R^2 = 0.81$

Population = 0.765CH + 2.698HB – 0.008SH [1]

Table 8.4 Summary of results.

	Population	estimated population / per-pixel procedure	estimated population / per-zone procedure
Total	334,294	268,397	307,839
mean per census unit	4,341	3,485	3,997
standard deviation per census unit	3,400	2,645	3,008
rms error		3,282	2,155

Per-pixel classification – Step 2
$R^2 = 0.80$
Population = 1.279DI + 0.249SH [2]

Per-zone classification – Step 1
$R^2 = 0.91$
Population = 1.235CH + 1.122HB + 0.200SH [3]

Per-zone classification – Step 2
$R^2 = 0.91$
Population = 1.166DI + 0.164SH [4]

The regression results were considered to be encouraging for further research.

.... DISCUSSION AND CONCLUSION

The spatial resolution of SPOT HRV XS and P data is adequate for the sampling of high-frequency land cover parcels characteristic of, for instance, typical North America urban environments (Welch, 1982; Treitz *et al.*, 1992). But in old European cities, where urban structures are quite different, the feasibility to integrate ancillary data could be considered as a convenient method for increasing the accuracy with which remotely sensed data could be classified. The per-zone classification method is considered as a viable means of increasing the accuracy with which remotely sensed data can be classified in an urban environment and thereby form the basis of urban population estimation programmes.

.... REFERENCES

ACSM-ASPRS, 1989, 'Multipurpose geographic database guidelines for Local Governments', *Photogrammetric Engineering and Remote Sensing*, 55: 1357–65

Anderson, D.E., Anderson, P.N., 1973, 'Population estimates by humans and machines', *Photogrammetric Engineering*, **39**: 147–54

Barnsley, M.J., Barr, S.L., Sadler, G.J., 1991, 'Spatial re-classification of remotely-sensed images for urban land-use monitoring', *Spatial Data 2000*, Remote Sensing Society, Nottingham: 106–17

Bertaud, M.A., 1989, *The use of satellite images for urban planning – A case study from Karachi, Pakistan*, World Bank, New York, 110pp.

Carbiener, R., 1986, 'Espaces verts urbains, peri-urbains et qualité de l'air', *Bulletin de la Société Industrielle de Mulhouse*: 111–20

Chrisman, N.R., 1987, 'The accuracy of map overlay: A reassessment', *Landscape and Urban Planning*, **14**: 427–39

Forster, B.C., 1985, 'An examination of some problems and solutions in monitoring urban areas from satellite platforms', *International Journal of Remote Sensing*, **6**: 139–51

Gong, P., Howarth, P.J., 1990, 'The use of structural information for improving land-cover classification accuracies at the rural-urban fringe', *Photogrammetric Engineering and Remote Sensing*, 56: 67–73

Haralick, R.M., 1979, 'Statistical and structural approaches to texture', *Proceedings IEEE*, **67**: 786–804

Holz, R.K., Huff, D.L., Mayfield, R.C., 1969, 'Urban spatial structure based on remote sensing imagery', *Proceedings, Sixth International Symposium on Remote Sensing of Environment*, University of Michigan, Ann Arbor, Vol. 2: 819–30

Jensen, J.R., 1981, 'Urban change detection mapping using Landsat data', *The American Cartographer*, **8**: 1237–47

Jensen, J.R., Toll, D.L., 1982, 'Detecting residential land-use development at the urban fringe', *Photogrammetric Engineering and Remote Sensing*, **48**: 629–43

Lo, C.P., Welch, R., 1977, 'Chinese population estimates', *Annals, Association of American Geographers*, **67**: 246–53

Lo, C.P., Shipmann, R.L., 1990, 'A GIS approach to land use change dynamics detection', *Photogrammetric Engineering and Remote Sensing*, **56**: 1483–91

Martin, L.R.G., Howarth, P.J., 1989, 'Change detection accuracy assessment using SPOT multispectral imagery of the urban fringe', *Remote Sensing of Environment*, **30**: 55–66

Manning, J., Evans, M., 1989, 'Revision of medium scale topographic maps using space imagery', *International Archives of Photogrammetry and Remote Sensing*, **27**: 233–45

Michel, A., 1990, *Stratification de l'espace urbain à partir d'images satellitaires pour réaliser un sondage à objectif démographique. Mise au point et évaluation des méthodes d'analyses des images SPOT et Landsat-TM en milieu urbain.* Thèse de Doctorat, Ecole des Hautes Etudes en Sciences Sociales, Paris

NRSC, 1989, *Monitoring Change at the Urban-Rural Boundary from Landsat and SPOT HRV Imagery*. WP56. National Remote Sensing Centre, Farnborough, 53pp.

Nicoloyanni, E., 1991, *Analyse des changements à partir des données sur Athènes et Meligalas*. Thèse de Doctorat, Université Louis Pasteur, Strasbourg

Paulsson, B., 1991, *Satellite remote sensing and GIS analysis for urban applications*, UNCHS (Habitat), World Bank, New York: 46pp.

Treitz, P.M., Howarth, P.J., Gong, P., 1992, 'Mapping at rural-urban fringe', *Photogrammetric Engineering and Remote Sensing*, **58**: 439–48

Verger, F., 1982, *L'Observation de la terre par les satellites*, PUF, France, 128pp.

Welch, R., 1982, 'Spatial resolution requirements for urban studies', *International Journal of Remote Sensing*, **3**: 139–46

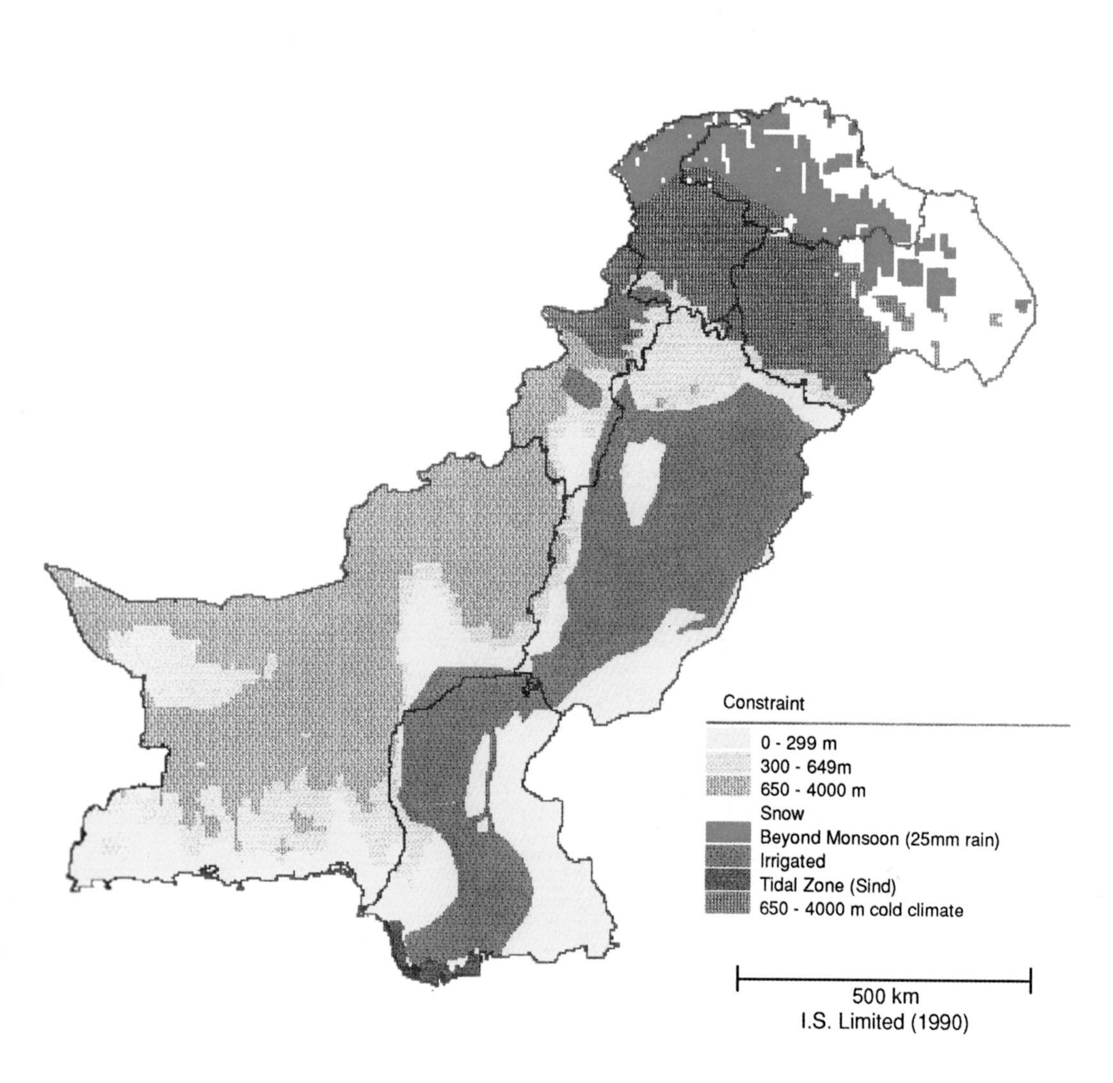

Plate 1. Pakistan: (**a**) Environmental constraints map.

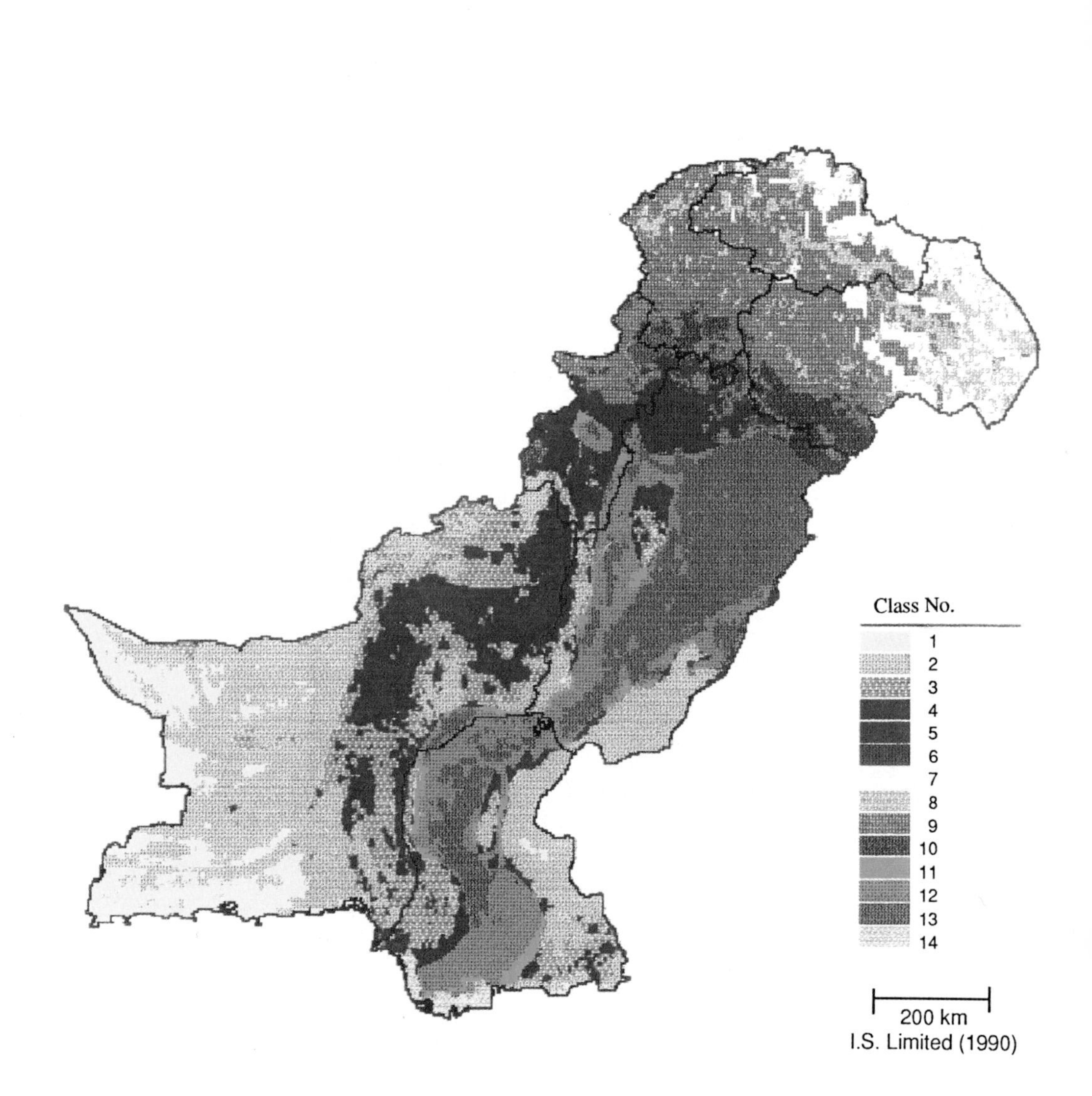

Plate 1. Pakistan: (**b**) Most frequent class (land cover class) map.

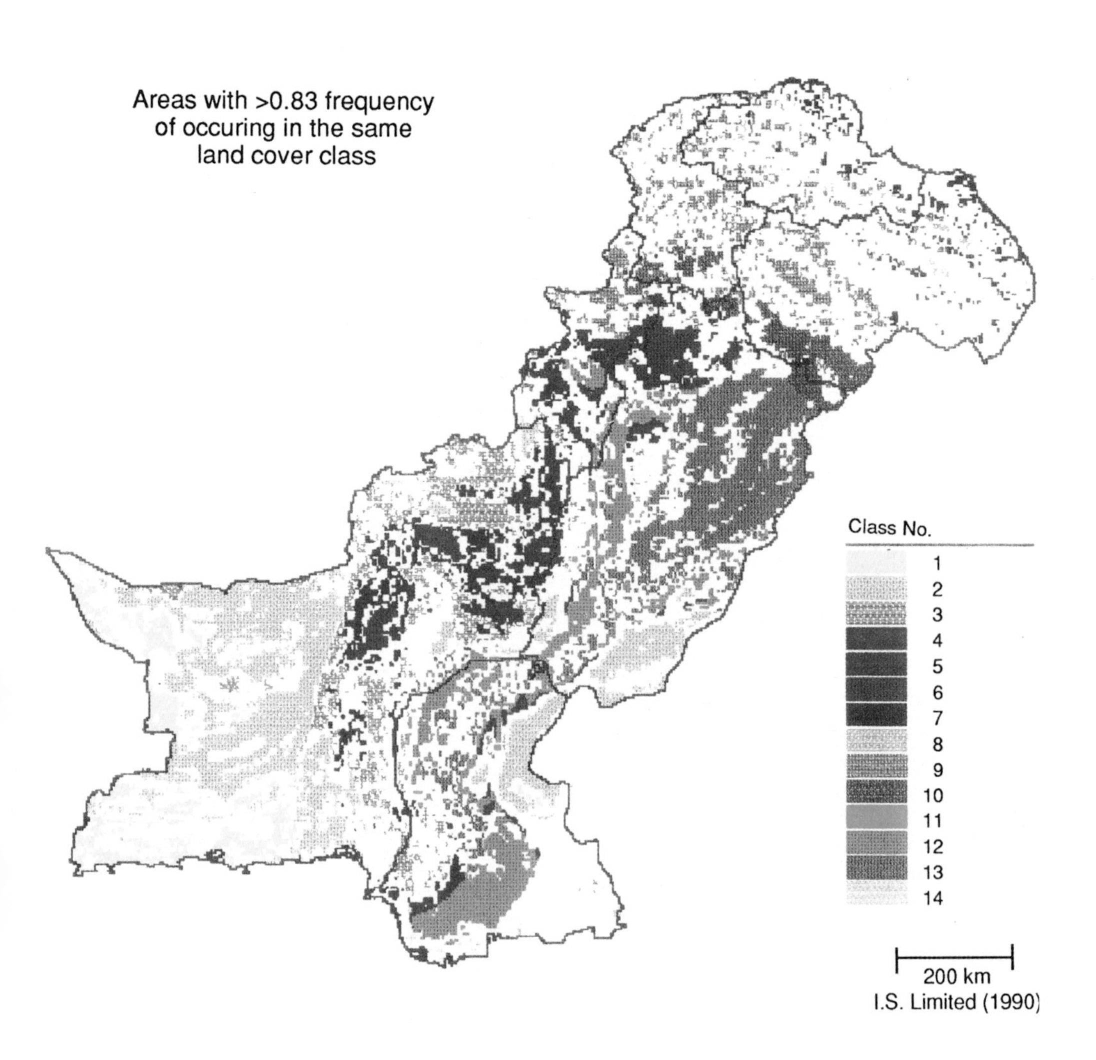

Plate 1. Pakistan: **(c)** Map of land cover core zones.

Plate 2. A NOAA AVHRR image of 27 February 1990 covering an area of approximately 710×10^3 km^2 and centred on the forests of south-western Ghana. The image is displayed at full spatial resolution, with approximate locations of national boundaries and 5° lines of latitude and longitude. The colour balance of the false colour comparile is adjusted to obtain the maximum contrast (Guo, 1991) and as a result the forest reserves are brown; cocoa/coffee plantations and woody scrub are light brown/red; cereal agriculture, irrigated agriculture and rubber plantations are red; bare soil, dry grassland and urban areas are blue/purple; clouds are white/yellow and smoke is pale yellow.

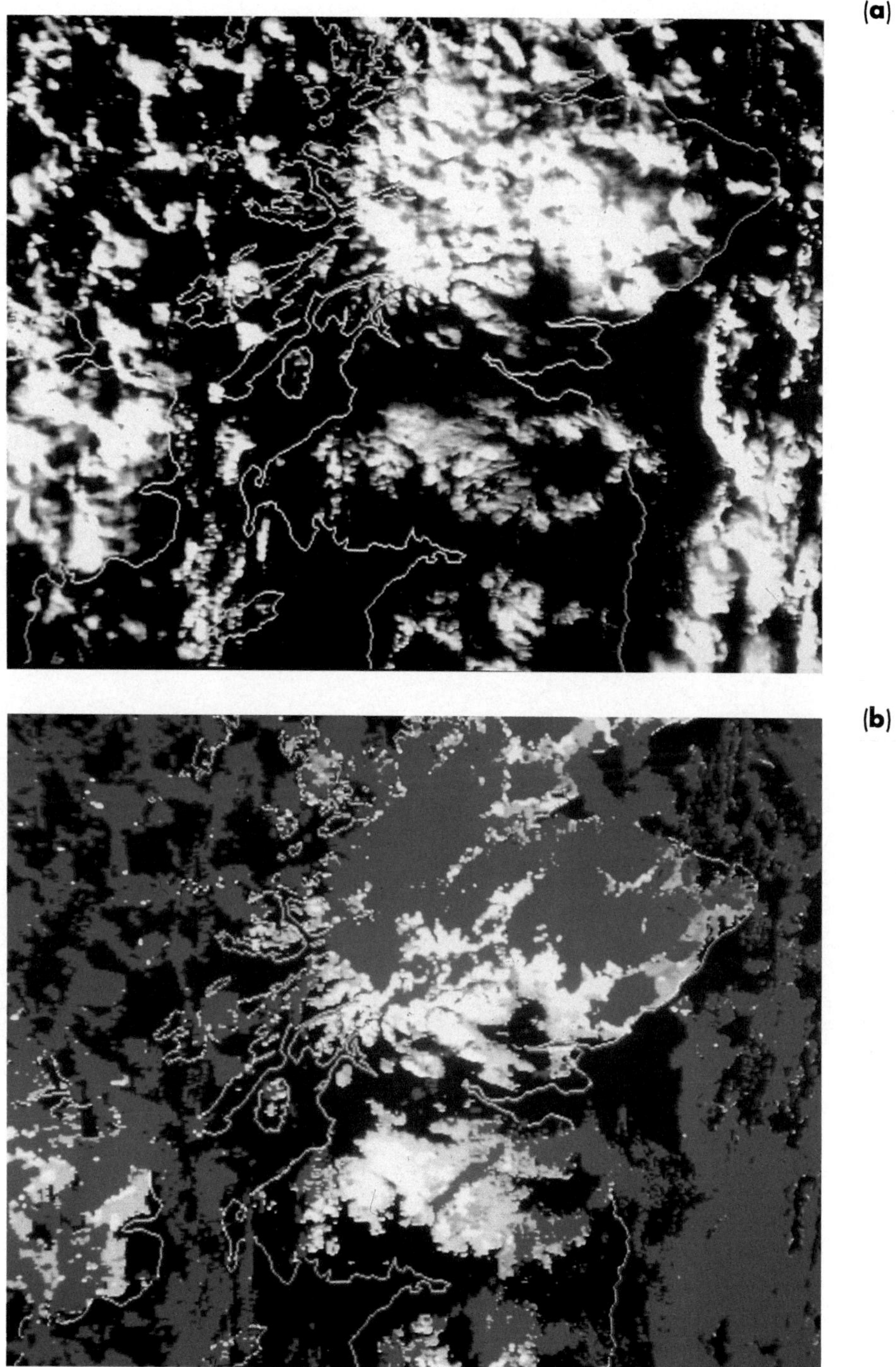

Plate 3. AVHRR imagery of northern Scotland (24 January, 1989). (**a**) Channel 1 image demonstrating an extreme case of cloud contamination within a snow-covered scene. (**b**) An unsupervised classification of the multispectoral imagery illustrating the availability of the algorithm for defining snow area (white) and detecting snow in forest (green).

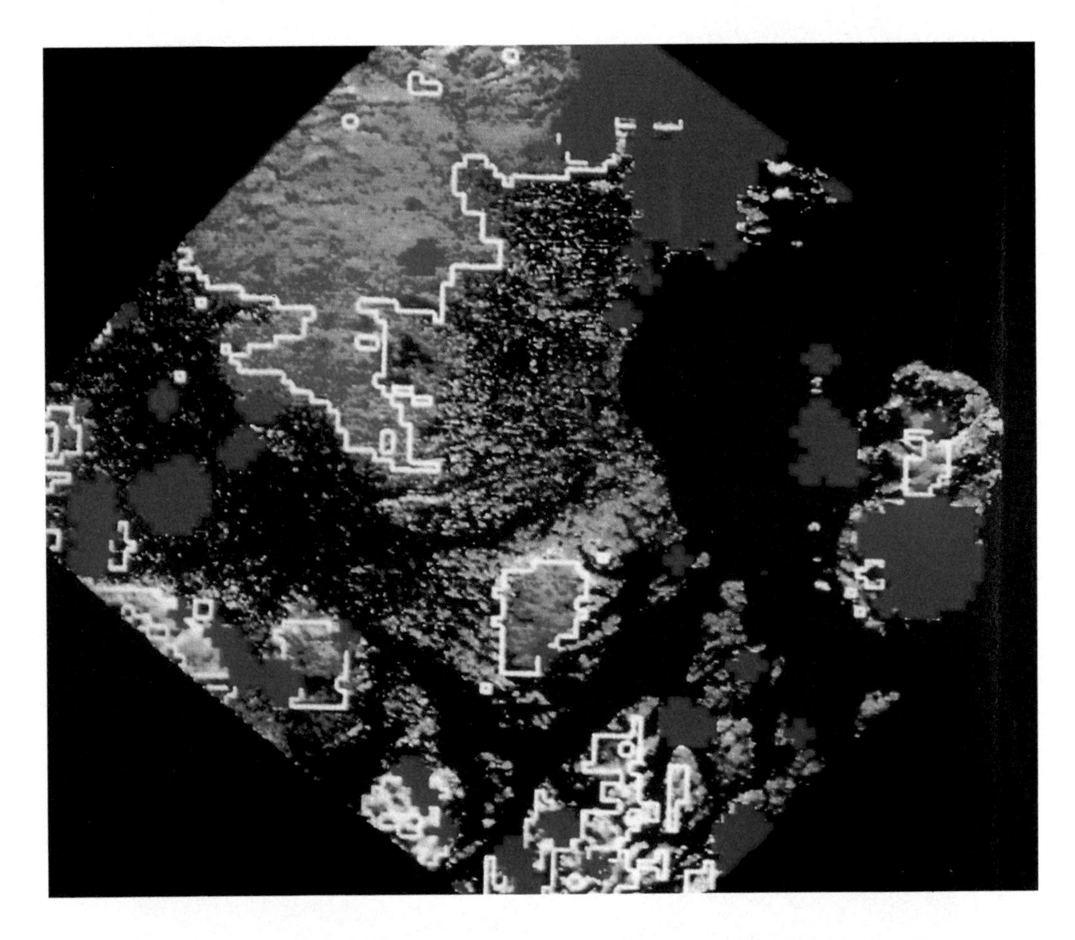

Plate 4. Landsat TM images of New Galloway showing areas of conifer forest, snow-covered and snow-free surfaces. The AVHRR-defined snow margin is displayed in white, and areas of cloud, identified by the algorithm, are shown in blue.

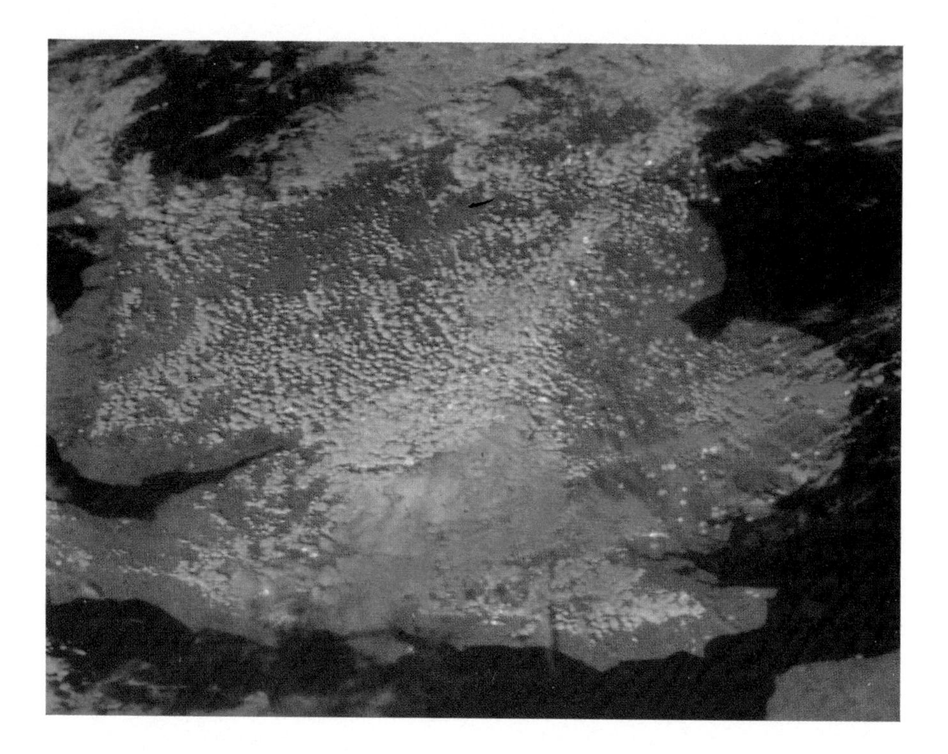

Plate 5. AVHRR colour composite for Friday 24 August, 1990.

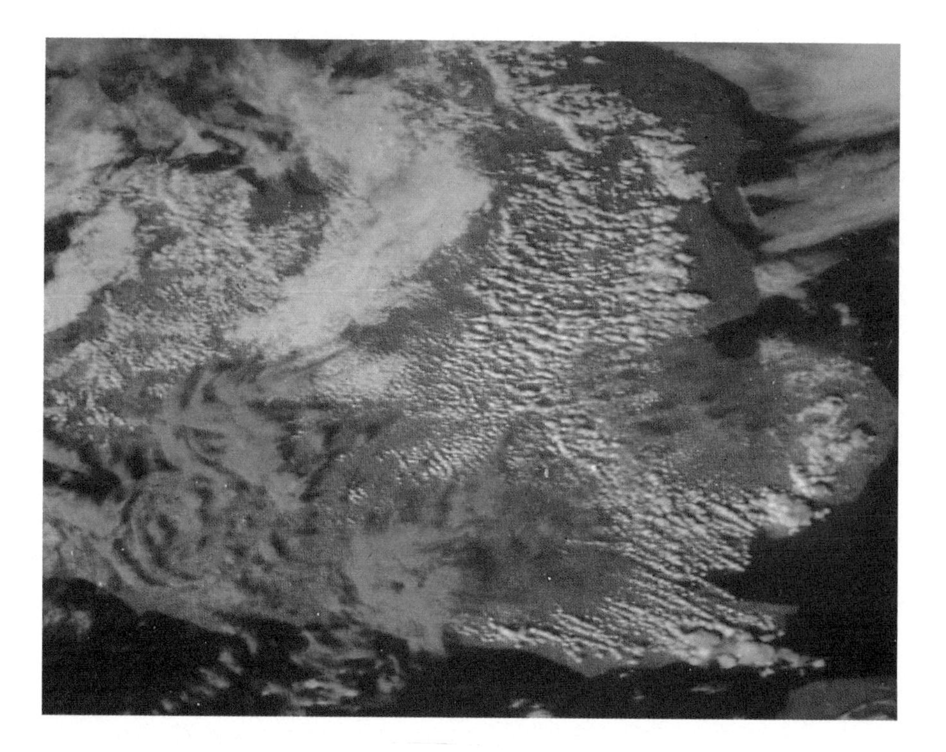

Plate 6. AVHRR colour composite for Saturday 25 August, 1990.

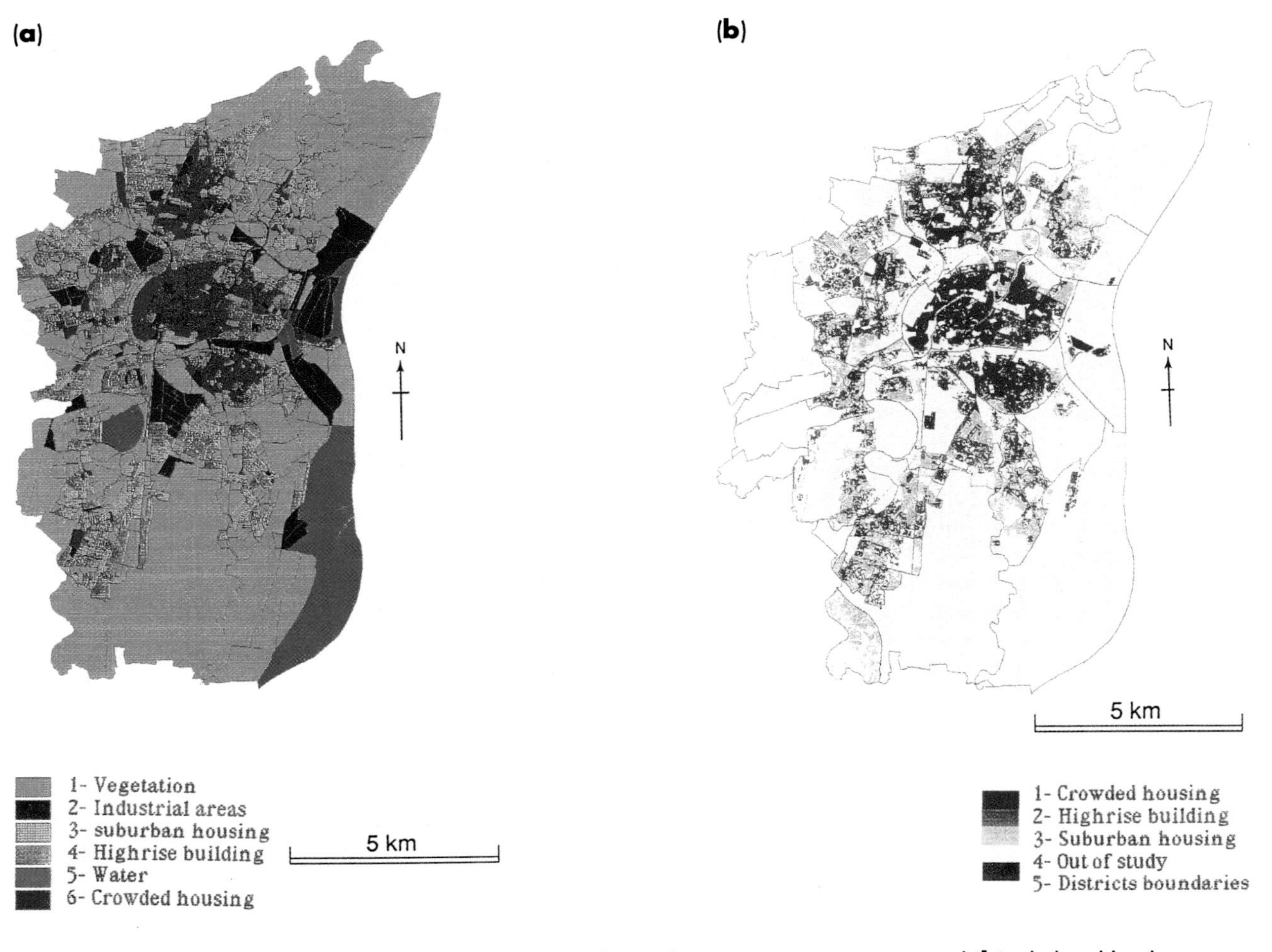

Plate 7. Strasbourg: **(a)** per-zone classification of Strasbourg from a SPOT HRV image, and **(b)** inhabited land cover.

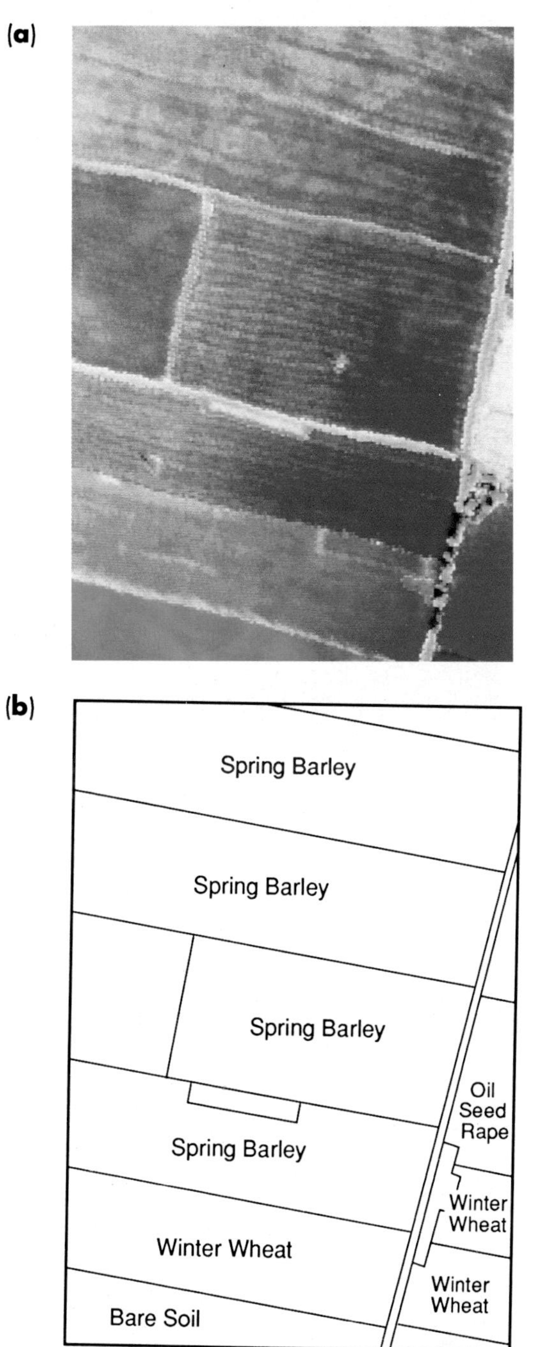

Plate 8. Diagram shows: (**a**) a standard false–colour composite produced using data recorded by an airborne multispectral scanner at near–IR, red and green wavelengths; (**b**) a sketch map of the principal land–types present in the study area (Gedney Hill, Lincolnshire, UK); (**c**) a false–colour composite of the same area produced using data recorded in a single spectral waveband (0.63–0.69μm, red) but at three separate sensor view angles; and (**d**) a multiple–view–angle composite produced using data recorded at near–infrared wavelengths (0.76–0.90μm).

(c)
(d)

9. Attempts to Drive Ecosystem Simulation Models at Local to Regional Scales

Paul J. Curran

'Geography is concerned with medium scale systems;
those lying somewhere between the scales
of the atom and the universe'
(Chorley and Kennedy, 1971, p. 4).

.... INTRODUCTION

Many of the ecological questions concerning the interplay between both *natural* and *anthropogenic* effects on our environment require accurate estimates of the net primary productivity (NPP) of terrestrial ecosystems, along with detailed information on its spatial variability, seasonality and annual variability (Dahlman, 1985). That we do not have such information has been recognised by many (NRC, 1986; Running, 1990), notably the International Geosphere-Biosphere Program (IGBP) which aims 'to quantify our knowledge of production and decomposition processes both regionally and globally, to determine the factors that control them and to understand their annual cycle and year-to-year variation' (Wickland, 1989, p. 697).

Ecologists have expended considerable effort in developing a quantitative understanding of short-term ecosystem processes and thereby productivity, at the scale of a plant, population and in some cases, community (Phillipson, 1966; Lieth and Whittaker, 1975). Much of this understanding has been achieved by modelling of some kind, ranging from relatively simple two-variable empirical models to more complex ecosystem simulation models (Jeffers, 1978). Applying these local scale models to ecosystems (at the regional scale) or the biosphere (at the global scale), and for time periods longer than those typical of a research grant, has proved to be difficult (Whittaker and Marks, 1975; Jeffers, 1988; Hobbs and Mooney, 1990). It is difficult because some models use 'scale-dependent' processes (Holling, 1992; Lathrop and Peterson, 1993) and empirical relationships (Begon *et al.*, 1990) but most importantly because the majority of these models are aspatial. To overcome these problems the models need to contain within them an understanding of the environment that is based on 'scale-independent' deterministic processes and to use data measured both on the ground and remotely to give spatial and temporal form to that understanding (Ehleringer and Field, 1993). These models can be given spatial form simply by applying them to geographic units. This can be done by determining the values of model-input parameters and variables for each of those geographic units by the use of remotely sensed data (Mather, 1992). In particular, vegetation can be located and identified via the classification of land cover, and then ecosystem variables that are both related to NPP and are input to the locally-based ecosystem simulation model can be estimated (IGBP, 1990; Wessman *et al.*, 1991). These ecosystem variables include, for example, the leaf area index (LAI) and the above ground biomass (O'Neill and DeAngelis, 1981; ESA, 1991).

The *potential* benefit of linking ecosystem simulation models with remotely sensed data is now well accepted (NASA, 1986; Schimel *et al.*, 1990; Goel and Norman, 1992).

> Two of the most important tools for extrapolating ecological understanding from local to regional scales are remote sensing and computer simulation modelling, two very synergistic technologies. Only a handful of the variables that terrestrial ecologists are interested in will ever be directly observable with optical satellite sensors. Most gas exchange processes such as photosynthesis, respiration, evapotranspiration and denitrification and soil processes such as decomposition and mineralization, cannot be detected by optical sensors. However, process-level simulation models that calculate the cycling of carbon, nitrogen and water through terrestrial ecosystems can simulate these invisible processes and are progressively being validated. While many of the earlier and current models require ground-based information (e.g., stem diameter) as driving variables, some recent efforts have been directed toward development of models that can utilize remote sensing data as input variables (Roughgarden *et al.*, 1991, pp. 1919–20).

This chapter aims to provide a primer to this research area by introducing ecosystem simulation models, remote sensing of vegetation and two studies involved with the input of remotely sensed data to ecosystem simulation models.

.... ECOSYSTEM SIMULATION MODELS

Ecosystem simulation models are simplified versions of reality (Stoddart, 1986) with many modes of construction (Thomas and Huggett, 1980). They can be divided into two broad and overlapping groups. The first group are the process-response models (Chorley and Kennedy, 1971) which are used primarily for environmental understanding and are based on the major functional interactions of carbon, water and nutrients. The second group are the morphological models (Chorley and Kennedy, 1971) which are used primarily for environmental management and are based on strong relationships between NPP, temperature, water, evapotranspiration and standing biomass (Aber and Melillo, 1991; Landsberg *et al.*, 1991). Within these two broad groups are a plethora of sub-groups (e.g., aggregated ecosystem, lumped parameter, gap-succession, physiological, nested-structural), each going part-way towards the eventual goal of a 'closed' control-system model (Chorley and Kennedy, 1971) in which all flows into, within and out of the system are accounted for (Luxmore *et al.*, 1986; Kaufmann and Landsberg,

1991). All of these ecosystem simulation models require data to parameterise them, drive them and, if possible, validate them. The model *parameters* are, by definition, *constant* in the case being considered and include, for instance, land cover and slope angle whereas the model *variables* used to drive the model *vary* in the case being considered and include, for instance, temperature and LAI (Webster and Oliver, 1990; Landsberg *et al.*, 1991).

By way of comparison, many of the current generation of *regional to global scale* soil-vegetation-atmosphere transfer (SVAT) models use land cover derived from remotely sensed data to parameterise and to infer the driving variables of albedo, biomass and roughness. Two of the earliest SVAT models designed to use remotely sensed data in this way were the Simple Biosphere (SiB) model (Sellers *et al.*, 1986) and the Biosphere-Atmosphere Transfer Scheme (BATS) model (Wilson *et al.*, 1987). Increasingly, attempts are being made to at least parameterise the *regional scale* global change models designed around commercially available geographical information systems (GIS). One example is the Modular Global Change Modelling System (MGCMS) (Aber *et al.*, 1993). This is a composite of three models: VEGIE that predicts NPP and nutrient uptake in relation to nutrients, water and light; MANE that models aqueous geochemistry and BROOK90 that is a general ecosystem modelling package developed for the study of the well known Hubbard Brook test site (Aber and Melillo, 1991). The remotely sensed databases in this GIS are used in parallel with those derived from cartographic sources.

However, the most demanding uses of remotely sensed data are to be found in the parameterisation and driving of *local to regional scale* ecosystem simulation models. These models tend to be based on dynamic ecosystem processes and to exploit the spatial, temporal and spectral capabilities of remotely sensed data (Asrar *et al.*, 1984; Waring *et al.*, 1993). Two such models developed specifically for forested landscapes are the FOREST-BGC model of Running and Coughlan (1988) and the labile carbon model of Cropper and Gholz (1993).

FOREST-BGC

The FOREST-BGC (Bio-Geo-Chemical cycling) model (Running and Coughlan, 1988) is a well-developed process-level simulation model that can be parameterised and driven using remotely sensed data. The model has two basic time resolutions with carbon and water exchange computed daily and nitrogen processes computed at least annually (Figure 9.1). To run the model requires the input of 20 state variables to define the initial conditions, 41 parameters and 15 driving variables (Running and Gower, 1991). Most of these inputs have species-specific defaults, thus putting the emphasis of the model on, first, the climatic variables of temperature, precipitation and radiation, which are available from local weather records, and second, LAI which can be estimated using remotely sensed data. Remotely sensed data could be used to estimate standing biomass and in the future remotely sensed data may be used to estimate both the state variable and parameter of leaf nitrogen concentration and the parameter of leaf lignin concentration (Curran, 1989).

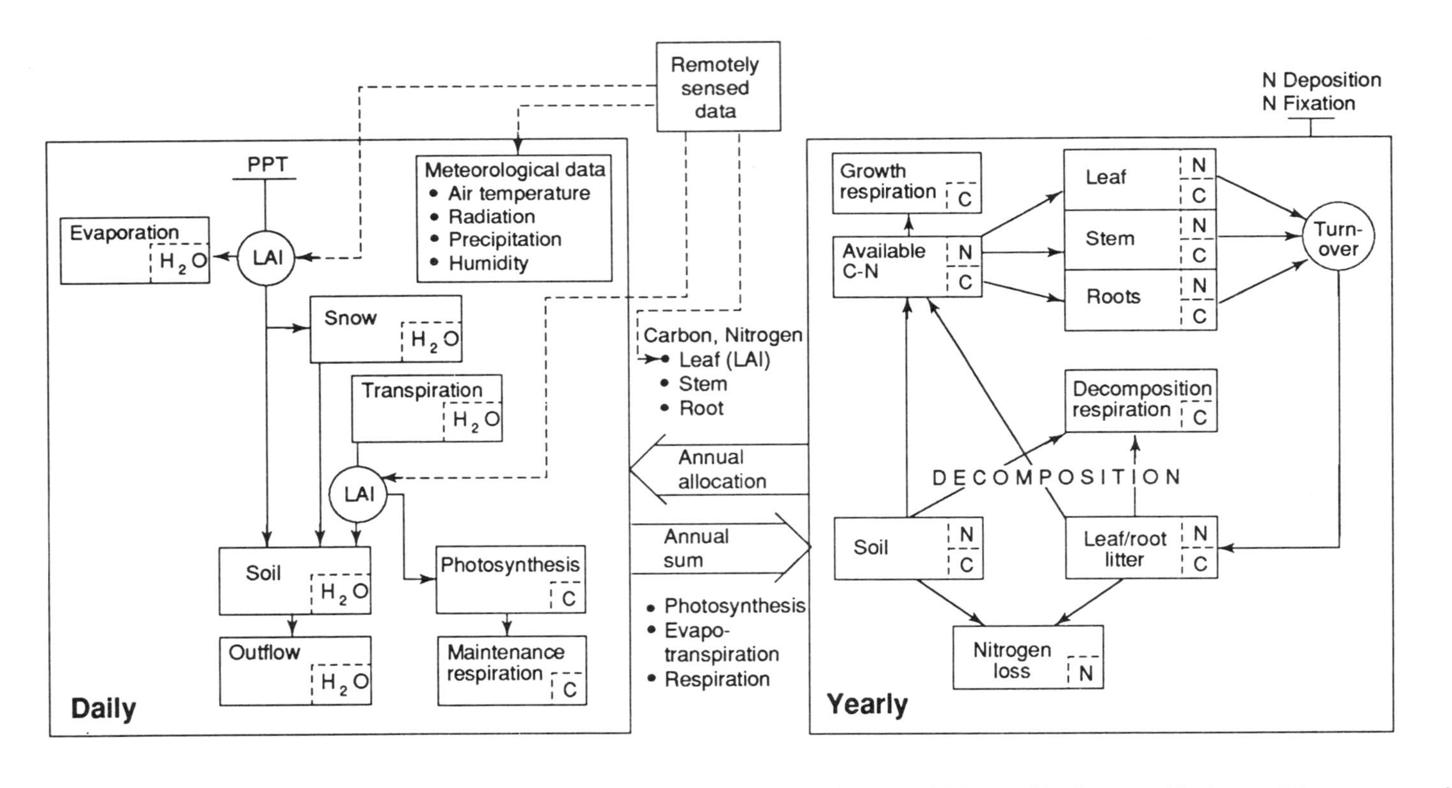

Figure 9.1 A flow diagram for FOREST-BGC (Bio-Geo-Chemical cycles); an ecosystem simulation model designed for forests and built around the movements of nitrogen (N), carbon (C), and water (H_2O). The two sub-models (left and right above) integrate daily and up to yearly movements. The major inputs to the model are meteorological data and LAI, and one of the outputs is NPP. (Modified from Running and Coughlan, 1988).

In addition to NPP the model provides estimates of snow melt, transpiration, soil water content and outflow, photosynthesis, maintenance and growth respiration, carbon allocation, litterfall, litter decomposition, nutrient uptake, nitrogen mineralisation and leaching. The performance of the model has been evaluated using both non-remote sensing inputs (McLeod and Running, 1988; Nemani and Running, 1989; Hunt *et al.*, 1991; Ryan, 1991) and remote sensing inputs (Running *et al.*, 1989). In addition, the annual NPP output from the model was found to be highly correlated with the annually integrated normalised difference vegetation index [NDVI = (near-infrared radiation) – (red radiation) / (near-infrared radiation) + (red radiation)] calculated using NOAA Advanced Very High Resolution Radiometer (AVHRR) imagery (Running and Nemani, 1988).

FOREST-BGC has now been used over a wide range of temporal and spatial scales and certainly has the flexibility of input required to produce a 'hierarchy of estimates of NPP for global ecological research' (Running, 1990, p. 87).

LABILE CARBON MODEL

The labile carbon model (Cropper and Gholz, 1990; 1993) is a model of stand level physiological processes in slash pine (*Pinus elliottii*). The model is constructed around the seasonal dynamics of labile carbon compounds (i.e., sugar and starch) at the level of the stand.

The model has been validated via physiological measurements on stands of slash pine (Cropper, 1988; Gholz *et al.*, 1991). Traditional sources are used to parameterise the model (e.g., forest age) and to drive part of the model (e.g., short-term meteorological variables). Remotely sensed data can be used to provide inputs such as meteorological data (either directly or indirectly), LAI and, in the future, foliar chemistry (Figure 9.2). There are several NPP-related outputs to the model, of which the three most important are net canopy carbon gain, maintenance respiration and finally 'growth potential' which is based on an estimate of the likely amount of carbon available for allocation to stems. This model is well developed but its use with remotely sensed data is still in its early stages.

Both FOREST-BGC and the labile carbon models require information on the location of the forest to parameterise the model, together with meteorological data, LAI and other NPP-related variables such as standing biomass and possibly canopy chemistry to drive the model. The challenge that this offers to current remote sensing instrumentation and procedures is discussed in the next section.

.... REMOTELY SENSED INPUTS TO ECOSYSTEM SIMULATION MODELS

Remotely sensed data are in essence a measure of radiation flux for a unit area of the Earth's surface. To parameterise or drive ecosystem simulation models, these radiation flux data need to be transformed into values that are related to an ecosystem component (e.g., area of forest) or process (e.g., Δ LAI) and are in some way superior to traditional sources of data.

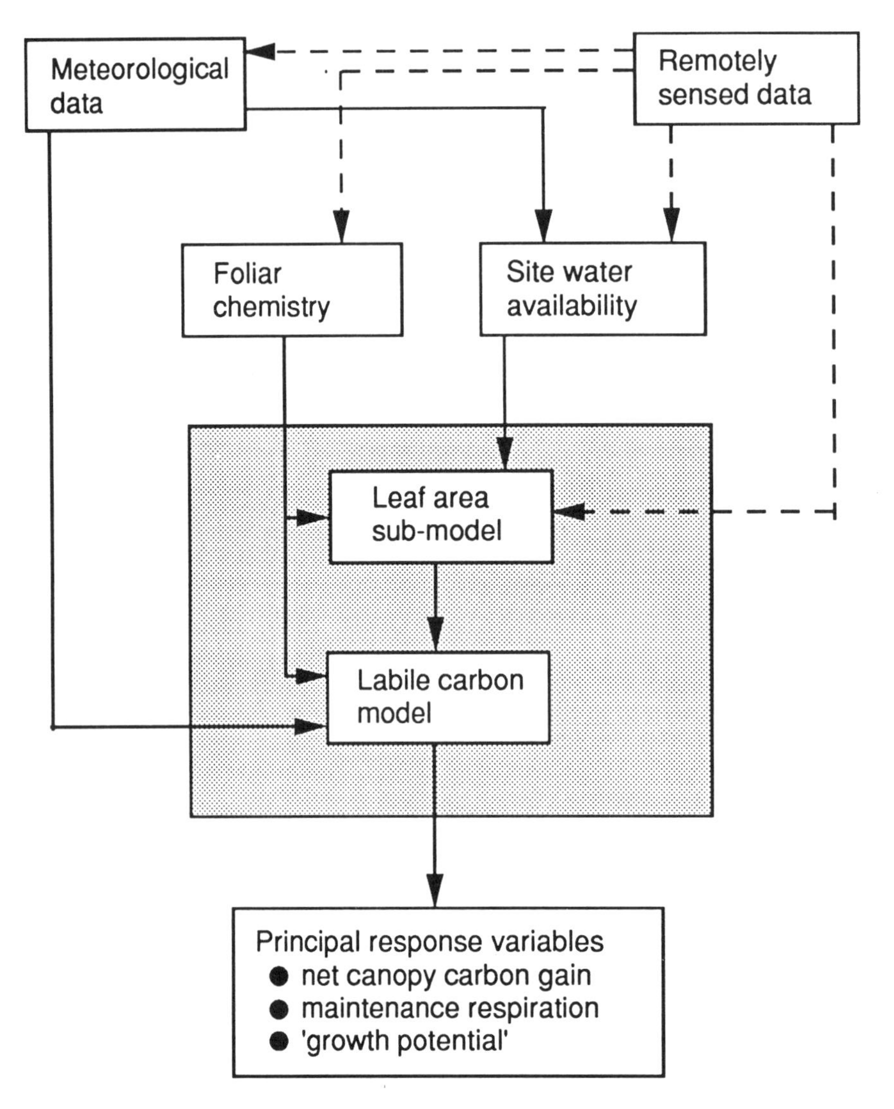

Figure 9.2 A flow diagram for the labile carbon model; an ecosystem simulation model designed for slash pine forests and built around the movement of the labile carbon compounds of sugar and starch. The major inputs to the model can be provided by traditional or remotely sensed data sources and all of the outputs are related to NPP.

Unfortunately, an ecological research sensor has never been launched (Wickland, 1991a) and so ecosystem simulation models must use sensors that are not designed or calibrated to record the subtle flux of radiation from a vegetation canopy (Roughgarden *et al.*, 1991). Despite this problem, data from existing sensors have been used with some success in environmental research. In particular, they have been used to provide nominal information (e.g., maps of geology, cloud cover and land cover) and interval or ratio data (e.g., sediment concentration in water, soil moisture, standing biomass, temperature, LAI and absorbed photosynthetically active radiation or APAR) (Curran, 1985; CEOS, 1992). However, only a few of these remotely sensed sources of data are reliable, accurate or available enough to parameterise or drive ecosystem simulation models. To date, attention has been focused upon land cover, LAI and standing biomass; *land cover* and *LAI* have been estimated using classifications and the NDVI transform on data from both the Landsat Thematic Mapper (TM) and NOAA AVHRR (ESA, 1991; Townshend, 1992; Vose *et al.*, 1993) and *standing biomass* has been estimated using SAR data (Wu and Sader, 1986; Wu, 1987; Kasischke and Christensen, 1990). These three remotely sensed estimates are of value to the study of ecological processes: land cover can be used to parameterise models, LAI is a key variable in determining the rates at which energy and matter are exchanged in ecosystems, while standing biomass can be used to both parameterise the models or to validate the NPP estimates of models.

The type of land cover and the amount of LAI and standing biomass each possess strong causal links with remotely sensed radiation flux. These causal links are respectively (i) a difference in the form and function, and thereby remotely sensed response, of each major land cover (Goward *et al.*, 1985; Graetz, 1990; Pedley and Curran, 1991); (ii) a negative relationship between red radiation and LAI, and a positive relationship between near-infrared radiation and LAI (Curran, 1983; Jensen, 1983; Danson and Curran, 1993); and (iii) a positive relationship between backscatter and standing biomass (Westman and Paris, 1987; Richards, 1990; Kasischke *et al.*, 1991). Recent research using airborne imaging spectrometers has pointed to the possibility of estimating those foliar chemicals that are related to ecosystem processes (Schimel *et al.*, 1990), notably chlorophyll that is related to the rate of photosynthesis, nitrogen that is related to the rate of carbon fixation and lignin that is related to the rate of decomposition (Wessman *et al.*, 1988; Curran, 1989; Wessman, 1990; 1992; Janetos *et al.*, 1992). Such work is still in its early stages (Wessman, 1990; Peterson, 1991; Curran, 1992) but could provide the fourth major input to ecosystem simulation models after land cover, LAI and standing biomass (Wessman, 1992; 1993). For example, foliar chemical data are possible inputs to both FOREST-BGC (Johnson *et al.*, 1993; Matson *et al.*, 1993) and the labile carbon model (Curran *et al.*, 1990a).

In recent years a number of sensors have been proposed for use on the Earth Observing System (EOS). Three sensors in particular were designed for ecological applications (Arvidson *et al.*, 1985; Wessman *et al.*, 1990; Dozier, 1991; Wickland, 1991a; 1991b) and could provide data to parameterise, drive and validate ecosystem simulation models (Ustin *et al.*, 1991). These sensors are the High Resolution Imaging Spectrometer (HIRIS), the Moderate

Table 9.1 Some of the biophysical measurements that could be derived from the High Resolution Imaging Spectrometer (HIRIS), the Moderate Resolution Imaging Spectrometer (MODIS) and the Synthetic Aperture Radar (SAR) for input to ecosystem simulation models. (Modified from Ustin *et al.*, 1991.) It is unfortunate from an ecological point of view that the HIRIS has not been selected for the Earth Observing System.

Vegetation parameter or variable	Sensor
Land cover class and area	HIRIS, SAR, MODIS
Fractional cover of soil and vegetation	HIRIS
Crown spacing and variance	HIRIS, SAR
Crown height	HIRIS, SAR
Green biomass	HIRIS, MODIS, SAR
Brown biomass	HIRIS, SAR
Phenologic state/seasonal growth	HIRIS, MODIS, SAR
Total chlorophyll concentration	HIRIS
Canopy cellulose concentration	HIRIS
Canopy lignin concentration	HIRIS
Foliar water content	HIRIS, SAR
Canopy water potential	SAR
Soil moisture	SAR
Vegetative APAR	HIRIS, MODIS
Non-vegetative APAR	HIRIS
Evapotranspiration	MODIS

Resolution Imaging Spectrometer (MODIS) and the Synthetic Aperture Radar (SAR) (Table 9.1). The two most developed of these sensors, MODIS and SAR, have now been selected for EOS (Moore and Dozier, 1992).

PREPARING TO DRIVE AN ECOSYSTEM SIMULATION MODEL WITH REMOTELY SENSED DATA

Few experiments have been performed that use remotely sensed data to parameterise, drive or validate ecosystem simulation models (Peterson and Waring, 1993). The two experiments outlined here are currently in progress and have a common aim, that of using remotely sensed data to drive an ecosystem simulation model for a forest (Curran *et al.*, 1990a; Danson *et al.*, 1992).

DRIVING FOREST-BGC, AN EXAMPLE IN WALES

The study site is a 28–35 year old (in 1992) Sitka spruce (*Picea sitchensis*) plantation in central southern Wales. The site is hilly, the climate is temperate and the nearly-closed canopy has no understorey (Curran, 1992). In 1991 thirteen relatively homogeneous stands were identified that spanned the range between well-drained with high productivity to less well-drained with low productivity (Figure 9.3). Within each of these stands, between three and five 50 × 50 m plots were established resulting in a sample of 51 plots.

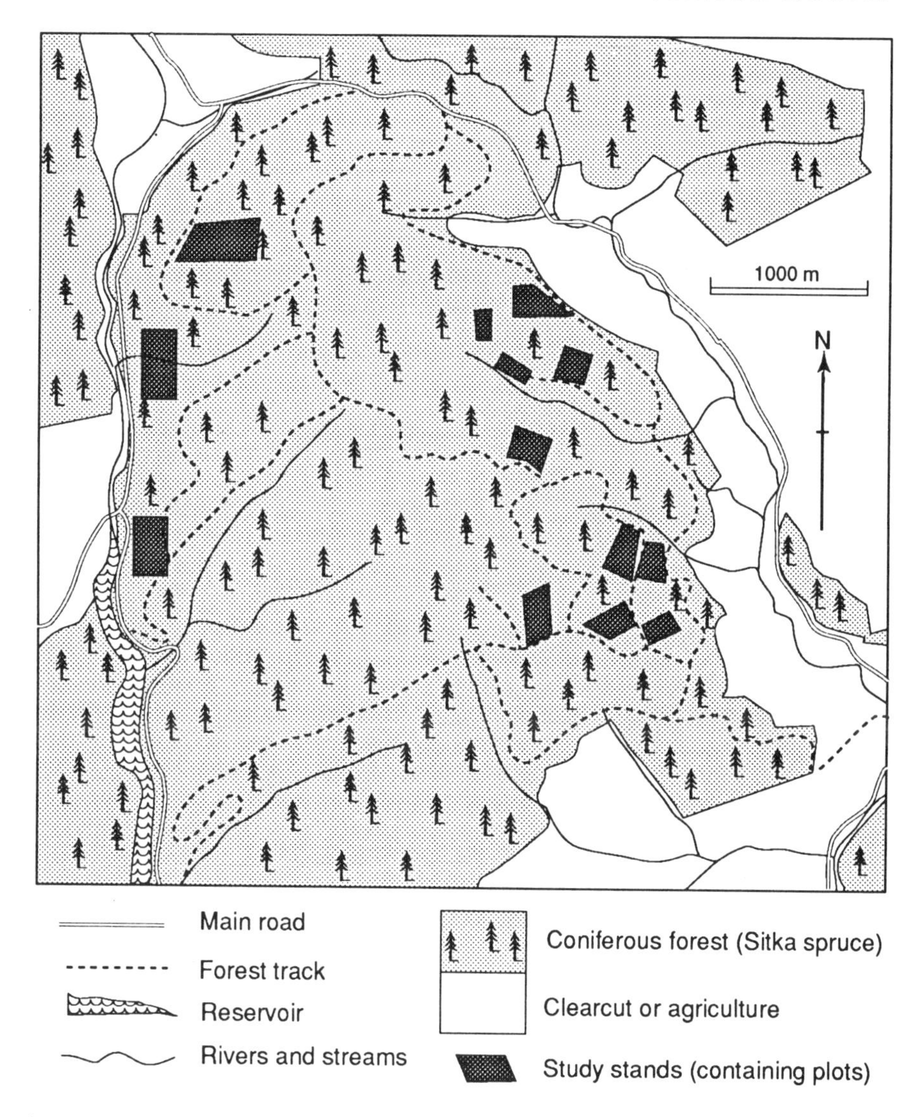

Figure 9.3 The study site at Llyn Brianne in Wales is a Sitka spruce plantation on very rugged terrain with 51 plots that vary between well-drained with high productivity to less well-drained with low productivity.

During 1991 the diameter at breast height (dbh) and tree height were measured for all of the trees within the central 1 ha of each plot. In July and August 1991, 11 trees were felled and leaves sampled. The relationship between dbh and LAI was very strong (R^2 = 0.97) and so the dbh was used to estimate the LAI of each plot. In addition the dbh and tree height were used to estimate the standing biomass of each plot (Danson *et al.*, 1992). In

Table 9.2 Sensors and aircraft used to collect remotely sensed data at Llyn Brianne, Wales, in 1991. The non-private aircraft owners are the National Aeronautics and Space Administration (NASA), USA and the Natural Environment Research Council (NERC), UK.

Sensor	Aircraft	Objectives
Airborne Visible/Infrared Imaging Spectrometer (AVIRIS)	NASA/ER2	Foliar chemistry
Daedalus 1268 Airborne Thematic Mapper (ATM)	NERC/Piper Chieftain	Leaf area index
Compact Airborne Spectrographic Imager (CASI)	NERC/Piper Chieftain	Chlorophyll concentration
Spectron spectroradiometer (SE590)	Private/Bell JetRanger	Point validation of ATM and CASI data
Synthetic Aperture Radar (AIRSAR)	NASA/DC8	Standing biomass

July 1991, 155 foliar samples were collected from the upper canopy. The concentrations of water, chlorophyll, lignin, nitrogen and cellulose in these samples were measured using standard wet laboratory analysis methods (Horwitz, 1970). There was little spatial variability in water and lignin but great spatial variability in the other foliar chemicals. For example, the foliar chlorophyll concentration on the well-drained and most productive plots was 11 per cent greater than on the less well-drained and less productive plots. Five types of digital remotely sensed data were collected during July and August 1991 (Table 9.2) as part of a joint National Aeronautics and Space Administration (NASA), USA/Natural Environment Research Council (NERC), UK project (Curran and Plummer, 1992).

Progress has been made towards the goals of estimating LAI with airborne thematic mapper (ATM) data, standing biomass with SAR data acquired by the AIRSAR system and chlorophyll concentration with data from the compact airborne spectrographic imager (CASI). These data in conjunction with climatic data will then form the drivers for FOREST-BGC.

DRIVING THE LABILE CARBON MODEL, AN EXAMPLE IN FLORIDA

The study site is a 27 year old (in 1992) slash pine (*Pinus elliottii*) plantation in northern Florida. The site is flat, the climate is sub-tropical and the open canopy has an understorey of shrubs and palmetto (Gholz and Fisher, 1982; Gholz *et al.*, 1991). In 1986, sixteen 50 $\times$ 50 m plots were established at the study site (Figure 9.4). Half of these plots were fertilised and the other half were controls.

The LAI of each plot was determined monthly from mid 1986 to early 1990. The procedures, detailed in Gholz *et al.* (1991), comprised annual LAI measurements using 20 to 30 trees and seasonal interpolation of these measurements using plot-level averages of needle growth and needle loss.

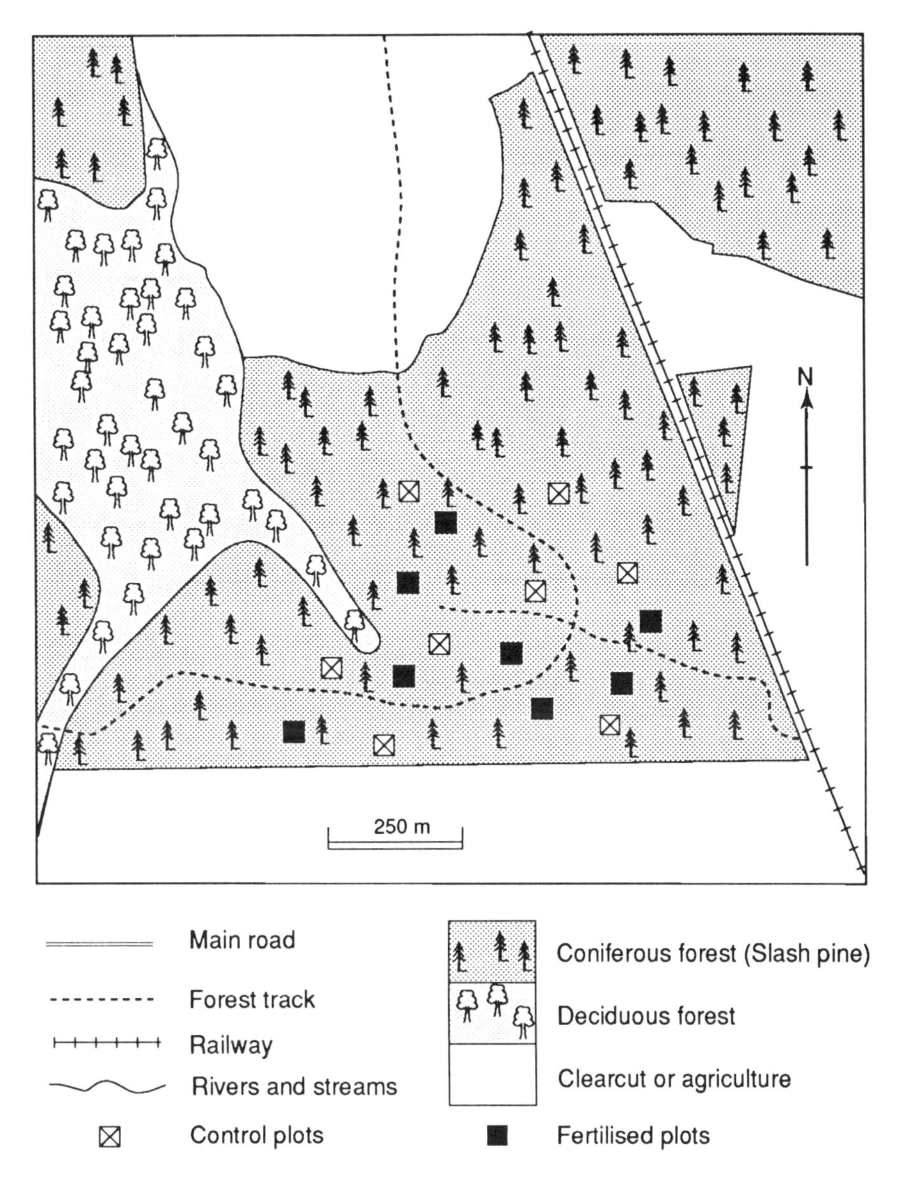

Figure 9.4 The study site near to Gainesville in Florida is a slash pine plantation on flat terrain with eight fertilised plots and eight control plots.

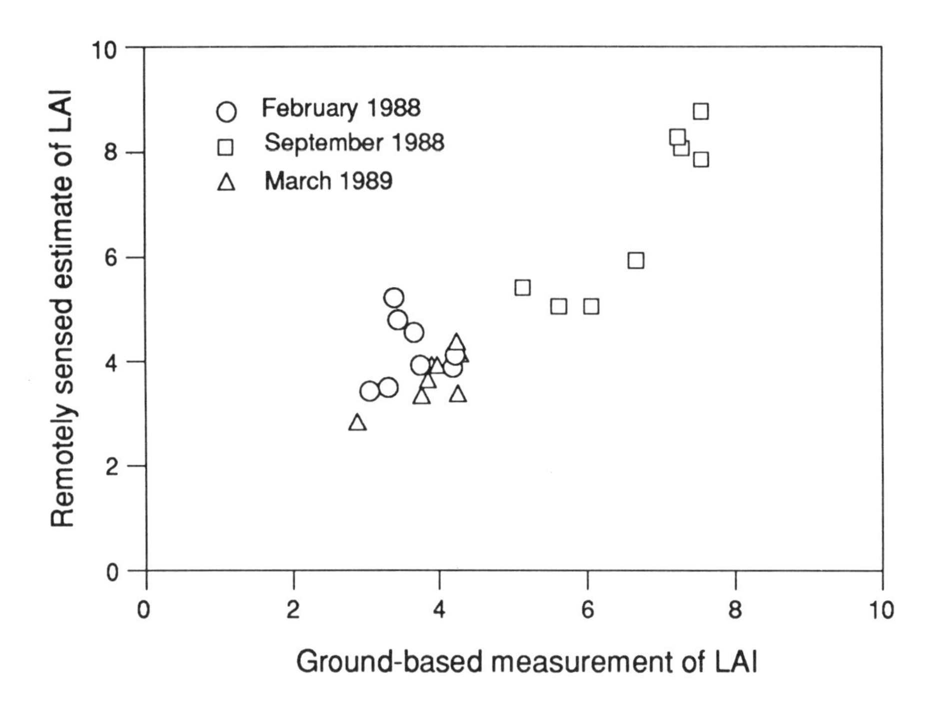

Figure 9.5 The relationship between the measured and estimated LAI for eight plots of *Pinus elliottii* in Florida, on three dates. The estimated LAI was derived using Landsat Thematic Mapper data and three predictive regression relationships, each based on data from a further eight plots on three dates.

Foliar samples were shot from the top of the canopy in each plot with 124, 162, 166, 160, 160, 320 and 360 samples being collected in February 1988, September 1988, July 1989, March 1990, September 1990, August 1991 and July 1992 respectively. The concentration of water, chlorophyll, lignin, nitrogen and cellulose in these samples is being measured using standard wet laboratory analysis methods (Horwitz, 1970). There was little spatial variability in water and lignin but great spatial variability in other foliar chemicals. For example, the foliar chlorophyll concentration on the fertilised plots was 66 per cent greater than on the control plots.

Two types of remotely sensed data were collected: Landsat TM in February 1988, September 1988 and March 1989 and Airborne Visible/Infrared Imaging Spectrometer (AVIRIS) data in March 1990, September 1990 and June 1992. All of the data were corrected radiometrically and atmospherically and, in addition, the Landsat TM data were standardised in terms of irradiance.

The Landsat TM data were used to calculate the NDVI which was then related to ground-based measurements of LAI for the 16 plots. There was a positive linear relationship between NDVI and LAI on all three dates with R^2 values of 0.35, 0.75 and 0.86 for February 1988, September 1988 and March 1989 respectively. The constants and NDVI coefficients varied as a result of a number of factors, the main one was hypothesised to be seasonal changes in understorey LAI.

Using data from half of the plots on each date three predictive relationships were used to estimate the LAI on the other plots (Figure 9.5), with a root-mean-square error of 0.74 LAI. These estimates were accurate to 15.6 per cent of the mean LAI and tracked the LAI increase from spring to autumn and the LAI decrease from autumn to spring (Curran *et al.*, 1992). This initial study demonstrated the potential of using Landsat TM imagery for estimating the stand-level LAI required for input to the labile carbon model.

Research on the estimation of foliar chemical concentration using AVIRIS data is underway (Janetos *et al.*, 1992; Curran, 1992). Preliminary observation of raw spectra indicate strong relationships between the slope of the spectra in the region of the 'red edge' (Curran *et al.*, 1990b) and foliar chlorophyll concentration. It is hoped that once the data set is complete both remotely sensed estimates of LAI and foliar chemistry will form drivers for the labile carbon model.

.... CONCLUDING COMMENTS

The linking of remotely sensed data with ecosystem simulation models is vital if we are to understand how our planet functions at regional to global scales.

> Satellites offer the potential of continuously viewing large segments of the earth's surface, thus documenting the changes that are occurring. The task, however, is not only to document global change, which will be an enormous job, but also to understand the significance of these changes to the biosphere. Effects on the biosphere may cover all spatial scales from global to local. The possibility of measuring biosphere function remotely and continuously from satellite imagery must be explored quickly and thoroughly in order to meet the challenge of understanding the consequences of global change. (Mooney and Hobbs, 1990, p. 1.)

The need for such an approach is now accepted (IGBP, 1990) and today there are several research groups around the World involved with:

> the modelling of terrestrial surface processes with the eventual aim of driving such models by remotely sensed data alone. An example of this approach is the Running model of forest ecosystems. This is intellectually the most challenging of uses of remotely sensed data, the one which is of greatest interest and potentially the most rewarding in terms of our understanding of the Earth's land surface. It is also the most difficult and least well advanced. (Briggs, 1991, p. 4.)

This chapter has been a brief introduction to this area of development in environmental remote sensing.

.... ACKNOWLEDGEMENTS

The research discussed in this chapter was supported by the Natural Environment Research Council (UK) and the National Aeronautics and Space Administrations' Earth Sciences Applications Division (USA). I am pleased to acknowledge the research collaboration of many individuals, notably

G. Foody, G. Smith, J. Kupiec, M. McCulloch, N. Lucas, R. Green, A. Pinar (University College of Swansea), D. Peterson, J. Dungan (NASA/Ames Research Center), H. Gholz, W. Cropper Jr. (University of Florida), S. Plummer (BNSC/NERC Remote Sensing Applications Development Unit), M. Danson, T. Powell (University of Salford), C. Milner (University of Cambridge), E. Milton, E. Rollin (University of Southampton), T. O'Neill (University of Wollongong) and those unsung heroes who helped with the collection of ground data at the field sites in Wales and Florida. Final thanks go to N. Lucas, J. Dungan and G. Foody for their useful comments on the manuscript.

.... REFERENCES

Aber, J.D., Melillo, J.M., 1991, *Terrestrial ecosystems*, Saunders College Publishing, Philadelphia

Aber, J.D., Driscoll, C., Federer, C.A., Lathrop, R., Lovett, G., Melillo, J.M., Steadler, P., Vogelman, J., 1993, 'A strategy for the regional analysis of the effects of physical and chemical climate change on biogeochemical cycles in Northeastern (U.S.) forests', *Ecological Modelling*, **67**: 37–47

Arvidson, R.E., Butler, D.M., Hartle, R.E., 1985, 'Eos: The Earth observing system of the 1990s', *Proceedings IEEE*, **73**: 1025–30

Asrar, G., Fuchs, M., Kanemasu, E.T., Hatfield, J.L., 1984, 'Estimating absorbed photosynthetic radiation and leaf area index from spectral reflectance in wheat', *Agronomy Journal*, **76**: 300–6

Begon, M., Harper, J.L., Townsend, C.R., 1990, *Ecology, individuals populations and communities* (Second Edition), Blackwell, Oxford

Briggs, S.A., 1991, *Note on the relevance of future Earth observation system options to terrestrial science*, Natural Environment Research Council/British National Space Centre, Remote Sensing Applications Development Unit, Abbots Ripton, 19pp.

CEOS, 1992, *The relevance of satellite missions to the study of the global environment*, Committee on Earth Observation Satellites, British National Space Centre, London

Chorley, R.J., Kennedy, B.A., 1971, *Physical geography: a systems approach*, Prentice Hall, London

Cropper, Jr., W.P., 1988, 'Labile carbon dynamics in a Florida slash pine plantation', *Forest growth modelling and prediction*, General Technical Report NC-120, USDA Forest Service, Washington, D.C.: 278–84

Cropper, Jr., W.P., Gholz, H.L., 1990, 'Modelling the labile carbon dynamics of a Florida slash pine plantation', *Silva Carelia*, **15**: 121–30

Cropper, Jr., W.P., Gholz, H.L., 1993, 'Simulation of the carbon dynamics of a Florida slash pine plantation', *Ecological Modelling*, **66**: 231–49

Curran, P.J., 1983, 'Multispectral remote sensing for the estimation of green leaf area index', *Philosophical Transactions of the Royal Society, Series A*, **309**: 257–70

Curran, P.J., 1985, *Principles of remote sensing*, Longman Scientific and Technical, Harlow

Curran, P.J., 1989, 'Remote sensing of foliar chemistry', *Remote Sensing of Environment*, **30**: 271–8

Curran, P.J., 1992, 'Estimating foliar chemical concentrations with the airborne visible/infrared imaging spectrometer (AVIRIS)', *International Archives of Photogrammetry and Remote Sensing*, Commission VII: 705–8

Curran, P.J., Plummer, S.E., 1992, 'Remote sensing of forest productivity', *NERC News*, **20**: 22–3

Curran, P.J., Gholz, H.L., Cropper, Jr., W.P., 1990a, 'Input of Landsat TM and AVIRIS data to a model of slash pine productivity', *International Congress of Ecology*, INTECOL, Yokohama, 189

Curran, P.J., Dungan, J.L., Gholz, H.L., 1990b, 'Exploring the relationship between reflectance red edge and chlorophyll content in slash pine', *Tree Physiology*, **7**: 33–48

Curran, P.J., Dungan, J.L., Gholz, H.L., 1992, 'Seasonal LAI in slash pine estimated with Landsat TM', *Remote Sensing of Environment*, **39**: 3–13

Dahlman, R.C., 1985, 'Modelling needs for predicting responses to CO_2 enrichment: plants, communities and ecosystems', *Ecological Modelling*, **29**: 77–106

Danson, F.M., Curran, P.J., 1993, 'Factors affecting the remotely sensed response of coniferous forest plantations', *Remote Sensing of Environment*, **43**: 55–65

Danson, F.M., Curran, P.J., Plummer, S.E., 1992, 'Remotely-sensed inputs to a forest ecosystem model', *Proceedings, 6th Australasian Remote Sensing Conference*, University of Christchurch, Christchurch, 130–9

Dozier, J., 1991, 'Recommended instruments for the restructured Earth Observing System (EOS)', *The Earth Observer*, **3**: 1–8

Ehleringer, J.R., Field, C.B. (editors), 1993, *Scaling physiological processes: leaf to globe*, Academic Press, San Diego

ESA, 1991, *Report of the Earth observation user consultation meeting*, European Space Agency SP-1143, Noordwijk

Gholz, H.L., Fisher, R.F., 1982, 'Organic matter production and distribution in slash pine (*Pinus elliottii*) plantations', *Ecology*, **63**: 1827–39

Gholz, H.L., Vogel, S.A., Cropper, Jr., W.P., McKelvey, K., Ewel, K.C., Tesky, R.O., Curran, P.J., 1991, 'Dynamics of canopy structure and light interception in *Pinus elliottii* stands, North Florida', *Ecological Monographs*, **61**: 33–51

Goel, N.S., Norman, J.M., 1992, 'Biospheric models, measurements and remote sensing of vegetation', *ISPRS Journal of Photogrammetry and Remote Sensing*, **47**: 163–88

Goward, S.N., Tucker, C.J., Dye, D.G., 1985, 'North American vegetation patterns observed with the NOAA-7 Advanced Very High Resolution Radiometer', *Vegetatio*, **64**: 3–14

Graetz, R.D., 1990, 'Remote sensing of terrestrial ecosystem structure: an ecologist's pragmatic view', in Hobbs, R.J., Mooney, H.A. (editors), *Remote Sensing of Biosphere Functioning*, Springer-Verlag, London: 5–30

Hobbs, R.J., Mooney, H.A. (editors), 1990, *Remote sensing of biosphere functioning*, Springer-Verlag, New York

Holling, C.S., 1992, 'Cross-scale morphology, geometry and dynamics of ecosystems', *Ecological Monographs*, **62**: 447–502

Horwitz, W., 1970, *Official methods of analysis* (Eleventh Edition), Association of Analytical Chemists, Washington, D.C.

Hunt, Jr. E.R., Martin, F.C., Running, S.W., 1991, 'Simulating the effects of climatic variation on stem carbon accumulation of a Ponderosa pine stand: comparison with annual growth increment data', *Tree Physiology*, **9**: 161–71

IGBP, 1990, *The International Geosphere-Biosphere Programme: A study of global change: the initial core projects*, International Geosphere-Biosphere Programme, Stockholm

Janetos, A.C., Aber, J., Wickland, D.E., 1992, *Workshop report: measuring canopy chemistry with high spectral resolution remote sensing data*, NASA White Paper, NASA Headquarters, Washington D.C.

Jeffers, J.N.R., 1978, *An introduction to systems analysis: with ecological applications*, Edward Arnold, London
Jeffers, J.N.R., 1988, 'Statistical and mathematical approaches to issues of scales in ecology', in Rosswall, T., Woodmansee, R.E., Risser, P.G. (editors), *Scales and global change*, Wiley, Chichester: 47–56
Jensen, J.R., 1983, 'Biophysical remote sensing', *Annals, Association of American Geographers*, 73: 111–32
Johnson, L.F., Hlavka, C.A., Peterson, D.L., 1993, 'Multivariate analysis of AVIRIS data for canopy biochemical determination along the Oregon Transect', *Remote Sensing of Environment*, in press
Kasischke, E.S., Christensen Jr., N.L., 1990, 'Connecting forest ecosystem and microwave backscatter models', *International Journal of Remote Sensing*, 11: 1277–98
Kasischke, E.S., Bougeau-Chavez, L.L., Christensen, Jr. N.L., Dobson, M.L., 1991, 'Relationship between aboveground biomass and radar back-scatter as observed on airborne SAR imagery', *Proceedings, third airborne synthetic aperture radar (AIRSAR) workshop*, National Aeronautics and Space Administration, Jet Propulsion Laboratory, Pasadena, California, 11–21
Kaufmann, M.R., Landsberg, J.J. (editors), 1991, *Advancing towards closed models of forest ecosystems* (Special Publication of *Tree Physiology*, 9), Heron Publishing, Victoria
Landsberg, J.J., Kaufmann, M.R., Binkley, D., Isebrands, J., Jarvis, P.G., 1991, 'Evaluating progress toward closed forest models based on fluxes of carbon, water and nutrients', *Tree Physiology*, 9: 1–15
Lathrop, R.G., Peterson, D.L., 1993, 'Identifying structural self-similarity in mountainous landscapes', *Landscape Ecology* (in press)
Lieth, H., Whittaker, R.H., 1975, *Primary productivity of the biosphere*, Springer-Verlag, New York
Luxmore, R.J., Landsburg, J.J., Kaufmann, M.R., 1986, *Coupling of carbon, water and nutrient interactions in woody plant soil systems* (Special Publication of *Tree Physiology*, 2), Heron Publishing, Victoria
Matson, P.A., Johnson, L.F., Miller, J.R., Billow, C.R., Pu, R., 1993, 'Seasonal changes in canopy chemistry across the Oregon Transect: patterns and spectral measurement with remote sensing,' *Ecological Applications* (in press)
Mather, P.M., 1992, 'Remote sensing and geographical information systems', in Mather, P.M. (editor), *TERRA-1: understanding the terrestrial environment*, Taylor and Francis, London: 211–19
McLeod, S.D., Running, S.W., 1988, 'Comparing site quality indices and productivity in Ponderosa pine stands of western Montana', *Canadian Journal of Forest Research*, 18: 346–52
Mooney, H.A., Hobbs, R.J., 1990, 'Introduction', in Hobbs, R.J., Mooney, H.A. (editors), *Remote sensing of biosphere functioning*, Springer-Verlag, New York: 1–4
Moore III, B., Dozier, J., 1992, 'Adapting the Earth Observing System to the projected $8 billion budget: Recommendations from the EOS investigators', *The Earth Observer*, 4: 3–10
NASA, 1986, *MODIS: moderate-resolution imaging spectrometer instrument panel report*, National Aeronautics and Space Administration, Washington, D.C.
Nemani, R.R., Running, S.W., 1989, 'Testing a theoretical climate-soil-leaf area hydrologic equilibrium of forests using satellite data and ecosystem simulation', *Agricultural and Forest Meteorology*, 44: 245–60
NRC, 1986, *Global change in the geosphere-biosphere: initial priorities for an*

IGBP, National Research Council, National Academy Press, Washington, D.C.

O'Neill, R.V., DeAngelis, D.L., 1981, Comparative productivity and biomass relations of forest ecosystems', in Reichle, D.E. (editor), *Dynamic properties of forest ecosystems*, Cambridge University Press, Cambridge: 411–49

Pedley, M.I., Curran, P.J., 1991, 'Per-field classification: an example using SPOT HRV imagery', *International Journal of Remote Sensing*, **12**: 2181–92

Peterson, D.L., 1991, *Report on the workshop, remote sensing of plant biochemical content: theoretical and empirical studies*, NASA White Paper, NASA Ames Research Center, California

Peterson, D.L., Waring, R.H., 1993, 'Overview of the Oregon Transect Ecosystem Research project', *Ecological Applications* (in press)

Phillipson, J., 1966, *Ecological energetics*, Edward Arnold, London

Richards, J.A., 1990, 'Radar backscatter modelling of forests: a review of current trends', *International Journal of Remote Sensing*, **11**: 1299–1312

Roughgarden, J., Running, S.W., Matson, P.A., 1991, 'What does remote sensing do for ecology?', *Ecology*, **72**: 1918–22

Running, S.W., 1990, 'Estimating terrestrial primary productivity by combining remote sensing and ecosystem simulation', in Hobbs, R.J., Mooney, H.A. (editors), *Remote sensing of biosphere functioning*, Springer-Verlag, New York: 65–86

Running, S.W., Nemani, R.R., 1988, 'Relating seasonal patterns of the AVHRR vegetation index to simulated photosynthesis and transpiration of forests in different climates', *Remote Sensing of Environment*, **24**: 347–67

Running, S.W., Coughlan, J.C., 1988, 'A general model of forest ecosystem process for regional applications, 1. Hydrologic balance, canopy gas exchange and primary production processes', *Ecological Modelling*, **42**: 125–54

Running, S.W., Gower, S.T., 1991, 'FOREST-BGC, A general model of forest ecosystem processes for regional applications, II. Dynamic carbon allocation and nitrogen budgets', *Tree Physiology*, **9**: 147–60

Running, S.W., Nemani, R.R., Peterson, D.L., Band, L.E., Potts, D.F., Pierce, L.L., Spanner, M.A., 1989, 'Mapping regional forest evapotranspiration and photosynthesis by coupling satellite data with ecosystem simulation', *Ecology*, **70**: 1090–101

Ryan, M.G., 1991, 'A simple method for estimating gross carbon budgets for vegetation in forest ecosystems', *Tree Physiology*, **9**: 255–66

Schimel, D., Wessman, C., Parton, W., Curtiss, B., Knapp, A., Seastedt, T., Goet, A., 1990, 'Modelling of terrestrial ecosystems with remotely sensed inputs', *Proceedings, IEEE Geosciences and Remote Sensing*, IEEE, New York: 557

Sellers, P.J., Mintz, Y., Sud, Y.C., and Dalcher, A., 1986, 'A simple biosphere model (SiB) for use with general circulation models', *Journal of Atmospheric Sciences*, **43**: 505–31

Stoddart, D.R., 1986, *On geography*, Basil Blackwell, Oxford

Thomas, R.W., Huggett, R.J., 1980, *Modelling in geography: a mathematical approach*, Harper and Row, London

Townshend, J.R.G. (editor), 1992, *Improved global data for land applications*, IGBP Report 20, International Geosphere-Biosphere Programme, Stockholm

Ustin, S.L., Wessman, C.A., Curtiss, B., Kasischke, E., Way, J., Vanderbilt, V.C., 1991, 'Opportunities for using the EOS imaging spectrometers and synthetic aperture radar in ecological models', *Ecology*, **72**: 1934–45

Vose, J., Dougherty, P.M., Long, J.N., Smith, F.W., Gholz, H., Curran, P.J., 1993, 'Factors influencing the amount and distribution of leaf area in pine stands', in Gholz, H.L., McMurtrie, R., Linder, S. (editors), *A comparative analysis of pine forest productivity*, Swedish Academy Press, Uppsala (in press)

Waring, R.H., Runyon, J., Goward, S.N., McCreight, R., Yoder, B., Ryan, M.G., 1993, 'Developing remote sensing techniques to estimate photosynthesis and annual forest growth across a steep climatic gradient in western Oregon, U.S.A.', *Scandinavian Journal of Forest Research* (in press)

Webster, R., Oliver, M.A., 1990, *Statistical methods in soil and land resource survey*, Oxford University Press, Oxford

Wessman, C.A., 1990, 'Evaluation of canopy biochemistry', in Hobbs, R.J., Mooney, H.A. (editors), *Remote Sensing of Biosphere Functioning*, Springer-Verlag, New York: 135–56

Wessman, C.A., 1992, 'Imaging spectrometry for remote sensing of ecosystem processes', *Advances in Space Research*, **12**: 361–68

Wessman, C.A., 1993, 'Remote sensing and the estimation of ecosystem parameters and functions', in Hill, J. (editor), *Imaging spectrometry as a tool for environmental observations*, Kluwer Academic, Dordrecht (in press)

Wessman, C.A., Aber, J.D., Peterson, D.L., Melillo, J.M., 1988, 'Remote sensing of canopy chemistry and nitrogen cycling in temperate forest ecosystems', *Nature*, **335**: 154–6

Wessman, C.A., Curtiss, B., Ustin, S.L., 1990, 'Large scale ecosystem modelling using parameters derived from imaging spectrometer data', *Imaging Spectrometry of the Terrestrial Environment*, Society of Photo-optical Instrumentation Engineers, Bellingham, Wa., **1298**: 164–70

Wessman, C.A., Ustin, S.L., Curtiss, B., Gao, B.C., 1991, 'A conceptual framework for ecosystem modelling using remotely sensed inputs', *Proceedings, 5th international colloquium – physical measurements and signatures in remote sensing*, European Space Agency SP-319, Noordwijk: 777–82

Westman, W.E., Paris, J.A., 1987, 'Detecting forest structure and biomass with C-band multipolarization radar: physical model and field tests', *Remote Sensing of Environment*, **22**: 249–69

Whittaker, R.H., Marks, P.L., 1975, 'Methods of assessing terrestrial productivity', in Lieth, H., Whittaker, R.H. (editors), *Primary productivity of the biosphere*, Springer-Verlag, New York: 55–118

Wickland, D.E., 1989, 'Future directions for remote sensing in terrestrial ecological research', in Asrar, G. (editor), *Theory and applications of optical remote sensing*, Wiley, New York: 691–724

Wickland, D.E., 1991a, 'Global ecology: the role of remote sensing', in Esser, G., Overdieck, D. (editors), *Modern ecology: basic and applied aspects*, Elsevier, Amsterdam: 725–49

Wickland, D.E., 1991b, 'Mission to planet Earth: the ecological perspective', *Ecology*, **72**: 1923–33

Wilson, M.F., Henderson-Sellers, A., Dickenson, R.E., Kennedy, P.J., 1987, 'Sensitivity of the Biosphere-Atmosphere Transfer Scheme (BATS) to the inclusion of variable soil characteristics', *Journal of Climatology and Applied Meteorology*, **26**: 341–62

Wu, S.T., 1987, 'Potential application of multipolarization SAR for pine plantation biomass estimation', *IEEE Transactions on Geoscience and Remote Sensing*, **25**: 403–9

Wu, S.T., Sader, S.A., 1986, 'Multipolarization SAR data for surface features delineation and forest vegetation characterization', *IEEE Transactions on Geoscience and Remote Sensing*, **25**: 67–76

10. Remote Sensing and the Determination of Geophysical Parameters for Input to Global Models

Arthur P. Cracknell

.... INTRODUCTION

A common experience of remote sensing is with the classification of multispectral imagery with the following characteristics:

(i) it is recorded in visible or near-infrared wavebands by sensors on Landsat (traditionally) or on one of several other satellites such as SPOT, IRS, MOS, etc.; and
(ii) it is be presented as a false-colour composite, either as hard copy for visual analysis and interpretation or on a screen for digital analysis and interpretation.

The information that one obtains from such data may be qualitative or it may be quantitative. However, even if it is quantitative, it is not usually quantitative in the sense to be discussed in this chapter. The kind of quantitative information that might be obtained is, say, the area planted with one particular crop. It is quantitative indeed, but what this chapter is concerned with is rather more fundamental from the physicist's point of view. Suppose that one is trying to model the atmospheric conditions for weather forecasting purposes. Then one looks at the basic physics of what is happening in the atmosphere and this leads to a set of differential equations that one has to try to solve. To do this one has to have the values of various physical parameters over time. These parameters include atmospheric temperature, pressure and humidity, sea surface temperature, etc. The conventional way to obtain such data is from meteorological stations, which are distributed around the surface of the Earth, augmented more recently by radiosonde balloon data and the occasional data from rocket soundings. Data of opportunity are also provided (i.e., at locations and with frequencies not of the meteorologists' choosing) from military and civilian ships and aircraft. In this chapter, the extent to which remotely-sensed data may either augment or replace such conventional sources of data is considered. One disadvantage of the conventional data sources is their uneven spatial distribution. The data come from locations clustered together in certain parts of the land area of the Earth's surface but are very sparse for other areas (e.g., deserts, Antarctica, etc.) and extremely sparse for many sea areas, particularly in the southern hemisphere. Satellite sensors, on the other hand, have the advantage of providing global coverage on a more-or-less uniform basis which is much more in line with the need to provide data as boundary conditions to a mathematical computer model. However, a satellite sensor has the disadvantage that it does not estimate the required parameters directly, but only indirectly and there is bound to be some long and complicated inversion process needed to estimate the required geophysical parameters, such as sea surface temperature or atmospheric humidity, from the data. There are also bound to be errors involved and these errors may be quite large. In this chapter some of the questions associated with the use of satellite sensor data to extract fundamental parameters related to the Earth's surface are examined.

It is convenient to separate the discussion under the three headings of air/atmosphere, sea/freshwater and the land. Basically these are listed in terms of increasing order of difficulty. It is relatively straightforward to extract information about atmospheric physical parameters from satellite sensor data, it is more difficult to extract physical parameters for the sea and it is almost impossible to extract fundamental parameters relating to the land from satellite sensor data.

.... THE ATMOSPHERE

The most obvious use of remote sensing data from satellite sensors in studying the atmosphere is the observation of weather systems and their movements. The image data are interpreted visually and this interpretation provides a valuable input to weather forecasts as well as a good basis for the public presentation of weather forecasts. But this does not count as fundamental parameters in the terms of this chapter. Three important meteorological parameters are the atmospheric temperature, pressure and humidity; these are obtained not from the image data but from the sounding data. One very important system is the TIROS Operational Vertical Sounder (TOVS), from which one can obtain vertical profiles of pressure, temperature and humidity (Olesen, 1992). This package of instruments is flown on the NOAA (National Oceanic and Atmospheric Administration) series of polar-orbiting satellites. Other sounders are flown on other satellites, including geostationary satellites. The TOVS actually consists of three separate instruments:

(i) the HIRS (High Resolution IR Spectrometer) which is a 20-channel instrument (19 infrared and one visible),
(ii) the SSU (Stratospheric Sounding Unit) with 3 channels, and
(iii) the MSU (Microwave Sounding Unit) with 4 microwave channels (actually 2 channels with 2 polarisations, vertical and horizontal, in each).

Table 10.1 Advantages and disadvantages of satellite sounding.

Advantages	Disadvantages
Even distribution of data points throughout the world	Indirect measurement of atmospheric parameters
Frequent and regular global coverage	Large amount of processing of data needed to estimate atmospheric parameters

Their scanning patterns are similar to that of the AVHRR (Advanced Very High Resolution Radiometer) but with a much coarser spatial resolution. Since these instruments are scanners one could represent the data they generate as images; but these images would be of such low spatial resolution as to convey little useful information. However, because they are scanners, these instruments do not measure directly the atmospheric parameters that one is trying to determine. There is an inversion process needed to extract the atmospheric parameters of pressure, temperature and humidity profiles from the sounding data. The theory involves the radiative transfer equation and is quite complicated. In practice, what one usually does is to obtain a TOVS software package and implement it on one's own computer. This is now done on a regular routine basis, though the accuracy of the results may not always be as great as one would like.

The main advantages and disadvantages of satellite sounding in relation to conventional methods of determining atmospheric profiles are summarised in Table 10.1.

.... THE SEA

BACKGROUND

The sea is much easier to handle than the land if one is trying to estimate fundamental physical parameters. With the sea there is a reasonable chance of being able to understand and analyse what is happening by using physical models because the systems are much simpler than those involving the reflection or emission of radiation by land surfaces. The land surface is covered by many different materials of various shapes and orientations, and having a wide range of spectral emissive and reflective properties. The sea and the atmosphere are, in principle, much more susceptible to modelling, although in practice the actual values of the parameters in those models may be rapidly changing and very hard to determine.

THE PROMISE

Few technological developments have captured the imagination of the marine research community to the extent of satellite remote sensing of the oceans; this is demonstrated by the performance of microwave, infrared and colour sensors carried on recent NASA (National Aeronautical and Space Administration) satellites and other current and proposed systems (Figure 10.1). For the first time it is possible to contemplate synoptic observations of geostrophic surface currents, wind stress and sea surface temperature over an ocean network approaching that which was previously quite unimagined. Previously one was only able to obtain data from limited sources comprising, typically, a few weather ships and moored buoys, merchant shipping and naval vessels, and oceanographic cruise ships. In terms of global coverage this was necessarily very patchy. What a satellite sensor offers is regular repetitive coverage on a regular grid of points over all the oceans of the world. What this means in terms of modelling, of weather systems or longer-term climate modelling, is that one now has the possibility of gathering data over a regular network of points and at regular intervals (i.e., one can obtain boundary values of important parameters at one very important boundary, namely the ocean-atmosphere interface). The disadvantage is that, as in the case of atmospheric sounding, one may either not measure the parameters that are needed for the model or one may not be able to determine their values accurately enough.

It should be noted that a satellite sensor is generally only able to observe the surface of the sea or a thin layer of the sea beneath the surface; we cannot usually obtain information about the sea below this surface layer (though there are some exceptions to this statement).

For small seas, coastal waters and estuaries, the spatial resolution achieved by a satellite sensor may be inadequate and one may have to resort to aircraft rather than satellites for remote sensing work.

WHAT CAN BE MEASURED?

One British oceanographer who has been a leading proponent of the use of space techniques is fond of saying that there are only four things which can be measured for the oceans by remote sensing from space (Allan, 1992). They are colour, roughness, slope and temperature. Of these, the temperature is directly useful but the others are not themselves useful geophysical parameters. The four parameters that can be estimated from the remotely sensed data may, however, be related to useful or usable parameters, or quantities, and these along with some others are indicated in Table 10.2. For each of the useful products in the left-hand column the crosses indicate which of the four parameters – colour, roughness, slope and temperature – are relevant.

The various different types of lines indicate the relation between the more fundamental parameters on the left and the more useful quantities on the right. The main sources of data through the 1970s and 1980s have been the following.

(i) The suite of microwave sensors carried on Seasat:
  Synthetic Aperture Radar (SAR)
  Scanning Multi-channel Microwave Radiometer (SMMR)
  Altimeter
  Scatterometer.

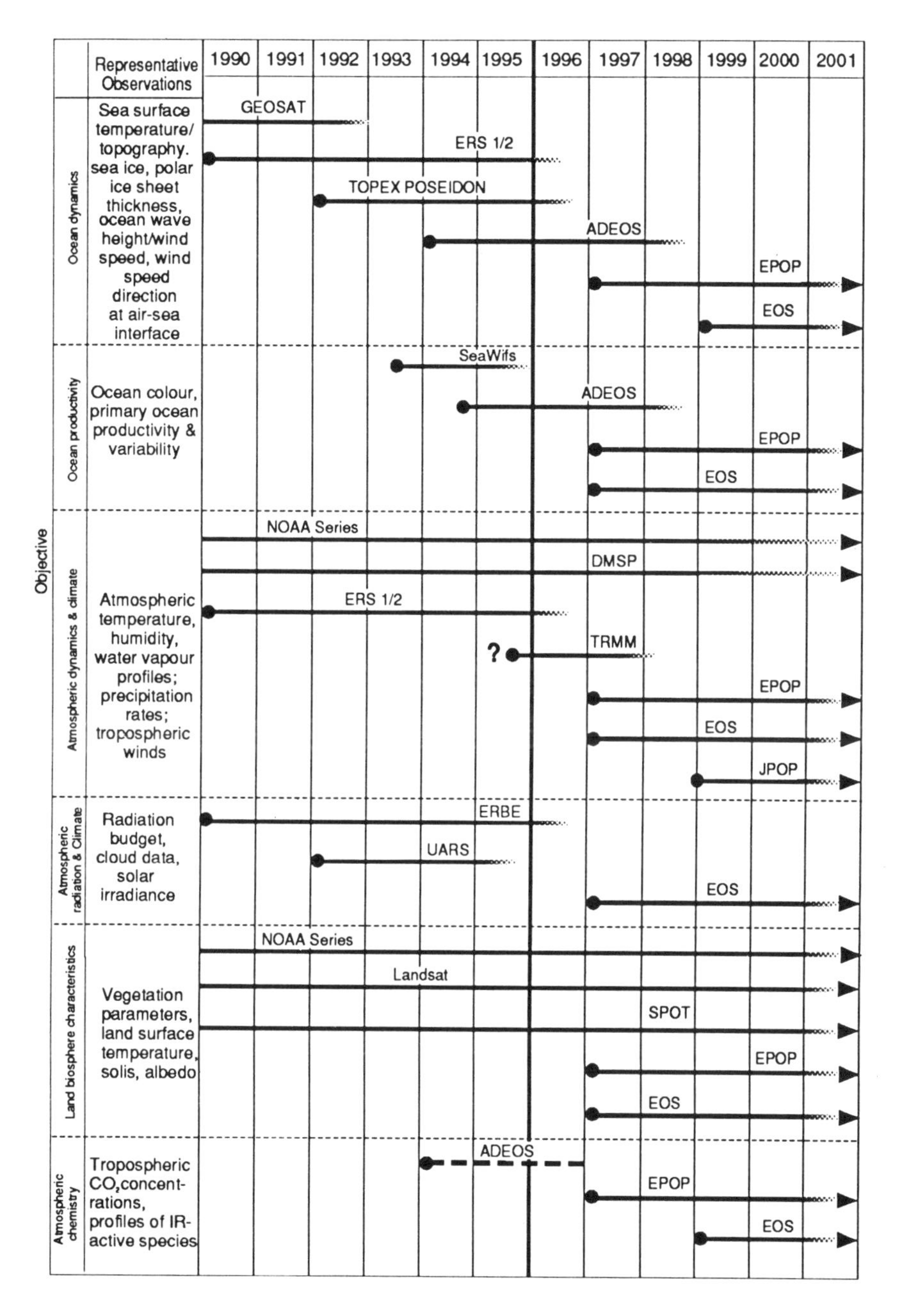

Figure 10.1 Some important measurements that may be made by satellite remote sensors (modified from Houghton *et al.*, 1992).

Table 10.2 The four parameters that can be remotely sensed and their potential areas of application.

	Remotely sensed parameter			
Application area	Colour	Roughness	Slope	Temperature
Ice charts	X	—	—	—
Suspended sediment concentration	X	—	—	—
Chlorophyll concentration	X	—	—	—
Bathymetry	X	X	X	—
Near-surface windspeeds	X	X	—	—
Pollution monitoring	X	—	—	X
Circulation patterns	X	—	—	X
Currents	X	X	X	—
Wave heights	—	X	—	—
Temperature charts	—	—	—	X

(ii) The instruments carried on Nimbus-7:
Scanning Multi-channel Microwave Radiometer (SMMR)
Coastal Zone Colour Scanner (CZCS).
(iii) The NOAA series of polar-orbiting satellites:
Advanced Very High Resolution Radiometer (AVHRR).
(iv) SPOT HRV, Landsat MSS and TM:
Visible, near-infrared and (on TM) thermal scanners.

There is not enough space to consider in detail all the things which one can do with the data; various conference proceedings, summer school proceedings and textbooks have been published on the subject (Cracknell, 1981; 1983; 1992; Robinson, 1985; Stewart, 1985; Vaughan, 1990). This chapter is necessarily selective in coverage and concentrates on some examples from research performed at the University of Dundee and in which quantitative geophysical parameters are obtained. These include chlorophyll and suspended sediment concentrations, near-surface wind speeds, sea-surface temperature charts and sea surface currents.

CHLOROPHYLL AND SUSPENDED SEDIMENT CONCENTRATIONS

From visible-band scanner data one can determine concentrations of chlorophyll and of suspended sediment. This involves a considerable amount of work on atmospheric correction of the data, and the use of sediment and chlorophyll algorithms that are, in general, empirical and therefore not wholly satisfactory.

The surface-leaving radiance from the sea surface is generally smaller than that from a land surface; therefore, for sea-surface observations the atmospheric effects are relatively more important than for land-surface observations. The question of atmospheric corrections to visible-band scanner data is therefore very important for work involving observations of the surface of the sea. In visible wavelengths, the total atmospheric contribution to the satellite-received signal over the sea is usually over 50 per cent and may approach 80–90 per cent. A useful introduction to the treatment of atmospheric corrections is given in Cracknell and Hayes (1991). Some discussions of recent work on chlorophyll algorithms will be found in papers by Sathyendranath *et al.* (1989), Bagheri and Dios (1990), Barale (1991), Eckstein and Simpson (1991), Hinton (1991), Parslow (1991), Ferrari and Tassan (1992) and Schiebe *et al.* (1992).

Over the years a considerable amount of work has been done by various research groups with CZCS (Coastal Zone Colour Scanner) data. However, it has, to some extent, ground to a halt following the demise of CZCS, though some people do attempt to estimate chlorophyll and suspended sediment concentrations with data acquired by systems such as the NOAA AVHRR, Landsat MSS, and TM or SPOT HRV data (e.g., Prangsma and Roozekrans 1989; Lopéz Garcia and Caselles 1990; Choubey and Subramanian 1992; Ekstrand 1992). However, for all of these the spectral bands are less suitable than those of CZCS for marine optical work.

NEAR SURFACE WINDSPEEDS

Windspeeds determined with satellite sensor data are most commonly obtained by using radar instruments of various kinds; for these the signal returned from the surface of the sea depends on the surface roughness which, in turn, depends on the windspeed. These instruments include the altimeter (Guymer, 1987; Rapley, 1990) and the scatterometer (Guymer, 1987; Brown, 1990; Offiler, 1983; 1990). Passive microwave scanners can also be used for the estimation of wind speed, but with much lower spatial resolution than the active microwave instruments (Guymer, 1987; Cracknell, 1992). However, there is another method which can be used in areas of sunglint and which will be mentioned briefly.

Sunglint occurs as a result of specular reflection of sunlight by a smooth, or nearly smooth, water surface. However, sometimes more complicated situations arise. The theory of the detailed analysis of sunglint patterns to determine windspeeds goes back to work by Cox and Munk (1954; 1956), though at that time they could not have foreseen the great scope for their theory with satellite sensor data.

When the sea surface is rough instead of smooth two things happen to a bright sunglint image:

(i) the intensity of the bright circular image is greatly reduced, and
(ii) the bright circular image is spread out over a larger area.

From the quantitative values of the intensities in various parts of the sunglint pattern one can calculate the windspeeds. The method is quite accurate but only works, of course, in areas of sunglint. For further details see, for example Cracknell (1990) and Khattak *et al.* (1991).

In connection with winds the very interesting work by Kuzmic (1991) on the testing of wind-induced current modelling using CZCS data for the northern Adriatic should also be noted.

SEA SURFACE TEMPERATURE CHARTS

Traditionally only very sparse data on sea surface temperatures was available. From reports from shipping etc., the Meteorological Office produces charts of sea surface temperature with general contours but lacking in fine spatial detail. Satellite thermal infrared sensor images show greater spatial detail. Many features of circulation patterns and of shelf fronts, which were only known about previously in rather general terms, can be studied in very great spatial detail (for an early reference see, for example, Simpson, 1981).

Detailed patterns are one thing but what is, perhaps, even more interesting is that one can convert these greyscale images (or the digital data behind them) into detailed charts or maps of sea surface temperature. The Advanced Very High Resolution Radiometer (AVHRR) was designed to estimate sea-surface temperature to an accuracy of about ±0.5 K. An early review of the determination of sea surface temperatures from satellite-flown infrared scanners was given by Robinson *et al.* (1984) and a good recent review is given by Fiúza (1992). The production of sea surface temperature charts from satellite thermal infrared sensor data is now done routinely, at a variety of scales and by a variety of organisations.

SEA SURFACE CURRENTS

One can make use of time/lapse sequences of sea surface temperature patterns in thermal infrared satellite sensor images to determine surface current vectors. This is an adaptation of a well-established procedure that is widely used with the half-hourly data from the geostationary satellites. By observing the difference between the positions of a cloud in two successive images and dividing this distance by the interval between the times of acquisition of the two images, one obtains the velocity of the cloud. With AVHRR images of temperature features, data from the same passes on successive days can be used to determine surface currents (Cracknell and Huang, 1988).

.... LAND

By contrast with the atmospheric and marine situations, the whole question of studying land surfaces to obtain simple or fundamental physical parameters is very rudimentary. The estimation of a simple parameter such as surface temperature is still a major problem for land surfaces whereas surface temperatures for the sea from thermal infrared satellite sensor data is a more or less routine operation. There are two, quite separate, problems.

(i) The emissivity of sea water is known and so the temperature can be calculated, whereas for the land surface the emissivity is not known and is spatially variable.
(ii) For the land surface it is far from clear exactly what we would measure for the land surface temperature even if we made a measurement on the ground with a thermometer. If there is vegetation cover, we might measure the soil temperature, or the temperature of the vegetation or the

temperature of the pockets of air trapped by the foliage of the vegetation. All these temperatures are likely all to be different from one another.

At the present time a start has been made on trying to determine land surface temperatures from thermal infrared data from satellite sensors but this work is still in its early stages.

.... GLOBAL MODELS

So far the problem of the determination of geophysical parameters from remotely-sensed data has been discussed, but this has been without any reference to the reasons why one might want to determine these parameters. The expression 'global models' was used in the title of this chapter to indicate that one important reason for determining these parameters is to provide input data to global models. These include weather forecast models, ocean circulation models and climate models. These models are operated on quite different timescales from one another but they all have a common requirement for the input of geophysical parameters over a regular grid of points to specify the initial conditions/boundary conditions for the model. Satellite systems have the advantages indicated in Table 10.1 in that they provide data over a uniform grid of points and at regular intervals of time. On the other hand, conventional atmospheric, marine or land-surface data-collection systems provide a very uneven spatial coverage and sometimes an irregular temporal coverage. Thus satellite sensors provide an important source of data for input to these various types of models and augment very considerably the data available from conventional environmental parameter measurements.

Considering weather forecast models briefly first. Over the last two decades there has been a transition from the traditional methods of weather forecasting to forecasts based on models run on computers (Houghton, 1991). The objective of a weather forecast model is to be able to predict the weather at a particular place and time. The idea is that one writes down the mathematical equations that describe the various physical processes that occur and tries to solve those equations for some future time. In an atmospheric model, the behaviour of the atmosphere is represented by the values of appropriate parameters specified on a three-dimensional grid of points. The UK Meteorological Office uses a global model operating with a grid with spacing of 90 km in the horizontal plane and also a limited area model with a grid spacing of about 40 km in the horizontal plane; in each model there are about 20 levels in the vertical plane. The parameters involved are indicated in Figure 10.2. The dynamics in this model include the horizontal momentum equations, the hydrostatic equation (any vertical acceleration is neglected) and the equation of continuity (i.e., matter is not destroyed or created). The other physics included in the model is the equation of state, the thermodynamic equation and parametric descriptions of processes such as evaporation, condensation, formation and dispersal of clouds, etc., of the radiative and convective processes within the atmosphere, and of the exchange of momentum, heat and water vapour with the Earth's surface. Further

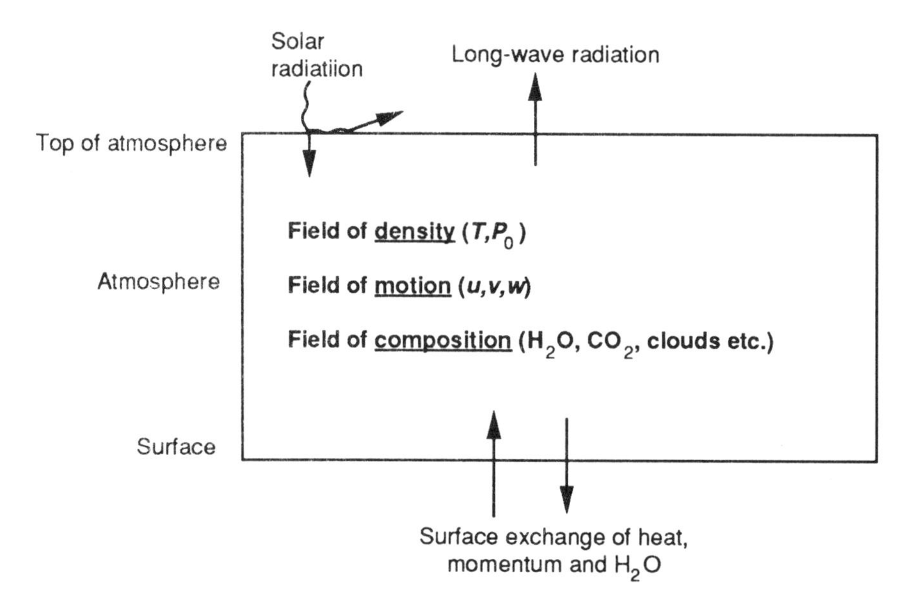

Figure 10.2 Schematic illustrating the parameters and physical processes involved in atmospheric models (modified from Houghton *et al.*, 1992).

allowance has to be made to account for motions which occur on scales different to the model's grid size. It is supposed that the results of meteorological observations, whether from conventional measurements or from satellite sensor data, allow the values of the atmospheric parameters to be specified at some starting time. The forecasting problem then is to solve the various equations describing the physical principles and processes mentioned for a subsequent time. That is, to determine new values of these atmospheric parameters at a subsequent time, in practice up to a few days ahead. Because there are several of these equations and, especially, because they are non-linear equations, the problem has to be tackled by numerical methods using a computer. Hence the terms 'computer model' or 'numerical model'.

The operational use of numerical models at the UK Meteorological Office began in 1965. Although their usefulness at that stage was limited, there have been rapid developments since then and within ten years the models were able to provide more accurate forecasts of the basic motion field than could be achieved by an unaided human forecaster. These improvements have come about for two reasons. First, the power of the computing facilities available has increased enormously and, second, the amount of data available to describe the present atmospheric conditions has been increased greatly. In addition to conventional surface measurements and radiosonde measurements, there are also satellite sensor observations, including atmospheric depth soundings and windspeeds obtained by cloud tracking from geostationary satellite sensors.

For weather forecasting purposes the surface temperature over the land and the sea provides a very important set of boundary conditions on the model.

Thus, the use of thermal infrared scanner data from satellite sensors to estimate sea surface temperatures provides a very important source of data for weather forecast models.

Because the timescale of changes in the oceans is very much longer than the timescale of atmospheric changes (i.e., of changes in the weather) one does not make elaborate provision in weather forecast models to take into account circulation, or changes, in the oceans. However, there are situations in which oceanic circulation is important; these include both scientific oceanographic studies and also climatological studies. In climatological models one is dealing with long periods of time and the circulation of water and of heat in the oceans is important. For climate modelling therefore one needs to use general circulation models for both the atmosphere and the oceans, and to allow these models to be coupled together. Because this is so much more complicated than using only an atmospheric global circulation model (GCM) for weather forecasting, the spatial resolution used in practice in climate modelling tends to be coarser than that used in weather forecasting. The typical oceanic GCM used for climate simulation studies follows basically the same set of equations as the atmosphere if the equation describing the water vapour balance in the atmosphere is replaced by one describing the salinity in the ocean. As with atmospheric GCMs, one specifies appropriate surface boundary conditions (i.e., fluxes of heat, momentum and fresh water), either from observations or from an atmospheric GCM. The vertical and horizontal exchange of temperature, momentum and salinity by diffusion or turbulent transfers is parametrised.

An extensive discussion of global climate change and climate modelling is given in the report of the IPCC (Intergovernmental Panel on Climate Change) (Houghton *et al.*, 1990; 1992). Towards the end of this there is an indication of the rôle to be played by space-borne instruments in climate modelling (see Figure 10.1).

.... CONCLUSIONS

In the early days of remote sensing (Earth observation) most work involved the qualitative interpretation of image data and in many situations this is still the way in which to proceed.

However, the quantitative use of remote sensing data is gaining very considerably in importance. This quantitative information is of two types. It may be geographical or statistical (the area of land planted with one particular crop, the estimated yield of a particular crop from a given area in a given season, etc.). Alternatively it may be geophysical, that is it may be possible to extract fundamental geophysical parameters for the air, sea or land from the data.

When concerned with gathering data for input into meteorological weather forecasting models, into oceanographic models or into climatological models then it is the second kind of information, giving us the values of geophysical parameters, which is required.

At the outset it was noted that along the sequence air, sea, and land there was a move from the fairly successful use of satellite sensor data for the extraction of geophysical parameters through to the situation for which it is very difficult. Remote sensing is already able to contribute now in some areas and it should be able to contribute even more in the future. In conclusion, therefore:

(i) it is now possible to extract some quantitative geophysical data from remote sensing observations;
(ii) some examples, but by no means a complete list, have been provided; and
(iii) there is an important future for this subject as newer and better satellite sensor systems and data-processing systems come on stream.

One important reason for seeking to determine geophysical parameters from space is to provide a regular and uniform source of data to supply boundary conditions for atmospheric and oceanic GCMs. The rôle that is expected to be played by future remote-sensing satellite systems in the next decade is indicated in Figure 10.1.

.... REFERENCES

Allan, T.D., 1992, 'An overview of satellite remote sensing in the marine environment', in Cracknell, A.P (editor), *Space oceanography*, World Scientific, Singapore: 1–12

Bagheri, S., Dios, R.A., 1990, 'Chlorophyll-a estimation in New Jersey's coastal waters using Thematic Mapper data', *International Journal of Remote Sensing*, **11**: 289–99

Barale, V., 1991, 'Sea surface colour in the field of biological oceanography', *International Journal of Remote Sensing*, **12**: 781–93

Brown, R.A., 1990, 'The scatterometer; data and applications', in Vaughan, R.A. (editor), *Microwave remote sensing for oceanographic and marine weather forecast models*, Kluwer, Dordrecht: 99–123

Choubey, V.K., Subramanian, V., 1992, 'Estimation of suspended solids using Indian Remote Sensing Satellite-1A data: a case study from Central India', *International Journal of Remote Sensing*, **13**: 1473–86

Cox, C., Munk, W., 1954, 'Measurement of the roughness of the sea surface from photographs of the sun glitter', *Journal of the Optical Society of America*, **44**: 838–50

Cox, C., Munk, W., 1956, 'Slopes of the sea surface deduced from photographs of sun glitter', *Bulletin of the Scripps Institute of Oceanography*, **6**: 401–88

Cracknell, A.P. (editor), 1981, *Remote sensing in meteorology, oceanography and hydrology*, Ellis Horwood, Chichester

Cracknell, A.P. (editor), 1983, *Remote sensing applications in marine science and technology*, Reidel, Dordrecht

Cracknell, A.P., 1990, 'Sunglint and the study of near-surface windspeeds over the oceans', in Vaughan, R.A. (editor), *Microwave remote sensing for oceanographic and marine weather forecast models*, Kluwer, Dordrecht: 125–39

Cracknell, A.P. (editor), 1992, *Space oceanography*, World Scientific, Singapore

Cracknell, A.P., Hayes, L.W.B., 1991, *Introduction to remote sensing*, Taylor and Francis, London

Cracknell, A.P., Huang, W.G., 1988, 'Surface currents off the west coast of Ireland studied from satellite images', *International Journal of Remote Sensing*, **9**: 439–46

Eckstein, B.A., Simpson, J.J., 1991, 'Cloud screening Coastal Zone Color Scanner Images using channel 5', *International Journal of Remote Sensing*, **12**: 2359–77

Ekstrand, S., 1992, 'Landsat TM based quantification of chlorophyll-a during algal blooms in coastal waters', *International Journal of Remote Sensing*, **13**: 1913–26

Ferrari, G.M., Tassan, S., 1992, 'Evaluation of the influence of yellow substance absorption on the remote sensing of water quality in the Gulf of Naples: a case study', *International Journal of Remote Sensing*, **13**: 2177–89

Fiúza, A.F.G., 1992, 'The measurement of sea surface temperature from satellites', in Cracknell, A.P. (editor), *Space oceanography*, World Scientific, Singapore: 197–279

Guymer, T.H., 1987, 'Remote sensing of sea-surface winds', in Vaughan, R.A. (editor), *Remote sensing applications in meteorology and climatology*, Reidel, Dordrecht: 327–57

Hinton, J.C., 1991, 'Application of eigenvector analysis to remote sensing of coastal water quality', *International Journal of Remote Sensing*, **12**: 1441–60

Houghton, J.T., 1991, 'The Bakerian Lecture 1991, the predictability of weather and climate', *Philosophical Transactions of the Royal Society of London*, **A337**: 521–72

Houghton, J.T., Callander, B.A., Varney, S.K., 1992, *Climate change 1992 – the supplementary report to the IPCC scientific assessment,* Cambridge University Press, Cambridge

Houghton, J.T., Jenkins, G.J., Ephraums, J.J., 1990, *Climate change – the IPCC scientific assessment*, Cambridge University Press, Cambridge

Khattak, S., Vaughan R.A., Cracknell, A.P., 1991, 'Sunglint and its observation in AVHRR data', *Remote Sensing of Environment*, **37**: 101–16

Kuzmic, M., 1991, 'Exploring the effects of bura over the northern Adriatic: CZCS imagery and a mathematical problem', *International Journal of Remote Sensing*, **12**: 207–14

López Garcia, M.J., Caselles, V., 1990, 'A multi-temporal study of chlorophyll-a concentration in the Albufera lagoon of Valencia, Spain, using Thematic Mapper data', *International Journal of Remote Sensing*, **11**: 301–11

Offiler, D., 1983, 'Surface wind measurements from satellites', in Cracknell, A.P. (editor), *Remote sensing applications in marine science and technology*, Reidel, Dordrecht: 169–82

Offiler, D., 1990, 'Wind fields and surface fluxes', in Vaughan, R.A. (editor), *Microwave remote sensing for oceanographic and marine weather forecast models*, Kluwer, Dordrecht: 355–74

Olesen, F.-S., 1992, 'Vertical sounding of the atmosphere and imaging radiometer', in Cracknell, A.P. (editor), *Space oceanography*, World Scientific, Singapore, 127–38

Parslow, J.S., 1991, 'An efficient algorithm for estimating chlorophyll from Coastal Zone Color Scanner data', *International Journal of Remote Sensing*, **12**: 2065–72

Prangsma, G.J., Roozekrans, J.N., 1989, 'Using NOAA AVHRR imagery in assessing water quality parameters', *International Journal of Remote Sensing*, **10**: 811–18

Rapley, C., 1990, 'Satellite radar altimeters', in Vaughan, R.A. (editor),

Microwave remote sensing for oceanographic and marine weather forecast models, Kluwer, Dordrecht: 45–63

Robinson, I.S., 1985, *Satellite oceanography: an introduction for oceanographers and remote sensing scientists*, Ellis Horwood, Chichester

Robinson, I.S., Wells, N.C., Charnock, H., 1984, 'The sea surface thermal boundary layer and its relevance to the measurement of sea surface temperature by airborne and spaceborne radiometers', *International Journal of Remote Sensing*, **5**: 19–45

Sathyendranath, S., Prieur, L., Morel, A., 1989, 'A three-component model of ocean colour and its application to remote sensing of phytoplankton pigments in coastal waters', *International Journal of Remote Sensing*, **10**: 1373–94

Schiebe, F.R., Harrington, J.A., Ritchie, J.C., 1992, 'Remote sensing of suspended sediments: the Lake Chicot, Arkansas, project', *International Journal of Remote Sensing*, **13**: 1487–1509

Simpson, J.H., 1981, 'Sea surface fronts and temperature', in Cracknell, A.P. (editor), *Remote sensing in meteorology, oceanography and hydrology*, Ellis Horwood, Chichester: 295–311

Stewart, R.H., 1985, *Methods of satellite oceanography*, University of California, Berkeley

Vaughan, R.A. (editor), 1990, *Microwave remote sensing for oceanographic and marine weather forecast models*, Kluwer, Dordrecht

11. Environmental Monitoring Using Multiple-view-angle (MVA) Remotely-sensed Data

M.J. Barnsley

INTRODUCTION

The detected reflectance of Earth surface materials, ρ, at one point on the Earth's surface can be expressed in terms of the following function:

$$\rho = f\,(\theta_i, \phi_i; \theta_r, \phi_r; \lambda; t; A)$$

where θ_i and ϕ_i are respectively the zenith and azimuth angles of the incident radiation (i.e., the illumination geometry), θ_r and ϕ_r are respectively the zenith and azimuth angles of the reflected radiation (i.e., the viewing geometry), λ is the wavelength of the reflected radiation, t is time, and A is the effect of the atmosphere.

Most remote sensing studies attempt to distinguish between different types of surface material – or to infer their fundamental physical, chemical and biological properties – on the basis of variations in ρ measured as a function of λ (i.e., using *multispectral* data) and, occasionally, t (i.e., using *multi-temporal* data). By comparison, very few studies have explored the potential of reflectance data recorded at several sensor view angles (i.e., as a function of θ_r and ϕ_r) for environmental monitoring. However, this situation may soon change, with the increasing availability of airborne and satellite sensors that are capable of acquiring such data; hereafter, referred to as *multiple-view-angle* or *MVA* data. These sensors offer a number of advantages with respect to conventional, nadir-viewing devices, including:

(i) the opportunity for more frequent observation of selected target areas (Pinter *et al.*, 1990; Barnsley *et al.*, 1993);
(ii) the ability to derive digital elevation data for terrain and cloud-tops from pairs of stereoscopic images (Lorenz, 1985; Day and Muller, 1988; Muller, 1989);

(iii) the capacity for Sun-glint avoidance over the oceans (NASA, 1986a; Salomonson *et al.*, 1989); and
(iv) the potential to infer unique information on Earth surface materials through an investigation of their angular reflectance properties (Barnsley and Muller, 1991; Barnsley *et al.*, 1993).

This chapter concentrates on the last of these four points. In particular, attention is focused on the potential use of MVA data for land-cover mapping, as well as for the derivation of improved estimates of Earth surface albedo (α) and the extraction of quantitative information on the three dimensional (3-D) structure of Earth surface materials. This is preceded by a brief review of recent research which highlights the close relationship between the angular reflectance properties and 3-D geometric structure of Earth surface materials and an examination of current and future sensors that are capable of acquiring MVA data.

THE ANGULAR REFLECTANCE PROPERTIES OF EARTH SURFACE MATERIALS

It has been known for some time that the detected reflectance of most Earth surface materials varies according to the angles at which they are illuminated and viewed (Salomonson and Marlatt, 1971; Suits, 1972; Kriebel, 1978; Kimes and Kirchner, 1982; Kimes, 1983; Barnsley, 1984; Otterman, 1985; Ranson *et al.*, 1991). However, recent research in this area suggests that these angular (or directional) reflectance properties are controlled by the 3-D geometric structure of the target, as well as its biophysical and biochemical characteristics (Deering and Eck, 1987; Otterman *et al.*, 1987; Ross and Marshak, 1988; Goel and Reynolds, 1989; Barnsley and Kay, 1990; Pinty *et al.*, 1990; Verstraete *et al.*, 1990). For example, vegetation canopies often exhibit a pronounced peak in reflectance in the backscatter direction (i.e., where the Sun and the sensor are at the same angular position relative to a given point on the Earth surface). This is known as the 'hot spot' (Suits, 1972) or 'opposition surge' (Hapke, 1986). Some studies have suggested that the amplitude and the angular width of this feature are closely related to specific canopy parameters. For simple vegetation canopies with continuous ground cover, such as cereal crops, the controlling factors are believed to include the average leaf size, the leaf-area index and the average leaf-inclination angle (Goel and Thompson, 1985; Gerstl and Simmer, 1986; Ross and Marshak, 1989; Brakke and Otterman, 1990; Pinty *et al.*, 1990; Verstraete *et al.*, 1990); while for canopies with discontinuous ground cover, such as woodland, the size, shape and density of individual tree crowns are thought to be critical (Li and Strahler, 1986; 1993; Strahler and Jupp, 1990; Jupp and Strahler, 1991).

The relationship between 3-D surface structure and angular reflectance has also been investigated for exposed soil (Otterman, 1985; Cierniewski, 1987; 1989; Deering *et al.*, 1989; 1990; Pinty *et al.*, 1989; Roujean *et al.*, 1989; Jackson *et al.*, 1990). Many of these studies note a strong peak in reflectance in the forward scatter direction (i.e., away from the Sun), particularly for smooth surfaces. This suggests that surface roughness plays a key rôle in determining the angular reflectance properties of soils.

The implication of these studies is that, all other things being equal, it may be possible to use MVA data to distinguish between land-cover types that differ in terms of 3-D geometric structure. It may also be possible to derive accurate, quantitative estimates of key structural parameters. This is directly analogous to the way in which multispectral data are used to map land cover and to infer information on the biophysical and biochemical properties of surface materials.

In reality, variation of the reflectance detected by an airborne or satellite sensor at different view angles is a function not only of the angular reflectance properties of the Earth surface, but also of atmospheric effects (Powers and Gerstl, 1988). For instance, the change in atmospheric path-length between nadir and off-nadir views will normally result in different degrees of atmospheric scattering and absorption (Tanré *et al.*, 1983). Furthermore, atmospheric scattering has a strong angular (or directional) component. This has been shown to contain useful information in its own right, relating to aerosol loading and optical depth (Martonchik and Diner, 1993). Indeed, several studies have suggested that it may be possible to estimate values for these parameters directly from MVA data (Diner and Martonchik, 1985; Martonchik and Diner, 1993).

Unfortunately, the radiance detected by an airborne or satellite-based sensor combines the angular scattering properties of *both* the ground *and* the atmosphere (Powers and Gerstl, 1988). Separation of these two components represents a major challenge, since neither is usually known *a priori*. In the long term, the optimum solution to this problem is to develop 'coupled' models of radiation interaction with the land surface and the atmosphere (Gerstl and Zardecki, 1985; Liang and Strahler, 1993; Martonchik and Diner, 1993). Meanwhile, ground-level angular reflectances must be estimated from exo-atmospheric measurements of radiance using existing methods for atmospheric correction of remotely-sensed images, such as 5S (Tanré *et al.*, 1990) or Lowtran 7 (Kneizys *et al.*, 1988), by assuming average regional values for key atmospheric parameters.

.... SENSOR DESIGN FOR MULTIPLE-VIEW-ANGLE (MVA) DATA ACQUISITION

In the previous section, it was indicated that an important prerequisite for full and effective use of MVA data in environmental monitoring is the need to deconvolve the angular scattering properties of the ground and the atmosphere. Without removing the atmospheric component, inferences made about the characteristics of Earth surface materials are likely to be inaccurate. The accuracy of such inferences will also be controlled by the number and range of view angles over which the data are acquired; or, in other words, by the design and operation of the sensor. This section examines a number of current and future sensors that can record images at several angles off-nadir, as well as vertically downwards.

CURRENT AND PROPOSED MVA SENSORS

There are several ways in which remote sensing devices can be designed to collect MVA data. For example:

- (i) by use of a wide field-of-view (FOV) in the across-track direction (Figure 11.1(a));
- (ii) by enabling a narrow FOV sensor to be tilted at different angles off-nadir in the across-track and/or along-track directions (Figures 11.1(b) and 11.1(c));
- (iii) by operating multiple sensors pointing at different (fixed) view angles in the along-track direction (Figure 11.1(d));
- (iv) by use of a wide FOV in *both* the along-track *and* the across-track directions (Figure 11.1(e));
- (v) by adopting a conical scanning pattern (Figure 11.1(f)).

Each of these strategies is used by sensors currently in operation or planned for launch in the near future (Barnsley *et al.*, 1993).

The first configuration is typified by the NOAA AVHRR (Advanced Very High Resolution Radiometer) sensors and a large number of airborne multispectral scanners (Barnsley, 1984). With a design such as this, the target area can be seen by the sensor at several different view angles from a series of parallel flight lines or a sequence of adjacent orbital overpasses (Figure 11.1(a)). The limitation of this approach is the length of time that it takes to acquire the MVA data set, particularly using Earth-orbiting satellites. Consequently, differences in the detected reflectance between individual images may result from changes in irradiance, atmospheric conditions and land-cover characteristics, as well as from angular reflectance effects.

The second configuration is used by the current generation of SPOT-HRV (High Resolution Visible) sensors. These are pointable up to ±27° in the across-track direction. The HRV instruments can be used to generate an MVA data set in much the same way as NOAA AVHRR (i.e., from a series of overpasses on separate days; Figure 11.1(b)). These sensors therefore suffer the same limitations for use in examining angular reflectance effects (Moran *et al.*, 1990).

Two sensors intended for launch as part of NASA's Earth Observing System (EOS), namely MODIS-T (Moderate Resolution Imaging Spectrometer – Tilt) and HIRIS (High Resolution Imaging Spectrometer), were to have an along-track tilting capability (NASA, 1986a; 1986b; Diner *et al.*, 1989; Goetz and Herring, 1989; Salomonson *et al.*, 1989; Ardanuy *et al.*, 1991). MODIS-T was intended to be tiltable up to ±50° in the along-track direction; while the specifications for HIRIS allowed the instrument to be pointed in both the along-track (up to 56° and 30°, fore and aft, respectively) and across-track (±45°) directions. The benefit of along-track tilting is that the study area can be observed at several different sensor view angles during a single orbital overpass, as the sensor platform approaches the target, overflies it and recedes from it (Figure 11.1(c)). Unfortunately, both MODIS-T and HIRIS have been de-selected recently from the EOS programme for budgetary reasons. An along-track tilting capability is, however, currently available on the Advanced Solid-state Array Spectro-

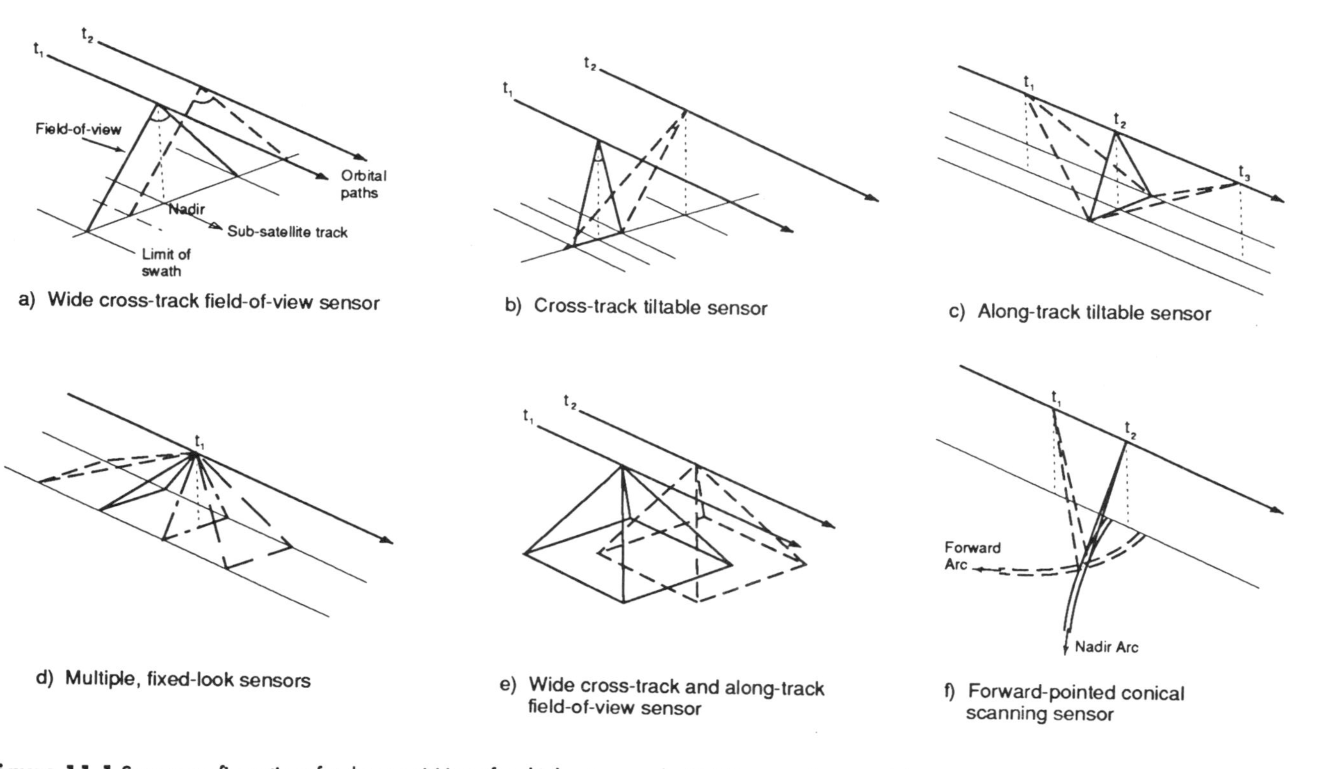

Figure 11.1 Sensor configurations for the acquisition of multiple-view-angle (MVA) remotely sensed data.

radiometer (ASAS), an airborne sensor developed by NASA (Irons *et al.*, 1987; 1991).

Despite the demise of MODIS-T and HIRIS, the EOS programme retains a third instrument capable of acquiring multiple images of the Earth surface at different sensor view angles within a single orbital overpass, namely the Multi-angle Imaging SpectroRadiometer, or MISR (Diner *et al.*, 1989). This sensor uses nine identical CCD cameras, each of which is pointed at a different (fixed) angle with respect to the Earth surface (e.g., Figure 11.1(d)). By clever use of different focal lengths for each camera, the resultant images have the same nominal spatial resolution (960 m or 1.96 km, depending on whether the instrument is operated in 'local' or 'global' mode). A similar multi-look capability is currently available from the CCD Airborne Experimental Scanner for Applications in Remote Sensing (CAESAR), developed by the Dutch National Remote Sensing Laboratory (Looyen *et al.*, 1990).

The fourth configuration will be used by the French sensor, POLDER (Polarisation and Directionality of the Earth's Reflectance), which is scheduled for launch on board ADEOS (Advanced Earth Observation Satellite) in 1996. This instrument will acquire data using a CCD array defining a rectangular field-of-view of approximately 86° (along-track) by 102° (across-track) (Deschamps *et al.*, 1990; Deuzé *et al.*, 1991). POLDER will acquire multispectral and multi-polarisation data at a spatial resolution of 6 km × 7 km. Its operation can be likened to that of an aerial survey camera, since overlapping images can be acquired during both a single orbital overpass and from adjacent orbital tracks (Figure 11.1(e)).

The last mode of operation is typified by the Along-Track Scanning Radiometer (ATSR) carried on board ERS-1 (Prata *et al.*, 1990). This sensor has an unusual conical scanning pattern which allows data to be recorded almost simultaneously along two scan lines, describing arcs across the Earth surface (Figure 11.1(f)). The axis of the conical scan lies at 23.45° forward from nadir in the along-track direction. As a result, the first arc is centred on the sub-satellite (i.e., nadir) point, while the second is tilted forward of the satellite at 46.9° along-track (equivalent to a 55° view zenith angle, when Earth surface curvature is taken into account). Consequently, any point on the ground falling within the nadir swath will be imaged twice during a single overpass.

Further discussion of the use of these and other sensors for making MVA measurements of Earth surface reflectance can be found in Barnsley *et al.* (1993)

ANGULAR SAMPLING CAPABILITIES OF CURRENT AND FUTURE SENSORS

The complete angular distribution of the radiation reflected by an object, for all illumination angles, is given by the Bidirectional Reflectance Distribution Function, or BRDF (Nicodemus *et al.*, 1977). Integration of the BRDF with respect to λ and (θ_r, ϕ_r), at a given illumination angle, yields the albedo (α) (Ranson *et al.*, 1991). Similarly, given knowledge of the BRDF and with the aid of an appropriate model of surface scattering, it should be possible to derive precise values for certain parameters describing the 3-D geometry of the target (Pinty *et al.*, 1990; Pinty and Verstraete, 1991a; 1991b; Verstraete *et al.*, 1990). Unfortunately, it is impossible to record reflectance data simultaneously for all angles of incidence and exitance. As a result, the BRDF and its related parameters must be *estimated* from a set of angular reflectance measurements recorded under a limited number of Sun-target-sensor geometries. Individual measurements of angular reflectance, such as these, may be thought of as *samples* drawn from the *population* of all such values (i.e., the BRDF). The accuracy with which the BRDF can be estimated will therefore be determined by both the size of the sample and its angular distribution within the viewing and illumination hemispheres.

The ability of a given sensor to sample the BRDF is determined not only by the geometry of the sensor itself, but also the time of day/year at which the data are collected (i.e., the illumination geometry) and the orientation of the flight path or orbital track. Clearly, in the case of an airborne sensor, the timing and direction of the overflights can, to a certain extent, be controlled to meet the needs of a specific sampling strategy. By contrast, a spaceborne sensor is much less flexible, in that the data collection times and the orientation of the overpass are conditioned by the orbital characteristics of the satellite on which it is based. The BRDF sampling capabilities of many of the satellite sensors discussed above have been simulated by Barnsley *et al.* (1993). The results obtained for two of these instruments, namely NOAA-10 AVHRR and the proposed EOS-MISR sensor, are reported below.

Figure 11.2 shows the capability of NOAA-10 AVHRR for sampling the BRDF over a 16-day period around the March equinox. In this diagram, the dot symbols represent individual occasions on which the sensor can see a fixed target area on the ground; in this case, a point at latitude 50°N. The angle at which the target is viewed on each occasion is represented by the location of the dot symbol within the polar plot. View zenith angle is plotted radially, in 10° increments, outwards from the centre (nadir); while the relative azimuth angle between the Sun and the sensor (ψ; where, by convention, $\psi = |\phi_r - \phi_i| + 180°$) is represented by the angular position of the dot symbol around the plot. Thus, $\psi = 180°$ denotes data collected in the backscatter direction. Finally, the solar zenith angle at the time of imaging is indicated by the shading pattern of each dot symbol.

Figure 11.2 illustrates that data from NOAA-10 AVHRR, with its early morning overpass, are sampled close to the solar principal plane ($\psi = 0/180°$) – that is, with the sensor scanning directly into and away from the Sun – and at large solar zenith angles. Maximum angular variation of detected reflectance is anticipated for many Earth surface materials under these conditions (Kimes, 1983). Closer examination of Figure 11.2 reveals that NOAA-10 AVHRR samples the BRDF near to the 'hot spot' – i.e., the pronounced peak in reflectance in the direction of the Sun ($\theta_i = \theta_r$ and $\psi = 180°$). All other things being equal, it may therefore be possible to use data obtained by NOAA-10 AVHRR to derive information on the surface biophysical parameters contained within the BRDF.

Although NOAA-10 AVHRR can acquire data in the solar principal plane, it is unable to sample within the remainder of the viewing hemisphere. A more complete sample could be obtained by combining data from the AVHRR on board NOAA-10 with that from similar sensors on NOAA-9 and -11. These satellites have the same nominal orbital parameters, but different equatorial crossing times (e.g., 7.30 a.m. and 7.30 p.m. for NOAA-10; 2.00 a.m. and 2.00 p.m. for NOAA-9 and -11). Consequently, a fixed point on the Earth

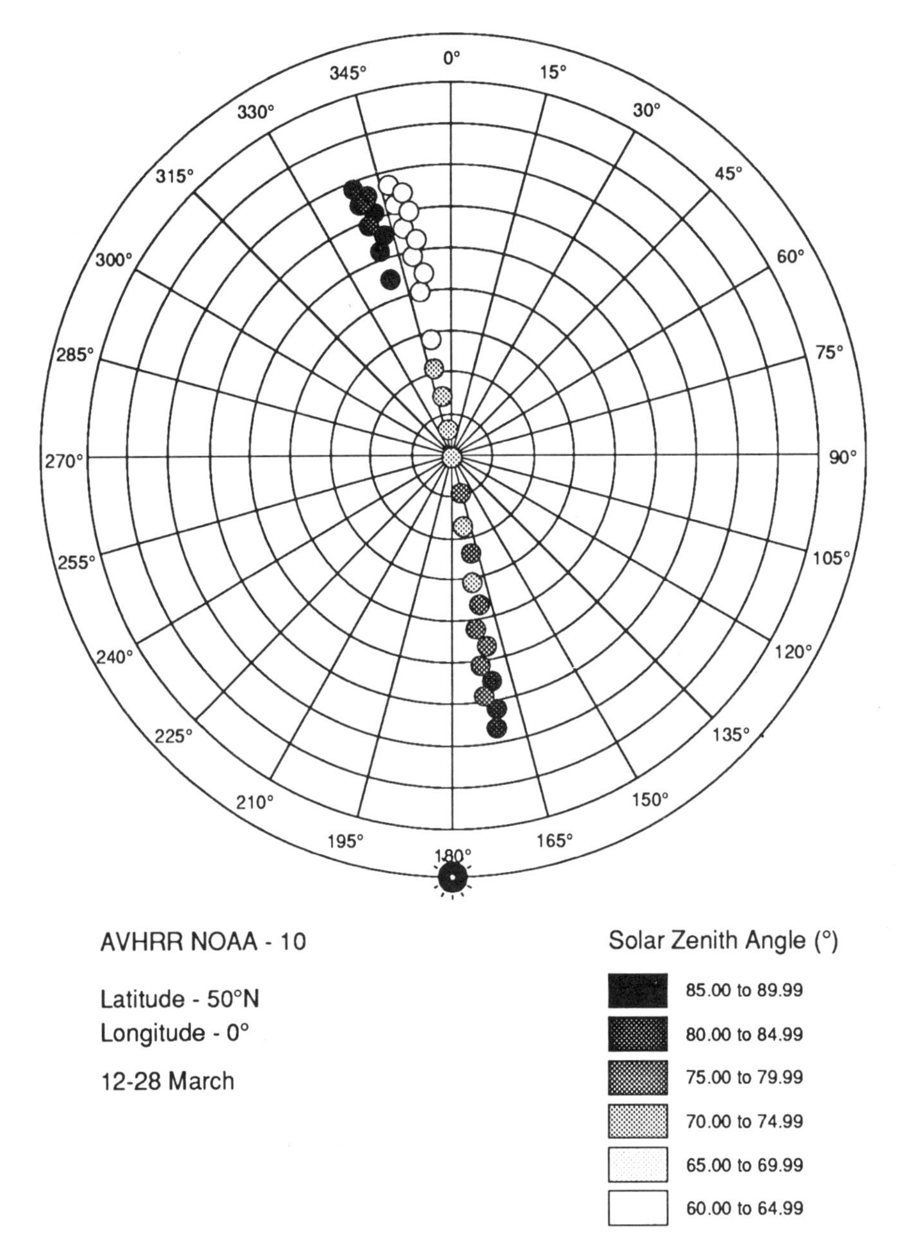

Figure 11.2 Angular sampling capability of NOAA-10 AVHRR over a sixteen-day period around the March equinox, for a site at latitude 50°N (after Barnsley *et al.*, 1993).

surface can usually be imaged on several occasions during a single day. As a result, the BRDF can be sampled over a much wider range of solar zenith and relative azimuth angles than is possible using a single instrument (Barnsley *et al.*, 1993).

An important consideration in the use of NOAA AVHRR data for angular reflectance studies is the extent to which measurements collected on different days can be regarded as samples of the *same* BRDF. This is difficult to determine *a priori* and will, inevitably, depend on the nature of the surface cover type, as well as any variations in atmospheric and illumination conditions during the period of data collection. Even over the period of a few days, ground conditions may change markedly as, for example, in the wetting and drying of soil during and after rainfall. Therefore, in practice, the optimum strategy for sampling the BRDF of Earth surface materials using NOAA AVHRR must be to balance the number and angular distribution of reflectance measurements with the time that it takes to acquire them (Barnsley *et al.*, 1993).

The second instrument considered here is the MISR. Currently, the MISR is proposed for launch as part of NASA's EOS programme in 1998. Unlike, NOAA AVHRR, the MISR has been designed specifically to acquire MVA data, using nine CCD cameras pointing at different (fixed) look angles along-track between ±58° (Diner *et al.*, 1989). This means that nine samples of the BRDF can be acquired during a single orbital overpass. Under most circumstances, ground, atmospheric and illumination conditions can be considered to be constant over this period of time. Thus, MISR offers considerable advantages for sampling the BRDF, by comparison with many other current or proposed satellite sensors. Further samples can be built up by observing the same target area from adjacent orbital tracks on other days, by virtue of the relatively wide cross-track FOV of MISR's CCD cameras (28°). Figure 11.3 illustrates that although the potential number of samples that MISR can obtain within a 16-day period is similar to that of the current generation of AVHRR sensors, the sample measurements are distributed more evenly within the viewing hemisphere.

.... LAND-COVER MAPPING USING MULTIPLE-VIEW-ANGLE DATA

It was noted above that the angular reflectance properties of Earth surface materials are controlled by their 3-D geometric structure. It was further suggested that, as a result of this, it might be possible to use MVA data to distinguish between certain land-cover types or vegetation communities. This has been demonstrated by Barnsley *et al.* (1990), using images acquired by an airborne multispectral scanner with a very wide (86°) field-of-view. In their study, Barnsley *et al.* (1990) constructed MVA data sets from a series of such images recorded on closely spaced, parallel flight lines. Due to the lateral displacement between these flight lines, a small test area on the ground was visible to the sensor at a different view angle in each of six images. MVA data were created by spatially co-registering these six sub-scenes. A separate MVA data set was produced in each of four spectral wavebands, namely 0.50–0.59μm (green), 0.63–0.69μm (red), 0.76–0.90μm (near-IR) and 1.55–1.75μm (middle-IR).

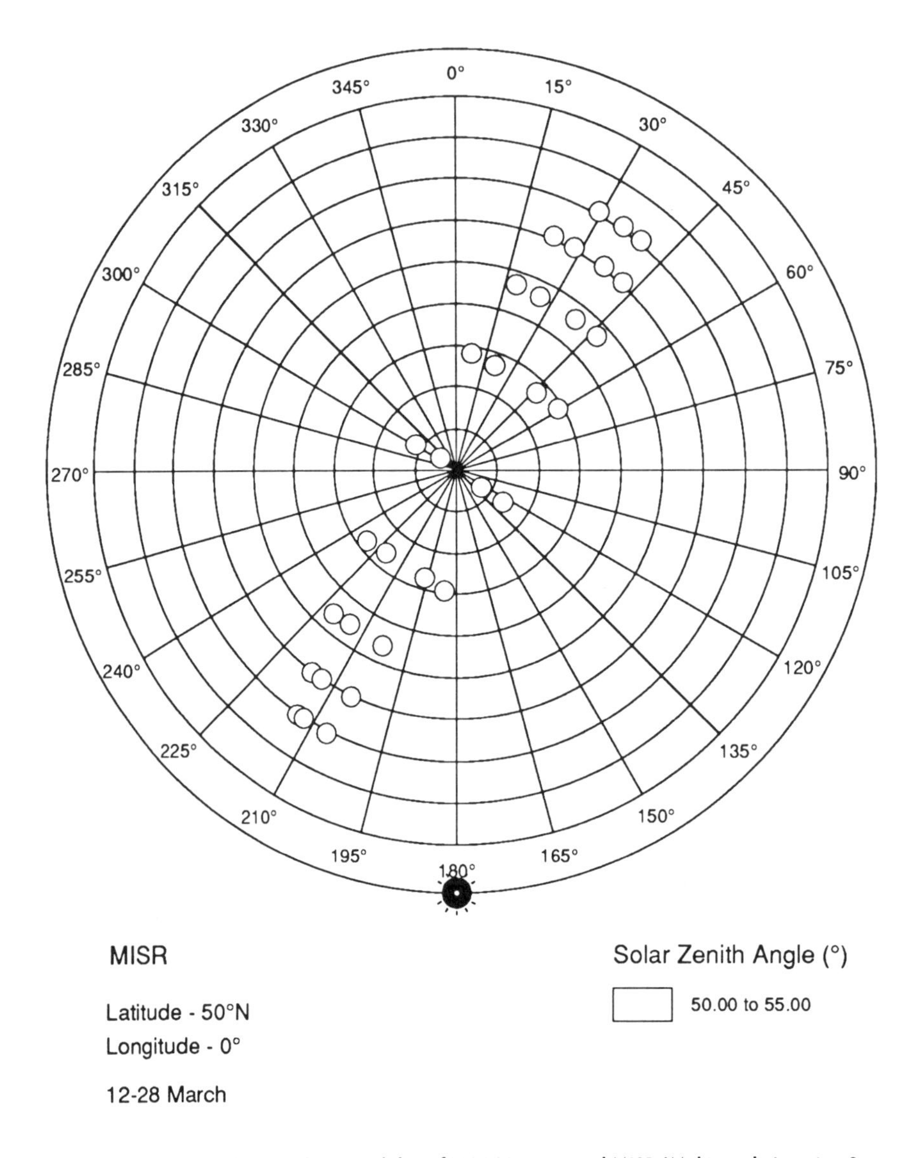

Figure 11.3 Angular sampling capability of NASA's proposed MISR (Multi-angle Imaging Spectro-Radiometer) instrument over a sixteen-day period around the March equinox, for a site at latitude 50°N (after Barnsley *et al.*, 1993).

Plate 8 shows a comparison between a standard false-colour composite (i.e., near-IR, red and green wavebands) and two MVA composites of the same study area, using the data produced by Barnsley *et al.* (1990). Each of the MVA data sets has been constructed from data acquired in a *single* spectral waveband, but at three different view angles (approximately 25° in the backscatter direction, 0°, and 30° in the forward-scatter direction). A small sketch map of the land-cover types present within the study area is also provided (Plate 8(b)). The most noticeable feature of this diagram is the remarkable similarity between the standard (nadir-viewing) false-colour composite (Plate 8(a)) and the MVA composite produced using data recorded solely in the red waveband (Plate 8(c)). In the latter, it is clearly possible to distinguish visually between the soil, oil-seed rape, spring barley and winter wheat. It is important to note that the colours evident in this MVA composite are a product of differences between the *angular* reflectance properties of the observed land-cover types which, in turn, are believed to be related to differences in their 3-D geometric structure.

In a single-band, single view-angle image, land-cover types that exhibit a difference in detected reflectance may be distinguished on the basis of the grey tone (or brightness) of the corresponding pixels. In MVA false-colour composites, land-cover types that have very weak angular reflectance properties (i.e., their reflectance does not vary significantly with sensor view angle) will also appear grey, since similar values will be displayed through each of the three colour guns on the VDU. The ability to distinguish between such cover types will, therefore, be reliant on differences in their *spectral* (as opposed to *angular*) reflectance, since this will condition their apparent brightness within the image. On the other hand, land-cover types which exhibit marked changes in reflectance at different view angles will appear to have some colour in the MVA composite, since unequal values will be displayed via the three colour guns. Moreover, if the angular reflectance properties of separate land-cover types differ, each cover type will appear to be a different colour within the MVA composite. This can be summarised as follows: when viewing a false-colour composite produced from single-waveband MVA data, the *brightness* of an individual land-cover type is controlled by its intrinsic *spectral* reflectance properties (i.e., a product of its biophysical and biochemical characteristics), while the *colour* (i.e., hue and saturation) is controlled by its *angular* reflectance properties which, in turn, are a product of its 3-D geometric structure.

With this in mind, an interesting comparison can be made between the MVA composites produced using data from the red and near-IR wavebands (Plates 8(c) and 8(d), respectively). The MVA composite for the near-IR waveband is essentially a grey-tone image. This is because the land-cover types present within the scene exhibit relatively weak angular reflectance properties in this waveband. For the vegetation canopies, in particular, this is probably a result of multiple scattering of radiation between plant elements, due to their high reflectance in this region of the electromagnetic spectrum, which reduces the angular reflectance anisotropy (Kimes, 1983). Further discussion of this is provided by Barnsley *et al.* (1994), who also considers the use of various techniques (such as RGB-IHS transformations), to isolate the *angular* (as opposed to the *spectral*) component of variation in MVA data sets.

Barnsley *et al.* (1990) use the same MVA data sets in a conventional, per-pixel, multispectral classification procedure. Their results indicated that the accuracy obtained from land-cover classification using single-band MVA data is slightly lower than that obtained using single view-angle, multispectral data. However, combined multispectral and MVA data yield a higher overall classification accuracy than either data set alone.

.... ESTIMATION OF EARTH SURFACE ALBEDO AND 3-D SURFACE STRUCTURE

Recent research into the angular reflectance properties of Earth surface materials has concentrated on the development of various mathematical models that can be used to re-construct the BRDF from a limited sample of angular reflectance measurements. These models can be used to generate accurate estimates of Earth surface albedo, a parameter of great importance in climate modelling (Dorman and Sellers, 1989; Saunders, 1990), by numerical or analytical integration (Walthall *et al.*, 1985; Barnsley and Muller, 1991; Ranson *et al.*, 1991). Moreover, some models can be inverted to derive quantitative estimates of various parameters describing the 3-D geometric structure of the target (Goel, 1988; Franklin and Strahler, 1988; Nilson and Kuusk, 1989; Pinty *et al.*, 1990; Pinty and Verstraete, 1991a; 1991b).

In general terms, two types of model are available. For the purpose of this discussion, these will be referred to as *functional* and *physical* models (Barnsley and Muller, 1991).

FUNCTIONAL MODELS

Functional models attempt to *describe* the angular distribution of the radiation reflected by an object in terms of a set of simple arithmetic and/or geometric functions. This allows the reflectance anisotropy to be represented by a small number of coefficients. The values of these coefficients can be determined by fitting the model to sample reflectance data, using a non-linear least-squares adjustment procedure. Examples of functional models include simple numerical interpolation (Ranson *et al.*, 1991), spherical harmonic functions (Barnsley and Muller, 1991; Sillion *et al.*, 1991) and the Walthall model (Walthall *et al.*, 1985). The principal strengths of these models are their computational simplicity and the absence of assumptions made about the nature of the target.

Functional models can also be used to derive accurate estimates of Earth surface albedo, through numerical or analytical integration with respect to θ_r and ϕ_r. Estimates of albedo derived in this way are inherently more accurate than those derived from nadir-viewing data alone, because they take into account angular anisotropy in the reflectance field (Irons *et al.*, 1988; Barnsley and Muller, 1991). In general, the accuracy with which albedo (or narrow-band hemispherical reflectance) can be estimated is expected to increase as more samples of angular reflectance are used to drive the functional model. Barnsley and Morris (1994) examine this relationship using the following procedure:

(i) generate a simulated BRDF for a surface with a given hemispherical reflectance;
(ii) sample the angular reflectance of that surface at different sensor view angles;
(iii) fit a functional model to these data;
(iv) derive the estimated hemispherical reflectance from the model using numerical or analytical integration; and
(v) compare actual and estimated hemispherical reflectances.

In their study, Barnsley and Morris (1994) generate a simulated BRDF for a soil surface with a hemispherical reflectance of 0.187, using the model developed by Pinty *et al.* (1989). Angular reflectance data for this target are derived by repeated sampling of the simulated BRDF at different sensor view angles. Various angular sampling strategies have been explored, including random, principal-plane only, and simulations of several current and future satellite sensors. A series of such data sets have been created, each containing a different number of angular samples. The results reported here relate to the random sampling strategy, using spherical harmonic functions to reconstruct the BRDF. Initial results indicate that both the mean error and the uncertainty (i.e., variance) in the estimate of hemispherical reflectance decrease very rapidly as the number of angular reflectance samples is increased (Figure 11.4). Indeed, the average error drops to < 5 per cent when 10 or more samples are used. By comparison, the error obtained using a single nadir-viewing measurement of the same surface is approximately 30 per cent. Similar results have also been obtained in studies by Kimes and co-workers (Kimes and Sellers, 1985; Kimes *et al.*, 1987).

One problem with most functional models is that a different set of coefficients must be calculated for each illumination geometry. Thus, representation of the complete BRDF requires $n \times m$ coefficients, where n is the number of coefficients used by the model and m is the number of illumination geometries considered; although Sillion *et al.* (1991) have used one-dimensional cubic splines to model variations in spherical harmonic coefficients as a function of illumination angle.

A further limitation of existing functional models is that they are generally poorly equipped to handle the 'hot spot'. This may lead to either under- or over-estimation of the albedo for many Earth surface materials. Finally, the values associated with the coefficients of many functional models cannot be directly interpreted in terms of key surface biophysical parameters. These relationships remain to be determined, either through further statistical analyses (by comparison with ground measurements of the surface cover types) or through the use of a physical model of surface scattering.

PHYSICAL MODELS OF THE BRDF

Physical models attempt not only to describe the BRDF, but also to account for the scattering of radiation at the Earth surface in terms of a number of physical parameters. A wide variety of physical models of the BRDF have been developed, ranging from geometric-optical models (Otterman and Weiss, 1984; Li and Strahler, 1986; 1993; Ross and Marshak, 1988) to various simplifications and abstractions of the complete radiative-transfer equation (Hapke, 1981; 1984; 1986; Hapke and Wells, 1981; Camillo, 1987; Pinty *et al.*, 1990; Verstraete *et al,* 1990; Pinty and Verstraete, 1991a; 1991b). Emphasis is commonly placed on the use of numerical inversion procedures to derive the model's parameters directly from a sparse set of angular reflectance measurements.

A comprehensive review of physical models of surface scattering is beyond

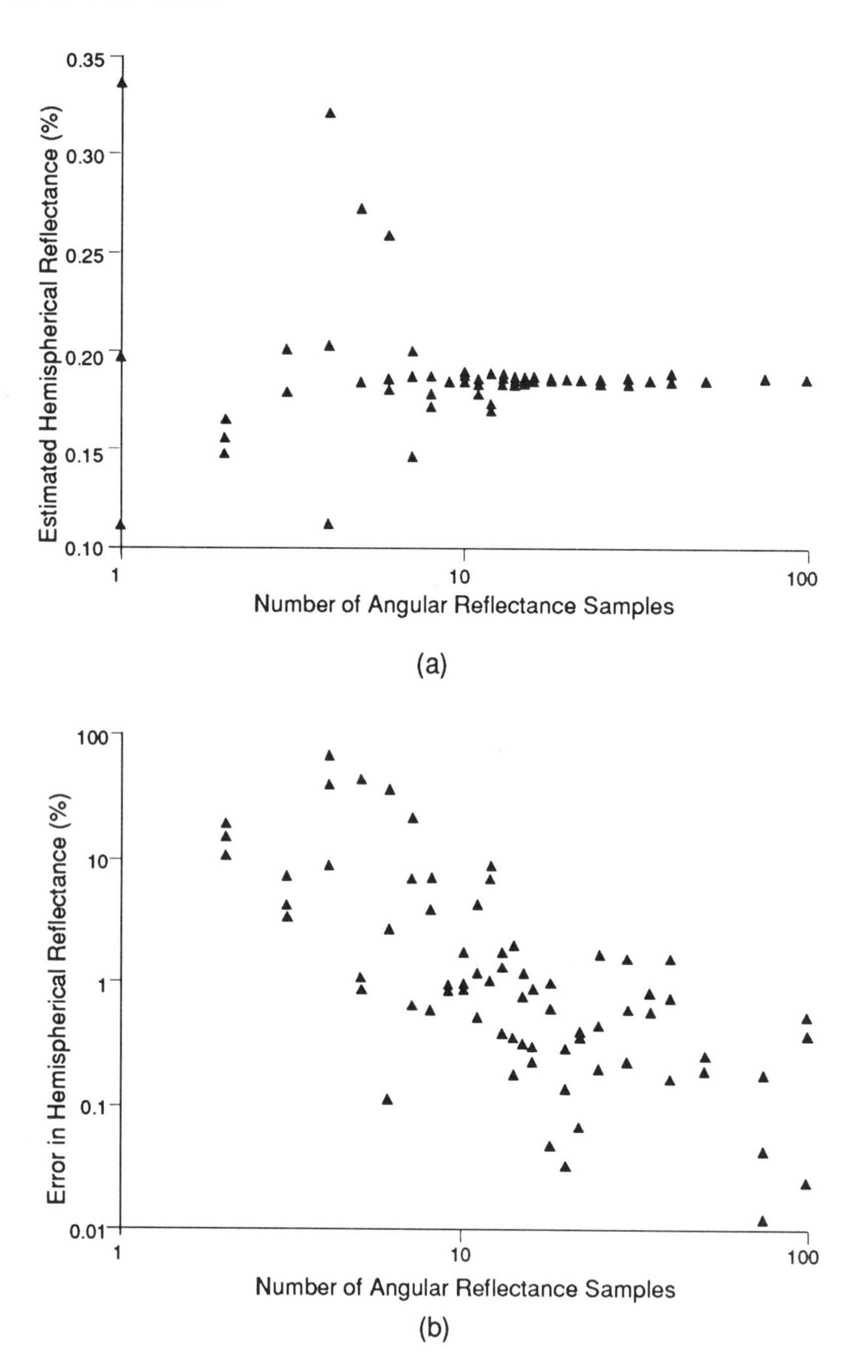

Figure 11.4 Hemispherical reflectance. (a) Estimated hemispherical reflectance produced by integration of spherical harmonic functions fitted to simulated angular reflectance measurements; (b) Percentage error in hemispherical reflectance estimates as a function of the number of angular reflectance samples.

the scope of this chapter; excellent accounts can, instead, be found in Goel (1988) and Myneni *et al.* (1989). However, a number of general observations can be made. First, physical models tend to be developed to represent radiation scattering within a *specific* type of vegetation canopy or surface material. For instance, Li and Strahler's geometric-optical approach has been developed to model the BRDF of vegetation comprised of spatially-discrete canopy elements, such as forests (Li and Strahler, 1986; 1993). By comparison, the model developed by Pinty, Verstraete and others, which is currently receiving considerable attention in the literature, has been developed to represent scattering within spatially-continuous, plane-parallel vegetation canopies, typified by cereal crops (Pinty *et al.*, 1990; Verstraete *et al.*, 1990; Pinty and Verstraete, 1991a; 1991b). This level of specificity is needed to extract accurate information on the detailed type of parameters required (e.g., crown size and shape, or average leaf size and leaf-area index). However, in an operational context, where MVA data from satellite sensors are used to generate information about all areas of the land surface, it will be necessary to develop some intelligent means for selecting the most appropriate model for a given area.

Calibration of physical models also presents significant problems. Principal among these is the need to derive accurate, detailed measurements of the 3-D geometric structure of surface materials, against which to compare values derived from the models. Traditional field-based techniques tend to be time-consuming, laborious and error-prone. Lewis has adopted an alternative approach, using photogrammetric techniques applied to close-range stereoscopic photographs – augmented, where necessary, by manual measurements of key structural parameters – to obtain a description of the complete 3-D geometry of selected plants (Lewis and Muller, 1990; Lewis *et al.*, 1991). These data have also been used, in conjunction with a Monte Carlo ray-tracing system, to simulate the angular reflectance properties of the measured surfaces (Lewis and Muller, 1992). This approach serves both to improve our understanding of the link between surface structure and angular reflectance, as well as providing test data sets against which to compare other physical models.

.... CONCLUSIONS

This chapter has examined the potential use of multiple-view-angle (MVA) remotely-sensed data for environmental monitoring. It has been demonstrated that MVA data contain useful information about the 3-D structure of Earth surface materials that cannot be readily derived from multispectral or multi-temporal analyses alone. The relationship between surface structure and angular reflectance can either be used *implicitly*, by means of conventional classification techniques applied to MVA for land-cover mapping, or *explicitly*, via models of radiation scattering, to derive improved estimates of Earth surface albedo and to infer the 3-D structure of surface materials.

Over the next few years, MVA images will become increasingly available through airborne sensors, such as ASAS and CAESAR, and satellite-based instruments, such as the ATSR, POLDER and MISR. Each of these sensors has a different spatial resolution. Consequently, attention now needs to be given to the effect of spatial scale on the measured angular reflectance trends, as well as the type of the information on 3-D surface structure that can be inferred at a given spatial resolution.

.... ACKNOWLEDGEMENTS

The author would like to thank the Natural Environment Research Council (NERC) for supporting this work through various research grants (GR3/7020, GR3/7901, TIGER III.1 and TIGER III.3) and for provision of the digital data through their annual Airborne Remote Sensing Campaigns. Thanks are also due to Prof. Peter Muller (Department of Photogrammetry and Surveying, UCL), Kevin Morris (currently at NERC IA Unit, University of Plymouth), Philip Lewis, David Allison, Richard Morris (Department of Geography, UCL) and Prof. Alan Strahler (Center for Remote Sensing, Boston University) for advice and assistance with various aspects of this work.

.... REFERENCES

Ardanuy, A., Han, D., Salamonson, V.V., 1991, 'The Moderate Resolution Imaging Spectrometer (MODIS) science and data system requirements', *IEEE Transactions on Geoscience and Remote Sensing*, **29**: 75–88

Barnsley, M.J., 1984, 'Effects of off-nadir view angles on the detected spectral response of vegetation canopies', *International Journal of Remote Sensing*, **5**: 715–28

Barnsley, M.J., Kay, S.A.W., 1990, 'The relationship between sensor geometry, vegetation-canopy geometry and image variance', *International Journal of Remote Sensing*, **11**: 1075–84

Barnsley, M.J., Morris, K.P., Reid, A., 1990, 'Preliminary analysis of a multiple-view-angle image data set', *Proceedings of the NERC 1989 Airborne Remote Sensing Campaign*, Natural Environmental Research Council, Swindon, UK: 49–68

Barnsley, M.J., Muller, J.P., 1991, 'Measurement, simulation and analysis of the directional reflectance properties of earth surface materials', *Proceedings of the 5th International Colloquium – Physical Measurements and Signatures in Remote Sensing*, Courchevel, France, European Space Agency SP-319, Noordwijk: 375–82

Barnsley, M.J., Strahler, A.H., Morris, K.P., Muller, J-P., 1993, 'Sampling the surface Bidirectional Reflectance Distribution Function (BRDF): 1. Evaluation of current and future satellite sensors', *Remote Sensing Reviews* (submitted)

Barnsley, M.J., Morris, K.P., 1994, 'Land-cover mapping using multiple-view-angle images' (in preparation)

Barnsley, M.J., Morris, K.P., Morris, R., 1994, 'Estimating hemispherical reflectance from a sparse set of angular reflectance measurements' (in preparation)

Brakke, T.W., Otterman, J., 1990, 'Canopy bidirectional reflectance dependence on leaf orientation', *International Journal of Remote Sensing*, **11**: 1023–32

Camillo, P., 1987, 'A canopy reflectance model based on an analytical solution to the multiple scattering equation', *Remote Sensing of Environment*, **23**: 453–77

Cierniewski, J., 1987, 'A model for soil surface roughness influence on the spectral response of bare soils in the visible and near-infrared range', *Remote Sensing of Environment*, **23**: 97–115

Cierniewski, J., 1989, 'The influence of the viewing geometry of bare rough soil surfaces on their spectral response in the visible and near-infrared range', *Remote Sensing of Environment*, **27**: 135–42

Day, T., Muller, J-P., 1988, 'Digital elevation model production by stereo-matching SPOT image-pairs: A comparison of algorithms', *Proceedings of the 4th Alvey Vision Conference*, University of Manchester, Manchester: 117–22

Deering, D.W., Eck, T.F., 1987, 'Atmospheric optical depth effects on angular anisotropy of plant canopy reflectance', *International Journal of Remote Sensing*, **8**: 893–916

Deering, D.W., Eck, T.F., Otterman, J., 1989, 'Bidirectional reflectances of three soil surfaces and their characterisation through model inversion', *Proceedings of IGARSS '89*, Vancouver, IEEE Publications, New York: 670–3

Deering, D.W., Eck, T.F., Otterman, J., 1990, 'Bidirectional reflectances of selected desert surfaces and their three-parameter soil characterization', *Agricultural and Forest Meteorology*, **52**: 71–90

Deschamps, P.Y., Herman, M., Podaire, A., Leroy, M., Laporte, M., Vermande, P., 1990, 'A spatial instrument for the observation of polarization and directionality of earth reflectances: POLDER', *Proceedings of IGARSS '90*, Washington, D.C.: 1769–74

Deuzé, J-L., Balois, J-Y., Devaux, C., Gonzalez, L., Herman, M., Lecomte, P., Verwaerde, C., 1991, 'Aircraft simulations of the POLDER experiment: First results', *Proceedings 5th International Colloquium – Physical Measurements and Signatures in Remote Sensing*, Courchevel, France, European Space Agency SP-319, Noordwijk: 393–6

Diner, D.J., Martonchik, J.V., 1985, 'Atmospheric transmittance from spacecraft using multiple view angle imagery', *Applied Optics*, **24**: 3503–11

Diner, D.J., Bruegge, C.J., Martonchik, J.V., Ackerman, T.P., Davies, R., Gerstl, S.A.W., Gordon, H.R., Sellers, P.J., Clark, J., Daniels, J.A., Danielson, E.D., Duval, V.G., Klaasen, K.P., Lilienthal, G.W., Nakamoto, D.I., Pagano, R.J., Reilly, T.H., 1989, 'MISR: A multiangle imaging spectroradiometer for geophysical and climatological research from EOS', *IEEE Transactions on Geoscience and Remote Sensing*, **27**: 200–14

Dorman, J.L., Sellers, P.J., 1989, 'A global climatology of albedo, roughness length and stomatal resistance for atmospheric General Circulation Models as represented by the Simple Biosphere Model (SiB)', *Journal of Applied Meteorology*, **28**: 833–55

Franklin, J., Strahler, A.H., 1988, 'Invertible canopy reflectance modelling of vegetation structure in semiarid woodland', *IEEE Transactions on Geoscience and Remote Sensing*, **26**: 809–25

Gerstl, S.A.W., Simmer, C., 1986, 'Radiation physics and modelling for non-Lambertian surfaces', *Remote Sensing of Environment*, **20**: 1–29

Gerstl, S.A.W., Zardecki, A., 1985, 'Coupled atmosphere/canopy model for remote sensing of plant reflectance features', *Applied Optics*, **24**: 94–103

Goel, N.S., 1988, 'Models of vegetation canopy reflectance and their use in the estimation of biophysical parameters from reflectance data', *Remote Sensing Reviews*, **4**: 1–222

Goel, N.S., Reynolds, N.E., 1989, 'Bidirectional canopy reflectance and its relationship to vegetation characteristics', *International Journal of Remote Sensing*, **10**: 107–32

Goel, N.S., Thompson, R.L., 1985, 'Optimal solar/viewing geometry for an accurate estimation of leaf area index and leaf angle distribution from bidirectional canopy reflectance data', *International Journal of Remote Sensing*, **6**: 1493–1515

Goetz, A.F.H., Herring, M., 1989, 'The High Resolution Imaging Spectrometer (HIRIS) for Eos', *IEEE Transactions on Geoscience and Remote Sensing*, **27**: 136–44

Hapke, B., 1981, 'Bidirectional reflectance spectroscopy, 1. Theory', *Journal of Geophysical Research*, **86**: 3039–54

Hapke, B., 1984, 'Bidirectional reflectance spectroscopy, 3. Correction for macroscopic roughness', *Icarus*, **59**: 41–59

Hapke, B., 1986, 'Bidirectional reflectance spectroscopy, 4. The extinction coefficient and the opposition effect', *Icarus*, **67**: 264–80

Hapke, B., Wells, E., 1981, 'Bidirectional reflectance spectroscopy, 2. Experiments and observations', *Journal of Geophysical Research*, **86**: 3055–60

Irons, J.R., Johnson, B.L., Jr., Linebaugh, G.H., 1987, 'Multiple-angle observations of reflectance anisotropy from an airborne linear array sensor', *IEEE Transactions on Geoscience and Remote Sensing*, **25**: 372–83

Irons, J.R., Ranson, K.J., Daughtry, C.S.T., 1988, 'Estimating big bluestem albedo from directional reflectance measurements', *Remote Sensing of Environment*, **25**: 185–99

Irons, J.R., Ranson, K.J., Williams, D.L., Irish, R.I., Huegel, F.G., 1991, 'An off-nadir pointing imaging spectroradiometer for terrestrial ecosystem studies', *IEEE Transactions on Geoscience and Remote Sensing*, **29**: 66–74

Jackson, R.D., Teillet, P.M., Slater, P.N., Fedosejevs, G., Jasinski, M.F., Aase, J.K., Moran, M.S., 1990, 'Bidirectional measurements of surface reflectance for view angle corrections of oblique imagery', *Remote Sensing of Environment*, **32**: 189–202

Jupp, D.L.B., Strahler, A.H., 1991, 'A hotspot model for leaf canopies', *Remote Sensing of Environment*, **38**: 193–210

Kimes, D.S., 1983, 'Dynamics of directional reflectance factor distributions for vegetation canopies', *Applied Optics*, **22**: 1364–72

Kimes, D.S., Kirchner, J.A., 1982, 'Irradiance measurement errors due to the assumption of a Lambertian reference panel', *Remote Sensing of Environment*, **12**: 141–9.

Kimes, D.S., Sellers, P.J., 1985, 'Inferring hemispherical reflectance of the earth's surface for global energy budgets from remotely-sensed nadir or directional radiance values', *Remote Sensing of Environment*, **18**: 205

Kimes, D.S., Sellers, P.J., Diner, D.J., 1987, 'Extraction of spectral hemispherical reflectance (albedo) of surfaces from nadir and directional reflectance data', *International Journal of Remote Sensing*, **8**: 1727–46

Kneizys, F.X., Anderson, G.P., Shettle, E.P., Gallery, W.O., Abreu, L.W., Selby, J.E.A., Chetwynd, J.H., Clough, S.A., 1988, *Users guide to LOWTRAN 7*, AFGL-TR-88-0177 Environmental Research Papers, No. 1010, Air Force Geophysics Laboratory

Kriebel, K.T., 1978, 'Measured spectral bidirectional reflection properties of four vegetated surfaces', *Applied Optics*, **17**: 253–8

Lewis, P., Muller, J-P., 1990, 'Botanical plant modelling for remote sensing simulation studies', *Proceedings of IGARSS '90*, Washington, D.C.: 1739–42

Lewis, P., Muller, J-P., Morris, K.P., 1991, 'Quality assessment of a botanical plant-modelling system for remote sensing simulation studies', *Proceedings of IGARSS '91*, Espoo, Finland: 1917–20

Lewis, P., Muller, J-P., 1992, 'The Advanced Radiometric Ray-Tracer: ARARAT for plant canopy reflectance simulation', *International Archives of Photogrammetry and Remote Sensing*, **29**: Commission VII: 26–34

Li, X., Strahler, A.H., 1986, 'Geometrical-optical bidirectional reflectance modelling of a conifer forest canopy', *IEEE Transactions on Geoscience and Remote Sensing*, **24**: 906–19

Li, X., Strahler, A.H., 1993, 'Geometric-optical bidirectional reflectance modelling of the discrete-crown vegetation canopy: Effect of crown shape and mutual shadowing', *IEEE Transactions on Geoscience and Remote Sensing* (in press)

Liang, S., Strahler, A.H., 1993, 'The computation of radiance for coupled atmosphere and canopy media using an improved Gauss-Seidel algorithm', *IEEE Transactions Geoscience and Remote Sensing* (in press)

Looyen, W.J., Verhoef, W., Clevers, J., Lamers, J., Boerma, J., 1990, 'Measurements on directional reflectance and acquisition of stereo information by CAESAR', *Proceedings of IGARSS '90*, Washington, D.C.: 1955–8

Lorenz, D., 1985, 'On the feasibility of cloud stereoscopy and wind determination with the ATSR', *International Journal of Remote Sensing*, **6**: 1445–61

Martonchik, J.V., Diner D.J., 1993, 'Retrieval of aerosol and land surface optical properties from multi-angle satellite imagery', *IEEE Transactions Geoscience and Remote Sensing* (in press)

Moran, M.S., Jackson, R.D., Hart, G.F., Slater, P.N., Bartell, R.J., Biggar, S.F., Gellman, D.I., Santer, R.P., 1990, 'Obtaining surface reflectance factors from atmospheric and view angle corrected SPOT-1 HRV data', *Remote Sensing of Environment*, **32**: 203–14

Muller, J-P., 1989, 'Key issues in image understanding in remote sensing', *Philosophical Transactions of the Royal Society Series A*, **324**: 381–95

Myneni, R.B., Ross, J., Asrar, G., 1989, 'A review of the theory of photon transport in leaf canopies', *Agricultural Forestry Meteorology*, **45**: 1–153

NASA, 1986a, *MODIS – Moderate Resolution Imaging Spectrometer*, 'Instrument Panel Report, Volume IIb, National Aeronautics and Space Administration', Washington, D.C.: 59pp.

NASA, 1986b, *HIRIS – High Resolution Imaging Spectrometer: Science Opportunities for the 1990s*, Instrument Panel Report, Volume IIc, National Aeronautics and Space Administration, Washington, D.C.: 74pp.

Nicodemus, F.E., Richmond, J.C., Hsia, J.J., Ginsberg, I.W., Limperis, T., 1977, *Geometrical Considerations and Nomenclature for Reflectance*, NBS Monograph 160, Institute for Basic Standards, Washington, D.C.

Nilson, T., Kuusk, A., 1989, 'A reflectance model for the homogeneous plant canopy and its inversion', *Remote Sensing of Environment*, **27**: 157–67

Otterman, J., 1985, 'Bidirectional and hemispheric reflectivities of a bright soil plane and a sparse dark canopy', *International Journal of Remote Sensing*, **6**: 897–902

Otterman, J., Strebel, D.E., Ranson, K.J., 1987, 'Inferring spectral reflectances of plant elements by simple inversion of bidirectional reflectance measurements', *Remote Sensing of Environment*, **21**: 215–28

Otterman, J., Weiss, G.H., 1984, 'Reflection from a field of randomly-located vertical protrusions', *Applied Optics*, **23**: 1931–6

Pinter, P.J., Jackson, R.D., Moran, M.S., 1990, 'Bidirectional reflectance factors of agricultural targets: A comparison of ground-, aircraft-, and satellite-based observations', *Remote Sensing of Environment*, **32**: 215–28

Pinty, B., Verstraete, M.M., Dickinson, R.E., 1989, 'A physical model for predicting bidirectional reflectances over bare soil', *Remote Sensing of Environment*, **27**: 273–88

Pinty, B., Verstraete, M.M., Dickinson, R.E., 1990, 'A physical model of the bidirectional reflectance of vegetation canopies, 2. Inversion and validation', *Journal of Geophysical Research*, **95**: 11767–75

Pinty, B., Verstraete, M.M., 1991a, 'Extracting information on surface properties from bidirectional reflectance measurements', *Journal of Geophysical Research*, **96**: 2865–74

Pinty, B., Verstraete, M.M., 1991b, 'Bidirectional reflectance and surface albedo: Physical modelling and inversion', *Proceedings of the 5th International Colloquim – Physical Measurements and Signatures in Remote Sensing*, Courchevel, France, European Space Agency SP-319, Noordwijk: 383–6

Powers, B.J., Gerstl, S.A.W., 1988, 'Modelling of atmospheric effects on the angular distribution of a backscatter peak', *IEEE Transactions on Geoscience and Remote Sensing*, **26**: 649–59

Prata, A.J., Cechet, R.P., Barton, I.J., Llewellyn-Jones, D.T., 1990, 'The Along-Track Scanning Radiometer for ERS-1: Scan geometry and data simulation', *IEEE Transactions on Geoscience and Remote Sensing*, **28**: 3–13

Ranson, K.J., Irons, J.R., Daughtry, C.S.T., 1991, 'Surface albedo from bidirectional reflectance', *Remote Sensing of Environment*, **35**: 201–11

Ross, J.K., Marshak, A.L., 1988, 'Calculation of canopy directional reflectance using the Monte Carlo Method', *Remote Sensing of Environment*, **24**: 213–25

Ross, J.K., Marshak, A.L., 1989, 'Influence of leaf orientation and the specular component of leaf reflectance on the canopy bidirectional reflectance', *Remote Sensing of Environment*, **27**: 251–60

Roujean, J.L., Leroy, M., Deschamps, P.Y., Podaire, A., 1990, 'A surface BRDF model to be used for the correction of directional effects in remote sensing multi-temporal data sets', *Proceedings of IGARSS '90*, Washington, D.C.: 1785–9

Salomonson, V.V., Marlatt, W.E., 1971, 'Airborne measurements of reflected solar radiation', *Remote Sensing of Environment*, **2**: 1–8

Salomonson, V.V., Barnes, W.L., Maymon, P.W., Montgomery, H.E., Ostrow, H., 1989, 'MODIS: Advanced facility instrument for studies of the earth as a system', *IEEE Transactions on Geoscience and Remote Sensing*, **27**: 145–53

Saunders, R.W., 1990, 'The determination of broad-band surface albedo from AVHRR visible and near-infrared radiances', *International Journal of Remote Sensing*, **11**: 49–67

Sillion, F.X., Arvo, J.R., Westin, S.H., Greenberg, D.P., 1991, 'A global illumination solution for general reflectance distributions', *Computer Graphics*, **25**: 187–96

Strahler, A.H., Jupp, D.L.B., 1990, 'Modelling bidirectional reflectance of forests and woodlands using Boolean models and geometric optics', *Remote Sensing of Environment*, **34**: 153–66

Suits, G.H., 1972, 'The calculation of the directional reflectance of a vegetation canopy', *Remote Sensing of Environment*, **2**: 117–25

Tanré, D., Herman, M., Deschamps, P.Y., 1983, 'Influence of the atmosphere on space measurements of directional properties', *Applied Optics*, **22**: 733–41

Tanré, D.C., Deroo, C., Duhaut, P., Herman, M., Morcrette, J.J., Perbos, J., Deschamps, P.Y., 1990, 'Description of a computer code to simulate the satellite signal in the solar spectrum: 5S code', *International Journal of Remote Sensing*, **11**: 659–68

Verstraete, M.M., Pinty, B., Dickinson, R.E., 1990, 'A physical model of the bidirectional reflectance of vegetation canopies. 1. Theory', *Journal of Geophysical Research*, **95**: 11755–65

Walthall, C.L., Norman, J.M., Welles, J.M., Campbell, G., Blad, B.L., 1985, 'Simple equation to approximate the bidirectional reflectance from vegetative canopies and bare soil surfaces', *Applied Optics*, **24**: 383–7

12. Earth Observation Data – or Information?

PAUL M. MATHER

INTRODUCTION

Satellite-borne sensors are observing routinely the Earth's land and sea surfaces in a variety of wavebands and at a number of spatial scales. The best-known of these systems, the multispectral scanning system (MSS) carried by Landsat, has been operating since 1972; more recently, the French Satellite Pour l'Observation de la Terre (SPOT), High Resolution Visible (HRV) and the Japanese Marine Observational Satellite (MOS) sensors have provided fine spatial resolution data in the visible and near infrared regions, while coarse spatial resolution imagery has continued to be supplied by the National Oceanic and Atmospheric Administration's (NOAA) Advanced Very High Resolution Radiometer (AVHRR) and the Meteosat radiometer and related systems. The launch in 1991 of the first European Remote Sensing Satellite (ERS-1) and the Japanese Fuyo-1 systems has made synthetic aperture radar (SAR) data widely available for the first time since Seasat in 1978 and short lived Space Shuttle programmes of the early 1980s. Plans for the next ten years include:

- ERS-2 (1993), which will carry a SAR similar to that on board ERS-1, and a visible and infrared narrow-band scanner called the Along Track Radiometer (ATSR-2);
- NOAA 'I'–'M' (1993–1996), carrying the AVHRR-2 and -3;
- Landsat-6 (1993), which will have an Enhanced Thematic Mapper (TM) instrument and Landsat-7 in 1998;
- SPOT-3 (1993), with the HRV sensor equivalent to those on SPOT-1 and -2;
- SPOT-4 (1996), carrying an enhanced HRV termed the HRVIV or High Resolution Visible and Infrared sensor, plus the Vegetation Sensor; and
- Radarsat (1995), which will have a SAR.

ESA is currently expected to launch the first Polar Orbit Earth Observation

Mission (POEM), called Envisat, in 1998. Two separate sensor packages are planned for POEM: the M-series, which are primarily meteorological, oceanographic and climatological, and the N-series oriented towards Earth resources and the atmosphere. The first European Polar Platform, EPOP-1, will carry the M-series sensor package, including VIRSR (similar to AVHRR), a SAR, MERIS (Medium Resolution Imaging Spectrometer for oceanographic applications), and AATSR (Advanced Along-Track Scanning Radiometer, designed primarily for sea-surface temperature studies), in addition to atmospheric sounding devices and a radar altimeter (Rast and Readings, 1992). NASA is planning a parallel system, called EOS (Earth Observing System) for launch in 1998. Plans for EOS and its associated Data and Information System, EOSDIS, are currently in a state of flux (Moore and Dozier, 1992).

The motivation underlying this remarkable planned expansion of Earth observation capabilities during the next ten years is primarily the perceived need to improve our understanding of the Earth system as a whole, including human activity. Rast and Readings (1992) list the four themes which underlie ESA's Earth observation programme, as follows:

- monitoring the Earth's environment on various scales, from local to regional to global;
- management and monitoring of the Earth's resources, both renewable and non-renewable;
- continuing provision of an operational meteorological system; and
- improving scientific knowledge of the structure and dynamics of the Earth's crust and interior.

They comment as follows: 'More organised information about the behaviour of the environment and factors influencing the Earth's natural resources can only be achieved on the basis of a better understanding of the Earth when viewed as a system ... Fundamental to this is the provision of data to identify processes and to validate models' (Rast and Readings, 1992, pp. 175–6). Clearly, these authors view the provision of Earth observation data in terms of the classical process of scientific explanation, which requires hypotheses or models to be established and subsequently tested against observational data. This procedure implies that observational data have the following characteristics:

- to be internally consistent;
- to relate to other observations in a known manner; and
- to be readily available to researchers.

Data consist of observations; in a scientific context these are generally numerical. Information is the combination of knowledge and observations. To be scientifically useful, Earth observation data must be combined with knowledge in order to improve our understanding of the Earth system, as envisaged by Wickland, who suggests that the goal of Earth System Science is: 'To obtain a scientific understanding of the entire Earth System on a global scale ...' (Wickland, 1989, p. 693). Such an aim is not achievable if the data that are combined with knowledge in the form of theories and models are inadequate. This is not a new observation; Monteith suggested that 'Most applications of remote sensing are still in the process of evolving through stages of development familiar in the experimental sciences ... Data banks, however comprehensive, cannot generate hypotheses spontaneously' (Monteith, 1990, p. 397).

The purpose of this chapter is to examine the obstacles that presently stand in the way of the use of Earth observation data in a scientific context and to propose mechanisms that will ensure that data of a sufficient quality are readily available to researchers. Given that the first two of ESA's research policy themes, listed above, are concerned with global problems, it is essential to ensure that researchers have access to global data sets.

OBSTACLES TO THE SCIENTIFIC USE OF EARTH OBSERVATION DATA

Earth observation (EO) data are provided to the user in the form of digital images, which are composed typically of registered arrays of numbers on a 0–255 or 0–1023 scale. These numbers are quantised counts, and are not immediately related to physical quantities such as radiant flux. Methods of processing these data, and extracting information from them, are widely known (Mather, 1987; Chuvieco, 1990). Generally speaking, such methods involve the enhancement of a single-date image using digital image processing techniques such as contrast improvement or filtering, or the derivation of map-like products using pattern recognition or classification techniques. It is important to realise that the data forming the image set are related only in a loose way to physical properties of the environment and that the quality of the information supplied is often far from the ideal. It is generally descriptive, nominal or ordinal-scale information such as that contained in land-cover maps that is derived, rather than estimates of physical or biophysical properties, such as evaporation, which are required as input to models. Thus, Monteith comments: 'I therefore believe that agronomists, like hydrologists and ecologists, are likely to have to wait for some years before remote sensing can provide them with estimates of evaporation better than what is now available from formulae that incorporate an informed guess about the magnitude of surface resistance' (Monteith, 1990, p. 399). Furthermore, such relationships between remotely-sensed observations and surface properties as exist may vary over the image, depending on topography and atmosphere. It is unlikely that the same relationships would hold at a different point in time. Earth observation data, in their raw form, are therefore internally and externally inconsistent, and consequently require preprocessing before they are suitable to input to predictive models (such as Running and Coughlan's (1988) forest ecosystem process model, the Biosphere-Atmosphere Transfer Scheme (Dickinson, 1984) and the simple biosphere model (Sellers and Dorman, 1987)).

The signal received by a sensor operating in the visible and near infrared wavebands of the electromagnetic spectrum is often represented in simple terms as the solar spectrum modified by interaction with the atmosphere and with the ground surface. However, these interactions are more complex, as outlined by Duggin and Robinove (1990). Their paper discusses eleven major assumptions that are implicit in the acquisition and analysis of passively-sensed digital image data; they identify the following factors, among others, as relating to the correlation between ground attributes in the real world and the corresponding properties inferred from remotely-sensed data:

- nature of the material forming the ground resolution elements;
- correlation of the attributes of this area with the upwelling spectral radiance field;
- effects of atmospheric attenuation and scattering;
- performance of the analogue-to-digital converter (affecting the response, calibration and linearity of the sensor);
- influence of any compression/decompression; and
- choice of image processing or pattern recognition procedures.

Other factors include the relationship between the sensor attributes, such as spatial resolution, spectral bands, and spectral response and the properties of the target of interest. Duggin and Robinove express the problem succinctly as follows: 'Fundamental implicit assumptions in image acquisition and analysis are that the radiance properties recorded on the image represent the optical properties of those features on the ground which are of interest and that the image, after processing, may be rectified and superimposed on a map to represent accurately the ground features which are of interest' (Duggin and Robinove, 1990, p. 1673).

To ensure that image data are internally coherent prior to analysis, preprocessing is necessary in order to remove, as far as possible, the 'external' effects recognised in the preceding paragraphs. Such preprocessing relates to the need for sensor calibration, correction for atmospheric, illumination and viewing geometry effects and geo-referencing. Geo-referencing is not strictly speaking a preprocessing operation, in that it is not expressly concerned with the removal or attenuation of a sensor-imposed property; rather, it refers to the registration of the image to an accepted map projection. In the sense that the operation is necessary if Earth observation data are to be used in conjunction with other spatial data sets, it will be considered at this point.

RADIOMETRIC CALIBRATION

Passive imaging sensors carried by orbiting satellites collect upwelling radiance from the Earth's surface in one or more spectral bands for each of a large number of ground resolution element(s). The output for each spectral band from the sensor is a voltage, which is related in an approximately linear manner to the spectral radiance. Asrar (1989) shows that the relationship relies on the gain and offset of the sensor system, which are found from pre-launch ground calibration and subsequently by in-flight calibration or by reference to ground targets of known reflectance. Gain and offset values are normally recorded in the header file of the data tapes. The quantised counts contained by the tape can be converted to radiances by reference to the gain and offset for the given spectral band. Since the sensor gains and offsets vary over the life of the sensor, it follows that the individual pixel values do not have any absolute or intrinsic physical meaning. It also follows that inferences regarding relative magnitudes of pixel values obtained by different sensors, such as HRV or TM, cannot be made unless the quantised counts are converted first to radiances. Remotely-sensed data in the form of digital counts cannot, therefore, be compared over time, or between sensors. However, they can be used for the extraction of information (such as a land-cover map) for which relative values only are required.

ATMOSPHERIC CORRECTION

Solar radiation incident upon the Earth's surface must first pass through the atmosphere. Similarly, reflected or emitted radiation passes upwards through the atmosphere to the sensor system. The spectral composition of the signal is modified by interactions with the atmosphere on both the upward and the downward journeys. The atmosphere is not, unfortunately, constant in its composition either in space or in time and, consequently, changes in the level of electromagnetic radiation recorded by the sensor for a given area cannot, even after radiometric correction, be presumed to indicate changes in the ground area being observed. Kaufman (1989) notes the following effects that are generated by the atmospheric effect: variation of the severity of the effect with the wavelength of the radiation being sensed, which may affect discrimination between, for instance, stressed and unstressed vegetation; alteration in the spatial distribution of reflected radiation, affecting the spatial resolution of the system; changes in the apparent brightness of a target, affecting measurements of albedo and reflection; and generation of spatial variations in apparent surface reflection through the effect of sub-pixel sized clouds. Correction for atmospheric effects is difficult, because the optical characteristics of the atmosphere (which are influenced by aerosol optical thickness, phase function, single-scattering albedo, gaseous absorption, and vertical profile) are rarely known in any detail at the time of imaging and over the area of the image. Atmospheric models such as LOWTRAN (Kneisys *et al.*, 1983) provide corrections based on theory, using either the observed or inferred characteristics of the atmosphere (the 'standard atmospheres', such as maritime temperate winter), but such approximations are correct only to the first order, at best. Hill and Sturm (1988) use a basic 'histogram minimum' method (USGS, 1979) to estimate atmospheric path radiance, followed by correction for absorption based on LOWTRAN functions. The removal of the effects on the spectral properties of the signal caused by atmospheric interactions remains one of the fundamental problems of remote sensing; without a reliable and satisfactory method of correction, the recovery of upwelling radiances, and hence the credible comparison of spectra across images, will remain problematical.

ILLUMINATION AND VIEWING GEOMETRY EFFECTS

Most natural terrestrial surfaces have reflectance spectra which depend upon the angle of view and the angle of illumination, relative to the ground surface. The distribution of reflected energy as a function of the angles of incidence and reflection is described by the hemispherical bidirectional reflectance function of the target, which is generally unknown. If the surface is assumed to be

Lambertian then a correction for solar illumination angle can be made, but this is only an approximation. Viewing angle depends on the properties of the satellite orbit and the characteristics of the sensor. For example, it would be realistic to suggest that the same object viewed by the AVHRR at either side of the swath (±56°) and at nadir would all produce different values, and a similar effect might be achieved by viewing the same area at different SPOT HRV tilt angles (HRV can be tilted up to 27° from nadir). It is therefore the case that a homogeneous cover, such as an agricultural crop growing on a flat area, will produce a variable spatial distribution of radiance in a specific spectral band. The effects of terrain must be added to the bidirectional reflectance characteristics of the target. It is well known that the topographic position of the ground cell, that is, its slope and aspect, have a significant effect on the characteristics of the signal emanating from that cell (for example, Holben and Justice (1980) conclude that the topographic effect can produce a wide variation in the radiances associated with a given land-cover type).

GEOREFERENCING

It is now widely accepted that Earth observation data alone cannot provide the information needed for the solution of many problems in the Earth and environmental sciences. In modelling applications, for example, Earth observation data represents only one of several inputs, and the location of the observation becomes an important property of that observation, allowing it to be related to other spatial data in a systematic fashion. At present it is not generally possible to fix the position of individual pixels from orbital data alone, though a reasonable approximation can be made for coarse spatial resolution NOAA AVHRR data. The solution to the problem of locating higher spatial resolution data of land areas is usually achieved through the well-known process of determining a bivariate polynomial transformation based upon ground control points that relates map and image coordinates using the method of least-squares, followed by resampling to generate a geo-coded image. If both image and map data are stored within a geographical information system (GIS), the operation becomes almost routine. Unfortunately, attention must be paid to the quality of the result, which cannot easily be ascertained through scrutiny of the residual errors at the individual ground control points. Nor is it wise to improve the apparent fit of the polynomial by eliminating from the analysis those ground control points which have high residual error values. All that these residual errors tell the user is the closeness of fit of the polynomial surface at the ground control points and nowhere else. Given that 'quality' is not an absolute term, depending as it does on the use to which the data are to be put, it is nevertheless important for the user to be aware of the adequacy of the transformation. Mather (1992) presents experimental results to show the variations in the shape of the bivariate polynomial surface that result from (i) altering the number of ground control points and (ii) changing their spatial distribution. In these experiments the true shape of the surface was known, which is not the case in real applications, but the conclusions drawn from the experimental results apply in practical applications. The number of ground control points should be significantly more than the minimum required by the computational algorithm, and the spatial distribution of control points should be well-scattered over the area of the image.

.... POSSIBLE SOLUTIONS

If users are unaware of the existence of data sets they will not use them. However, in order to make appropriate and proper use of data the user must be aware of the characteristics of the data, as well as the methods that are available to extract the required information from the data. Quality of data, which relates to the context in which the data are to be used as well as to the internal characteristics of the data set, is an important consideration. Quality is also a characteristic of the algorithms that are used in information extraction – some algorithms may produce lower-quality results than others. The first of the two interlinked proposals that are made here with the intention of reducing or eliminating obstacles to the scientific use of Earth observation data, as discussed above, is the development of a metadata network with sufficient intelligence available to assist the user in the choice of data set and processing technique. The second proposal relates to the fact that, for many users, it is the information that is extracted from the data rather than the data themselves (and, by implication, the data processing methods that are used) that is of interest. An analogy may be made between data usage and the more widespread use of computers. Many scientists began to use computers only when they became accessible in the form of work-stations and personal computers. The majority, though, became computer users only when the 'user interface' was of a quality sufficient to allow the use of the computer without the necessity of becoming an expert. The interface (such as Microsoft Windows or the Mackintosh graphical interface, both examples of graphical user interfaces incorporating 'point and click' approaches) embodied the expertise that was previously within the realm of the computer specialist. In a similar way, image processing software and GIS will need to become more accessible to Earth and environmental scientists if they are to become standard tools.

METADATA NETWORKS

Metadata means descriptions of data. A metadata network can be conceived of as an interlinked system containing summary information about data sets, including information about the nature of the data, access, coverage, spatial resolution, and other relevant characteristics. The network may be distributed or centralised; the former would consist of a number of nodes, each of which would contain some metadata. A node located at a data centre would include the characteristics listed above, whereas a user node would hold information about specific applications of particular data sets. Clients access metadata through the issuing of queries which are passed around the network, and matching details are sent to the client's workstation. A centralised system would hold all metadata in standard form at a single location, and all queries would be answered through a connection with this central point. The NASA Master Directory or the Earthnet system could be cited as examples of a centralised network. The UK's GENIE project, begun in mid-1992 at the Universities of Loughborough and Nottingham, is intended to be a distributed and intelligent metadata system.

GENIE (Global Environmental Network for Information Exchange) is funded by the UK Economic and Social Research Council (ESRC) on behalf of the Inter-Agency Committee for Global Environmental Change (IACGEC). It will initially allow users to locate metadata relating to specific applications. Some metadata are to be entered by individuals and will refer to their use of the data; the main 'descriptive' metadata sets will be entered by data centres covering the areas of social and economic science, hydrology, Earth observation, ecology, oceanography and glaciology. The system is intended to be flexible, requiring no fixed-format metadata entry and allowing free-format queries. It will also be capable of establishing linkages between different metadata sets (based on users' descriptions of their requirements and on the metadata they select to meet these requirements).

A second aspect of GENIE is its attempt to establish some measure or indication of data quality. For Earth observation data this measure will include descriptions of straightforward intrinsic characteristics of the data set, such as the presence and extent of cloud cover or of banding in the imagery. For derived products, mainly biophysical or geophysical parameters such as sea-surface temperature maps, quality will be expressed in terms of the accuracy and precision of the estimates. Locational accuracy is an important attribute of spatial data and metadata will also provide estimates of this parameter for georeferenced data. Finally, it is accepted that a proportion of users will have little or no background or training in the use of Earth observation data. For some users, the reference provided by GENIE will be their first contact with remotely-sensed images. GENIE will provide initial help on descriptive topics as an aid to the beginner. At a later stage, metadata will include advice on the choice of data processing methods (based on users' experiences as reported to the system) and estimates of the likely errors resulting from the application of these methods. For example, the use of supervised image classification procedures provides estimates of classification accuracies; such information is of considerable significance to the non-expert. The aim is to provide access to all relevant properties of a data set, not merely to produce a catalogue of data sets and expect the user to discover independently those properties.

At a later stage, it may become feasible to consider the development of a library of algorithms for processing remotely-sensed data. It is possible to draw a parallel with the Numerical Algorithms Group library (NAG) which began in the early 1970s as the Nottingham Algorithms Group, and provided access to standard algorithms in numerical analysis. Such algorithms were not generally available, and many researchers had to learn programming in order to carry out relatively simple statistical analyses of their data. Subsequently NAG has expanded to include a wide range of algorithms, some of which are supplied by workstation vendors. A similar pattern can be seen in the provision of software systems for remote sensing, in that many of the methods (and, indeed, the code) emerged from university and other research groups. However, there is always a considerable time-lag between the development of new methods by researchers and their general availability via standard software packages. No mechanism for the transfer of new techniques from researcher to applications scientist has yet been established. While GENIE is not yet funded to provide that mechanism, it may be possible to extend its capabilities into the provision of computer software for the processing of data sets that are referenced by GENIE.

The aim of a metadata network is primarily to inform potential users of the characteristics of data sets which may be of relevance in their studies. Software directories such as 'Archie' provide an analogous service for programmers, but they lack several vital properties that GENIE is intended to incorporate, principally relating to the quality of the data, the uses to which the data have already been put, and methods for the processing and extraction of information from the data. Current plans are for the introduction of a limited service on GENIE at the end of 1993.

EXPERT SYSTEMS

Many – perhaps the majority – of users of remotely-sensed imagery are specialists in a particular field of Earth or environmental science. A major obstacle facing them is the apparent complexity of the techniques of data acquisition (including the extraction of data from magnetic tapes), processing and validation. Learning a new system, or even remembering how to use an existing system, is a significant deterrent to the new or occasional user. Yet, as remote sensing becomes more sophisticated – for example, the development of methods of processing that are integrated within a GIS – little thought has been given to this aspect of data use. The ecologist, geologist or botanist is expected to become proficient in the theory and basis of remote sensing, to learn the characteristics of image data and to become proficient in the use of image processing or GIS software. It is akin to an initiation rite; if the potential user cannot learn these things then he/she is unlikely to use the information properly anyway. Such an attitude is reminiscent of that adopted by a minority of statisticians during the 1960s and 1970s when techniques of statistical analysis found their way into the methodology of (then) non-numerate disciplines such as geography and the social sciences. Left unaided, the geographer could be expected to misuse the methods and produce meaningless, or at best suspicious, results while the professional statistician looked on. Nowadays it is possible to encapsulate the knowledge of the expert in a well-defined and understood field and to make that knowledge available to a novice or occasional user through the medium of an expert system.

Applications of expert systems to remote sensing problems are described by Wilkinson and Fisher (1984), Goodenough *et al.* (1987), Heard *et al.* (1992), M^c^Keown (1987), Ripple and Ulshoefer (1987), Schowengerdt and Wang (1989), Skidmore (1989) and Kimes *et al.* (1991). These studies suggest that the need for continuing research in the area of expert systems is a pressing one; as Kimes *et al.* remark, 'The EOS era will challenge remote sensing research. A key issue will be how to integrate knowledge intelligently' (Kimes *et al.*, 1991, p. 1987). The complexity of image processing and analysis methods is such that currently there is little prospect of a successful all-embracing expert system. It is more realistic to seek out individual aspects of the process of information extraction and data integration that are more amenable to a knowledge-based approach. These operations are likely to lie in the general area of image processing rather than in the more complex and less clearly-understood areas of image analysis (extraction of objects, their descriptions and spatial relationships). This observation is reinforced by the fact that conventional image processing operations are well-defined, whereas research into image analysis problems is being pursued actively. An expert system can only be built where the system concerned is stable, well-understood and generally accepted. Surveys of the applications of artificial intelligence techniques in other subject areas have reached a similar conclusion, namely that knowledge-based systems are capable of producing results in specific, limited applications areas, such as the analysis of images of airports (e.g., McKeown *et al.*, 1985; Matsuyama, 1987).

Together with the facilities provided by a comprehensive metadata system, such as GENIE, the availability of expert system-based processing systems will release researchers from the need to acquire knowledge about the technology and methods of remote sensing beyond that required for the intelligent use of the data. As GIS packages become more sophisticated and demanding of the user, it is essential that consideration is given to the consequences of such developments.

.... CONCLUSIONS

Data from orbiting space platforms provide the only means of observing global change on the surface of the Earth on a regular and consistent basis. Effort is required to ensure that these data are made as widely available as possible to the multidisciplinary and interdisciplinary global change research community. Questions of data quality, which relate to fitness for the specific purpose, must be quickly resolved if the information contained in remotely sensed data is to be used properly. Access to data, and to procedures for their correction, is essential if these data are to be used outside a narrow specialist community. This chapter described mechanisms, based on the concept of the metadata network, that can meet such requirements and provide the global environmental change research community with access to a reliable and consistent source of information.

.... REFERENCES

Asrar, G., 1989, 'Introduction', in Asrar, G. (editor), *Theory and applications of optical remote sensing*, Wiley, New York: 1–13

Chuvieco, E., 1990, *Fundamentos de teledeteccion espacial*, Ediciones Rialp, S.A. Madrid

Dickinson, R.E., 1984, 'Modelling evapotranspiration for three-dimensional global climate models', *Geophysics Monographs, American Geophysical Union*, **29**: 58–72

Duggin, M.J., Robinove, C.J., 1990, 'Assumptions implicit in remote sensing data acquisition and analysis', *International Journal of Remote Sensing*, **11**: 1669–94

Goodenough, D.G., Goldberg, M., Plunkett, G., Zelek, J., 1987, 'An expert system for remote sensing', *IEEE Transactions on Geoscience and Remote Sensing*, **25**: 349–59

Heard, M.I., Mather, P.M., Higgins, C., 1992, 'GERES: a prototype expert system for the geometric rectification of remotely-sensed images', *International Journal of Remote Sensing*, **13**: 3381–5

Hill, J., Sturm, B., 1988, 'Radiometric normalisation of multi-temporal Thematic Mapper data for the use of greenness profiles in agricultural landcover classification and vegetation monitoring', *Proceedings of the EARSeL 8th Symposium: Alpine and Mediterranean Areas*, Capri, Italy, European Space Agency, Noordwijk: 21–40

Holben, B.N., Justice, C.O., 1980, 'The topographic effect on spectral response of nadir-pointing sensors', *Photogrammetric Engineering and Remote Sensing*, **46**: 1191–1200

Kaufman, Y., 1989, 'The atmospheric effect on remote sensing and its correction', in Asrar, G. (editor), *Theory and applications of optical remote sensing*, Wiley, New York, 336–428

Kimes, D.S., Harrison, P.R., Ratcliffe, P.A., 1991, 'A knowledge-based expert system for inferring vegetation characteristics', *International Journal of Remote Sensing*, **12**: 1987–2020

Kneisys, F.X., Shettle, E.P., Gallery, W.O., Chetwynd, J.H., Abreu, L.W., Selby, J.E.A., Clough, S.A., Fenn, R.W., 1983, *Atmospheric transmittance/radiance: computer code LOWTRAN 6*, AFGL TR-83-0187, Air Force Geophysics Laboratory, Hanscom Air Force Base, Massachusetts

McKeown, D.M., 1987, 'The role of artificial intelligence in the integration of remotely sensed data with geographic information systems', *IEEE Transactions on Geoscience and Remote Sensing*, **25**: 330–48

McKeown, D.M., Harvey, W.A., McDermott, J., 1985, 'Rule-based interpretation of aerial imagery', *IEEE Transactions on Pattern Analysis and Machine Intelligence*, **7**: 570–85

Mather, P.M., 1987, *Computer processing of remotely-sensed images: an introduction*, Wiley, Chichester

Mather, P.M., 1992, 'Geometric correction using least squares: a sensitivity analysis', *Remote sensing: from research to operation*, Remote Sensing Society, Nottingham: 560–70

Matsuyama, T., 1987, 'Knowledge-based aerial image understanding systems and expert systems for image processing', *IEEE Transactions on Geoscience and Remote Sensing*, **25**: 305–16

Monteith, J.L., 1990, 'Remote sensing in agriculture: progress and prospects', in Steven, M.D., Clark, J.A. (editors), *Applications of remote sensing in agriculture*, Butterworths, London: 397–402

Moore III, B., Dozier, J., 1992, 'Adapting the Earth Observing System to the projected $8 billion budget: Recommendations from the IEOS investigators', *The Earth Observer*, **4**: 3–10

Rast, M., Readings, C.J., 1992, 'The ESA Earth Observation Polar Platform Programme', in Mather, P.M. (editor), *TERRA-1: Understanding the terrestrial environment – the role of Earth observation from space*, Taylor and Francis, London: 175–84

Ripple, W.J., Ulshoefer, V.S., 1987, 'Expert systems and spatial data models for efficient geographic data handling', *Photogrammetric Engineering and Remote Sensing*, **53**: 1435–41

Running, S.W., Coughlan, J.C., 1988, 'A general model of forest ecosystem processes for regional applications, I: Hydrological balance, canopy gas exchange and primary production processes', *Ecological Modelling*, **42**: 125–54

Schowengerdt, R.A., Wang, H.-T., 1989, 'A general-purpose expert system for image processing', *Photogrammetric Engineering and Remote Sensing*, **55**: 1277–84

Sellers, P.J., Dorman, J.L., 1987, 'Testing the simple biosphere model (SiB) using point micrometeorological and biophysical data', *Journal of Climatology and Applied Meteorology*, **26**: 622–51
Skidmore, A.K., 1989, 'An expert system classifies Eucalypt forest types using Thematic Mapper data and a digital terrain model', *Photogrammetric Engineering and Remote Sensing*, **55**: 1449–64
USGS, 1979, *Landsat data users handbook*, Eros Data Center, Sioux Falls, South Dakota, USA
Wickland, D.E., 1989, 'Future directions for remote sensing in terrestrial ecological research', in Asrar, G. (editor), *Theory and applications of optical remote sensing*, Wiley, New York: 691–724
Wilkinson, G.G., Fisher, P.F., 1984, 'The impact of expert systems in future operational remote sensing', *Satellite remote sensing – review and preview*, Remote Sensing Society, Nottingham: 53–60

13. Spatial Data: Data Types, Data Applications and Reasons for Partial Adoption and Non-integration

J.A. Allan

.... Introduction

Most users of spatial information feel intuitively that the spatial data are part of some seamless unified body of information that should be readily integrated into a comprehensive system. That the history of map making, including its very recent history, is marked by conflicts concerning method, the pace of innovation and the priority of provision contradicts this intuition. The purpose of the chapter is to identify how the many types of data recorded on maps and in geographical information systems differ and why they have been difficult to integrate. The review will require an examination of the goals of major users such as military and government departments, as well as those requiring maps for environmental science in its widest sense, and those professionals requiring maps and spatial data for resource evaluation purposes.

Information which is to be shown on maps has to be acquired by survey techniques, and has to be recorded and stored. To be useful it has to be geometrically registered to a desired projection and manipulated in a variety of ways to enable the production of a cartographic product. All of these processes have been greatly affected by new technologies and especially by computing systems. Such technologies have transformed the context of map making in the past two decades and the map making industry is very much in the midst of the revolution, precipitated by the increasingly widespread adoption of new technologies deployed in the space, ground and user segments.

Associated with each stage of the map production process is a large professional body located in the military, the public and the private sectors. These surveyors, cartographers and printers have practices developed mainly in the past two hundred years. Driven by the perceived needs of the military and other users, a system of symbolising 'permanent' features was evolved by means of geometrically coordinated points, lines and symbols together with text.

The development of new technologies has affected all stages of the map making process. Aircraft (by 1910) and satellites (by the 1960s) have provided new platforms from which to record the position of surface features. The camera (mid-nineteenth century) was the first sensor which provided a permanent record and it has been joined in the mid-twentieth century by a suite of sensors able to record, electronically and in hard copy, information outside the visible spectrum with fine spectral resolution. On-the-ground techniques such as photogrammetry have reduced survey costs and enabled the mapping of difficult and inaccessible terrain. All digital manipulation has been speeded immeasurably by the contribution of computing, and computers have widened immensely and made more flexible the display of spatial information.

.... THE STRUCTURE OF SPATIALLY DISTRIBUTED DATA: DEEP AND IMPOSED STRUCTURES

If one looks at an aerial image of a tract of land and the image has the necessary spatial resolution to record both the land surface cover as well as the artefacts introduced by humans, it is clear that there are two major 'structural' elements in the landscape. First there is the 'deep structure'; in other words, the natural landforms, soils, soil moisture and, where people have not interfered, vegetation. In most tracts, and especially those where populations are gaining an economic return from the tract, there is also an 'imposed structure' which includes the boundaries of agricultural fields, roads, dwellings, drainage channels, villages, towns and the built environment generally.

The users of maps have, throughout history, been mainly concerned with the 'imposed structures'. These were perceived to be permanent, intrinsically valuable, legally significant, economically significant, useful navigationally and significant militarily.

The areas within parcel-defining boundaries were not of concern to those using the maps for conventional purposes. Variations within parcels were not regarded as significant by such users except with respect to the 'trafficability' of terrain and, since the representation of such information on a map would add to the expense of the maps as well as making them valuable to an 'enemy', these types of data were never shown on publicly available maps.

The unrecorded 'deep structure' of the landscape has always been of importance, however, because, for example, landforms, surface texture and soil moisture determine the amelioration, improvement, protection and other management practices in which a farmer would have to invest to gain returns from farming activity and thereafter increase productivity. The natural landscape is characterised by spatial variability of properties such as slope, elevation, soil, including soil moisture and vegetation, and temporal variability of properties such as soil moisture, vegetation, phenology, livestock and wildlife. It is the temporal variability which so very much affects the attitude of map makers, surveyors and users as to whether or not to map a particular tract of land and, if so, with what accuracy to record the surface phenomena.

Since the majority of early modern maps (in the nineteenth century) were produced for military organisations, the properties and needs of the military community played a very significant rôle in determining the range of information on topographic maps. The features were the permanent ones, useful for navigation and to the artillery.

The portrayal of topography was necessary, however, it was not necessary to record the elevations accurately. Since the portrayal of topography was difficult cartographically it was not surprising that an acceptable specification for the accuracy of contour lines was the very relaxed one of 'half the contour interval'. That it would have been extremely expensive to increase this accuracy was another reason for accepting it. In the very flat areas, where the most important economic and military activity has always taken place, the problem of providing an economically viable, dense network of accurate height information is only just being solved as the result of the deployment of Earth observation systems. These new systems enable a very dense spatial sampling of terrain elevation. That there were many potential users with the need for spatial and height data at great levels of accuracy was a problem never previously addressed, despite the major economic importance of the relatively flat tracts of the Earth's surface for agricultural and other important economic uses.

.... SPATIAL AND TEMPORAL RESOLUTION OF MAPS: THEIR RELEVANCE TO MAP MAKERS AND MAP USERS (ALIENS, 'VISITORS' AND REMOTE USERS) *VERSUS* LOCAL USERS AND LOCAL MANAGERS

THE EVOLUTION OF THE MAP-USING COMMUNITY AND ITS INSTITUTIONS

Comprehensive large-scale mapping of extensive areas has been, and will continue to be, carried out for institutions which can mobilise significant resources. Systematic mapping has always been relatively expensive and has required the mobilisation of advanced technology and specialised human resources and *advanced* mapping has only been possible since the emergence, in the nineteenth century, of the nation/sovereign-state along with the panoply of government institutions which characterise these states. In the recent past other economic entities with the economic power equivalent to that of a substantial nation-state could also aspire to produce advanced mapping. For example, major transnational companies, especially in the oil industry, have the need and the economic competence to pay for the survey and cartographic processes necessary to assemble relevant spatial information systems as well as to engage in map production themselves. New technologies are making it increasingly possible for minor bodies to make accurate maps of remote small areas and data acquired from satellite sensors also enables the production of previously unobtainable global small scale mapping at high temporal frequency, based on comprehensive coverage of a selection of the essential variables for global environmental modelling. There has been a revolution in the expectations of the minor user through the availability of the raster data sets provided by Earth observation and the acceleration in the power of the computer workstation.

Meanwhile, the surplus technical competence in the industrialised countries, in both the public and the private sectors, is available to contract for the production of mapping in countries lacking, for the moment, the necessary technical competence for map production. International agencies such as those of the United Nations, and many multi-lateral and bi-lateral bodies can mobilise these technological resources to produce mapping for general and specialist development purposes.

SECURITY AND RELATED IMPEDIMENTS TO SPATIAL DATA USE

Map making has always been a sensitive security issue. This sensitivity is a strong indicator of the attitudes of the various communities concerned with initiating map making, as well as of the map making professionals and of those whose land, artefacts, economies and socio-political institutions are recorded. Map making is a 'top-down' activity and it serves those with central power. In the past it served those with imperial influence, just as in the post-imperial era it has served the super-powers and their strategic allies. Advancing the technologies of mapping has always been, and remains, a prime aim of the defence establishments of the major industrialised economies.

Spatial data can be seen, therefore, as useful to those officials and institutions with a legitimate interest in unfamiliar and remote locations within the boundaries of a nation-state. Where the central government is not seen to represent the interests of regional and other groupings, then the act of mapping by, for example, an organisation seen to be identified with urban and/or military interests, will not be viewed as benign by alienated ethnic or tribal groups, or by poor and exploited social and economic sectors (for example, rural communities).

Spatial data are also useful to officials and institutions with a strategic/military interest in the territory and resources of other countries. Mapping and imperial control have always been closely associated and the methods of surveying large areas were advanced substantially in the process of surveying the, often very large, colonised areas in South Asia, South-East Asia and Africa. As it happens, the aircraft and the techniques of photogrammetry came just as modern imperialism was coming to an end, and the acquisition of data by satellite sensors is a post-imperial phenomenon and one which has been mightily exploited by the post-World War II super-powers. Most of the products of surveillance by satellite sensors have, thus far, remained in the military domain.

THE CONCERNS OF LOCAL COMMUNITIES

From the point of view of a local community, who know the terrain, the soils, the soil moisture and the vegetation of their locality intimately, mapping is completed for aliens and 'visitors', by the agents of aliens and 'visitors', and is normally produced at a scale which is irrelevant to the detailed management activity of individual farmers and local rural institutions. Conventional mapping has neither been detailed enough, nor current enough, to be relevant for local scale decision making nor does it relate to the ephemeral surface features of day-to-day concern to farmers and managers of livestock. The term 'visitor' is intended to embrace that group of past, present and future civil and military professionals who aspire to understand, evaluate and manage land, and govern and administer communities and resources.

.... TYPES OF USER APPLICATIONS AND THEIR RELATIONSHIPS WITH THE UNDERLYING AND IMPOSED STRUCTURES

Some of the needs of the 'visitors' have been addressed by the national mapping departments. Since the 'visitors' in the past have been concerned only with the *imposed structure*, their needs have been served by the selection and portrayal of a narrow range of surface features by means of points, lines and conventional signs surveyed, stored and plotted according to the conventions of coordinate geometry. This apparently comprehensive cartographic procedure fits extremely well with the new vector systems of digital mapping. It is becoming increasingly clear that it also accords well with the administrative conventions of government which require the segmentation of national space hierarchically, with systems of provinces or counties, districts and/or boroughs, constituencies, wards, parishes, regional health authorities, river authorities, etc. Sometimes these hierarchies are harmonised but their separate genesis is such that normally they are not in harmony.

Both the map makers and the statistical departments of national, regional and local government have invested a great deal in organising the recording of data on a number of grossly simplified assumptions. In so far as these relate to space, these procedures have misleadingly assumed that the polygons into which information is fitted are uniform within an individual segment for the particular characteristics being mapped. Meanwhile, the census makers have made a virtue of recording the data in simple arrays or tables which are compellingly attractive in their compatibility with spatial information systems of the vector type and especially because of their palpable economy in recording large volumes of quantified information on a large number of variables relating to polygons or points. The information is apparently spatially structured as all the data refer to points and polygons with a geometrically definable spatial address. That the *spatial segmentations* are not in harmony is a profound problem, especially in comprehensive and complex spatial data bases. For example, the segmentation based on hydrological units rarely coincides with administrative or natural boundaries; the public utilities, public transport and other infrastructures are rarely spatially organised according to consistent concepts; and the same is true of health services, the police, census units, electoral units etc.

While there will always be a large number of 'visitors' who will be perfectly happy with maps showing points, lines and polygons and a small range of thematic features, there has at the same time been a large community of potential map users who have never been provided with the spatial data they have required. These are the scientists and professionals requiring information on the deep structure; on the terrain, on soil qualities, soil moisture, hydrology, vegetation and land use. All these features, except landforms, change within temporal cycles, seasonal and longer term, which are of great significance to land resource managers. The features of concern are only rarely shown on map sets and frequently they are only recorded if the scientists themselves have mobilised dedicated surveys to assemble the information.

This community of land resource orientated professionals has been particularly impressed by the data provided since the early 1970s by orbiting Earth resource satellites in that for the first time image data showing the extent of surface phenomena of concern have been available. More important, the satellite system's coverage was capable of sampling adequately, for the first time, the dynamic character of the underlying structures (as opposed to the imposed structures) of the land surface, with which environmental scientists and land resource managers at all levels have always been concerned. The commitment of the land resources professionals to satellite sensor imagery is strong and increasing. Image data from satellite sensors are recorded typically as digital arrays, known as raster systems, and since the areas covered can be very extensive the volumes of data involved are huge, even at the relatively coarse spatial resolutions of the current systems.

.... VECTOR AND RASTER MAPPING: PROFESSIONAL AND INSTITUTIONAL FACTORS

Two applications communities concerned with spatial information can be identified. They are characterised by differing applications, by different survey traditions, by differing information requirements in terms of the way they approach the sampling of spatial information, by different preferences in terms of the way data are structured in their preferred system, and, possibly most important, by differing levels of economic significance of their application field and the overriding political significance of the domain in which they play a rôle (Table 13.1). These approaches and applications can be summarised and shown diagrammatically in Figure 13.1. The raster and vector approaches are seen to be very different in a number of major respects and especially with regard to the provenance of the data used, and in terms of the volume of activity as measured either by the numbers of users and customers or by financial turnover (Figure 13.1).

Because the vector products are designed to portray segmented space rather than variations in the underlying structure those who support them, through a long tradition of deploying conventional survey systems of the vector type, make a virtue out of the compatibility of the vector system with the data management procedures of powerful bodies, such as central government departments and local government departments. These bodies record information in relation to regions, districts and other administrative segmentation procedures. At the same time the political influence and buying power of long established public institutions such as national mapping organisations, which include the even more powerful defence mapping bodies as well as a large group of government statistical organisations and land registries, is very strong. Meanwhile, the bodies concerned with raster data are all relatively uninfluential and have only recently had available comprehensive multispectral image coverage which has transformed their access to relevant spatial information.

While a number of professionals, for example, the meteorologists, the

Table 13.1 The two approaches to the use of spatial information of the terrestrial environment: applications, data needs, data structures, economic and security aspects and significance.

The vector approach	*The raster approach*
Applications	**Applications**
Military and defence, Civil topographic mapping	Renewable natural resource inventory, monitoring, evaluation and management
Planning for civil government at the national and regional levels and some aspects of local government	Local level resource management
Leisure and tourism	
Data requirements	**Data requirements**
Fine spatial resolution	Coarse spatial resolution except for military surveillance
Coarse temporal resolution	Fine temporal resolution
Preferred data structures	**Preferred data structures**
Vector systems	Raster systems
Economic and security significance	**Economic and security significance**
In the defence domain	*In the defence domain*
Military survey Surveillance products	Fine resolution products of prime importance - not necessarily processed beyond images, not in the public domain
In the civil domain	*In the civil domain*
Long established Government statistical services	Users small in number, specialist and not influential
National mapping sevices	
Land registration	
Leisure and tourism	

climatologists, the oceanographers and the hydrologists, have recognised the importance of the Earth observing systems in relation to their professional concerns, many have not. Vegetation scientists have been particularly blind to the advantages of using image data from space (Allan, 1991). Possibly the most important impediment to the rapid adoption of the raster-type image product is that the only really mighty proponent, the defence and intelligence community, does not operate in the public domain and does not share its fine spatial resolution products with those interested in the natural environment. The older vector-type product which also originated in the military mapping agencies, but two centuries ago, has been assimilated gradually and totally by the civil mapping organisations. Presumably, in due course, the raster products will likewise be universally adopted, but in the meantime it will be necessary for there to be an unlikely change of attitude

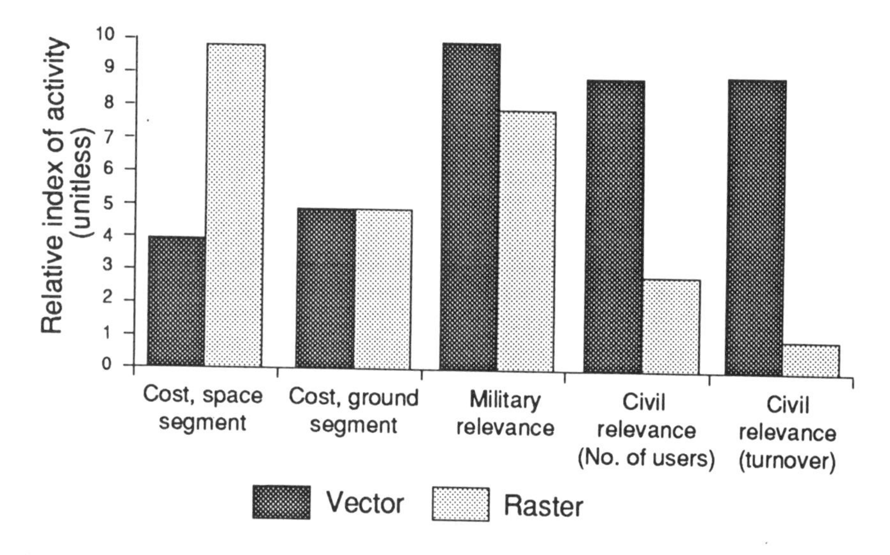

Figure 13.1 Comparison of the space and ground segment costs, and of user relevance of the vector and raster approaches.

on the part of the military and intelligence agencies and of their political masters.

Meanwhile, the needs of those operating at the local level and concerned with 'the deep structure' so relevant to natural resource management have not been addressed. For the moment their data needs are not being articulated and their ability to purchase data is very limited. Neither they nor the institutions which should be capable of mounting and presenting the arguments, the national level ministries and departments concerned with environmental management and protection, have adopted the economic theory which would underpin the relevant arguments for utilising data derived from Earth observation (Johansson, 1987; Conrad and Clark, 1987; Neher, 1990). These bodies are guilty of not calculating the cost of denying themselves the data by not deploying Earth observation systems and environmental geographical information systems (GIS).

.... FUTURE NEEDS AND LIKELY DEVELOPMENTS

The final question to answer is whether the argument being posed here reflects an unchangeable situation or whether the attitudes and approaches of users and those who provide them with spatial information will change. Changes, including in those in attitudes, will certainly take place. The development of systems and software to accelerate the processing of spatial information and transform the delivery of visual products via extensive networks is already well underway. The integration of vector and raster data is receiving considerable research and development attention, and if surveyors and users do not contribute to the solution of problems then the computer scientists certainly will. The absence of a contribution by surveyors and applied scientists will only make more likely the omission of procedures which take into account the need to register data spatially; the specification and adoption of relational data base procedures will become industry standard. It is important that advanced research is carried out on how to develop and deploy systems with, for example, the intuitive relevance of quadtree procedures (Rhind and Green, 1988).

But the profile of perceived relevance and the expression of demand for spatial data shown in Figure 13.1 will, for the foreseeable future, determine the political context for the allocation of resources to the various branches of spatial data capture, data processing and spatial information provision. In other words, those wanting to see the level of use of raster data increase will have to do a considerable amount of research in unfamiliar areas of the political economy of spatial information (Friday and Laskey, 1991). It is only when the scale of activity in the raster domain approximates to that in the vector domain in terms of the number of users and professionals involved, but especially in terms of financial turnover, that integration will increase. For the moment, the vector community will dominate in all except meteorological and oceanographic applications both in the commercial and the research sectors.

.... REFERENCES

Allan, J.A., 1991, 'Global biogeography: who is responsible?', *Journal of Biogeography*, 18: 121–2

Conrad, J.M., Clark, C.W., 1987, *Natural resource economics*, Cambridge University Press, Cambridge

Friday, L.E., Laskey, R.A., 1991, *The fragile Earth*, Cambridge University Press, Cambridge

Johansson, P., 1987, *The economic theory and measurement of environmental benefits*, Cambridge University Press, Cambridge

Neher, P.A., 1990, *Natural resource economics*, Cambridge University Press, Cambridge

Rhind, D.W, Green, N.P.A., 1988, 'Design of a geographical information system for a heterogenous scientific community', *International Journal of Geographical Information Systems*, 2: 171–89

14. Scale and Environmental Remote Sensing

Giles M. Foody and Paul J. Curran

.... INTRODUCTION

Scale is a key issue in the spatial and environmental sciences. It is widely recognised that many environmental processes and patterns are scale-dependent (NRC, 1985; Graetz, 1990; Baker, 1989; Milne *et al.*, 1989; Haines-Young, 1991; Bian and Walsh, 1993; Field and Ehleringer, 1993). Consequently, it is essential that appropriate scales of study are adopted and that the Earth be studied at a range of scales and levels of detail (NRC, 1986; Wickland, 1991a). The critical variable of scale has not received explicit attention in the preceding chapters, therefore the aim of this chapter will be to review briefly what is meant by the term 'scale' and then to discuss some of the main scale-related conclusions reached elsewhere in this book.

.... SCALE AND ENVIRONMENTAL REMOTE SENSING

The term scale is often used in a loose fashion. For those with an interest in the spatial sciences it may be natural to think of scale in its cartographical context, where its meaning is unambiguous. In cartography the scale of a map, S, is the most important mathematical property of the map (Maling, 1989) and it relates distance on a map to actual distance on the ground via the equation,

$$1/S = d/D$$

where D and d are respectively the distances between two locations on the ground and their representations on the map. Therefore, a small scale map (e.g., 1:1,000,000) represents a large area but with little spatial detail, whereas a large scale map (e.g., 1:10,000) provides coverage of a smaller area but in considerably more spatial detail (Robinson *et al.*, 1984; Lam and Quattrochi, 1992). Unfortunately, however, the term scale is often used differently in general usage and interpreted to mean the reverse of the cartographical definition. This may be considered as being 'geographical scale', where the

Table 14.1 Generalised spatial and temporal characteristics of five environmental remote sensing systems.

Sensing system	Spatial resolution (m)	Temporal resolution (days)	Image size (km)[1]
SPOT HRV (P)	10	26[2]	60
SPOT HRV (XS)	20	26[2]	60
Landsat MSS	80	16[3]	185
Landsat TM	30[4]	16	185
NOAA AVHRR	1100[5]	1	3000

Notes: 1 – expressed as swath width; 2 – the SPOT HRV system is steerable and so data may be acquired at a finer temporal resolution than that quoted; 3 – for Landsats 1–3 the temporal resolution was 18 days; 4 – the spatial resolution of data acquired in the thermal waveband (band 6) is 120m; 5 – some NOAA AVHRR data products are available at coarser spatial resolutions.

term denotes the spatial extent of the region of interest (Golley, 1989; Lam and Quattrochi, 1992) with large scale referring to large regions (e.g., continental/global scale) and small scale referring to small regions (e.g., individual agricultural fields). Throughout this book, the scale of an investigation has been discussed in its cartographical context with emphasis on studies at medium–small and thereby regional–global scales.

In environmental remote sensing an issue related to scale is that of spatial resolution (Woodcock and Strahler, 1987). In general, and for the purposes of this book, the latter may be considered to be a measure of the spatial detail that can be observed with a remotely sensed data set (Townshend, 1981). A fine spatial resolution system such as the SPOT HRV provides spatially detailed data on a small region whereas a coarse spatial resolution image, such as a NOAA AVHRR image, provides less spatially detailed data of a much larger area (Table 14.1). In this context NOAA AVHRR data are most appropriate for regional to global scale investigations whereas SPOT HRV data are more appropriate for local scale investigations.

Further to the use of scale in a spatial context, scale may also be used in a temporal context. This is important in many remote sensing investigations where an ability to monitor change over time periods that range from hours (e.g., in meteorology) through years (e.g., in urban growth) to centuries (e.g., soil development) is vital. For a variety of reasons, notably practical constraints of data handling and manipulation, the spatial and temporal aspects of scale need to be considered together in many remote sensing studies (Allan, 1984).

It is important, however, that scale be recognised as a *variable*, with many environmental features and processes operating at characteristic spatial and temporal scales (Figure 14.1). Thus, for example, processes apparent at one scale may not be apparent at another. Results of studies are therefore scale-specific (Johnston, 1984) and, for instance, the relationships between variables can vary markedly in strength and direction with changes in scale (Openshaw, 1983). Since the effect of scale changes can be complex and non-linear (IGBP, 1992) inferences drawn from results of studies performed at one scale and applied to another may be incorrect. This applies to both scaling-up and

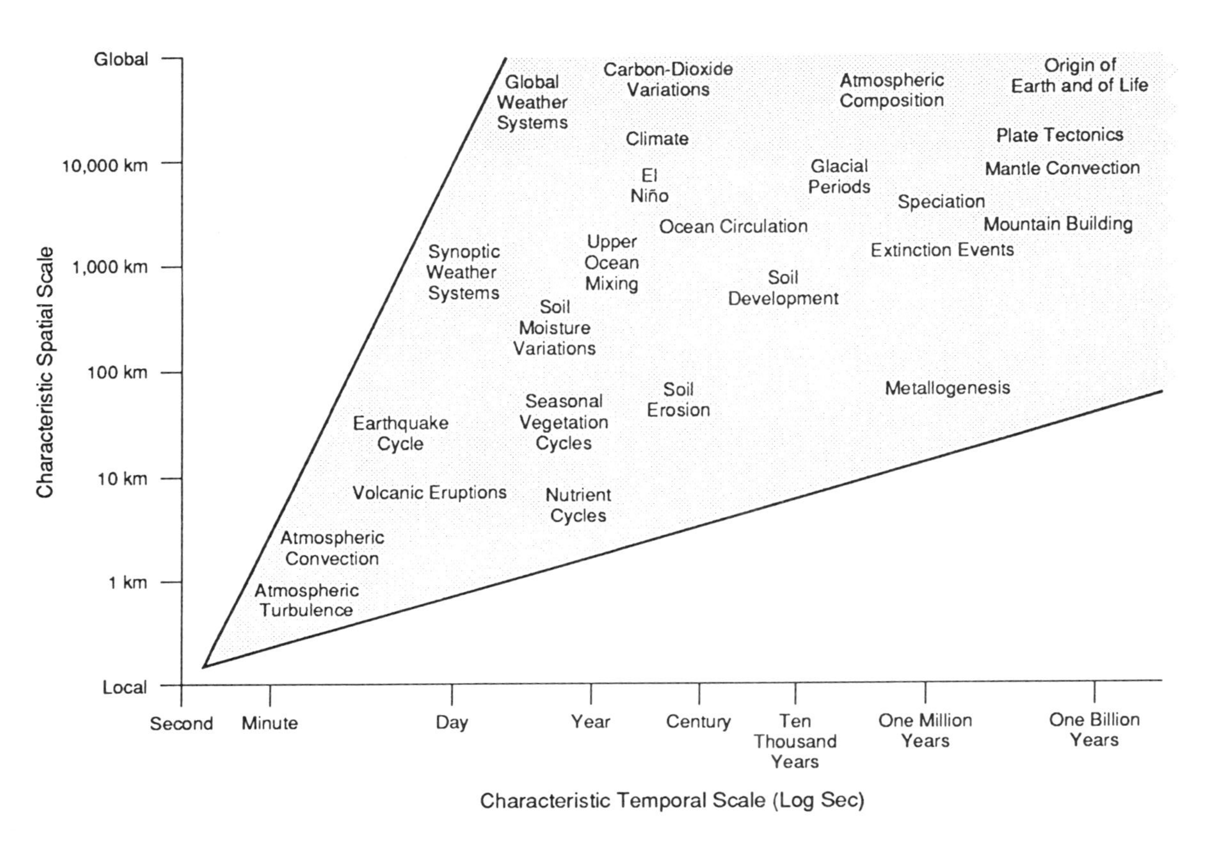

Figure 14.1 Characteristic spatial and temporal scales of some key Earth system processes (modified from NASA, 1988).

scaling-down. With the latter, attempts to infer the characteristics of individuals from aggregate population data are typically made. These may suffer from the widely known problem of the ecological fallacy where incorrect conclusions about individuals are drawn from grouped data (Johnston, 1981; 1984). Of more concern here though are the problems of scaling-up detailed knowledge on the environment derived from local scale investigations to large regions. This is because the complexity of the effect of scale variations rules out the use of simple generalisations (NRC, 1985; IGBP, 1992). Our ability to scale-up is also limited by a lack of information and theory on scale effects as well as the limitations of experience (Gleick, 1988). Undoubtedly, however, the only practical approach to observe the environment at a range of scales, and particularly at regional to global scales, is through remote sensing and many of the environmental problems we currently face are at these small scales (NRC, 1985; Graetz, 1990; IGBP, 1992). However, since the results obtained at one scale may have little or no validity at other scales, it is important that an appropriate scale be selected for a particular application (Woodcock and Strahler, 1987; NASA, 1988; Atkinson and Curran, 1993) and that the analysis of the data be undertaken with regard to appropriate scene models which account for scale effects (Graetz, 1990). This requires consideration of the sensing system's spatial resolution with respect to the scale of investigation and recognition of the hierarchy of scales that may exist (Woodcock and Harwood, 1992). Furthermore, some investigations may benefit from an analysis of data acquired at a nested hierarchy of scales (Figure 14.2) and remote sensing is unique in offering an ability to study the environment over a wide range of scales (Table 14.1) (Justice *et al.*, 1991). This is particularly valuable since many environmental problems that are evident at one scale may have influences at others. Thus a global problem may be manifest at a particular local region and a local action can have global ramifications (Mannion and Bowlby, 1992). Information at a range of scales is therefore required and remote sensing has the potential to provide environmental information at local to global scales. Thus, while remotely sensed data have considerable potential for the provision of data for studies of the Earth's atmosphere, oceans and land surfaces (e.g., Allan, 1992; Saunders and Seguin, 1992; Townshend, 1992; Cracknell, 1993), the appropriate scale(s) of observation and study must be selected.

SCALE AND ENVIRONMENTAL REMOTE SENSING ISSUES ADDRESSED IN THIS BOOK

Throughout this book the term scale has been used in its cartographical context, often loosely expressed in relation to the local–global scale continuum. It is relatively easy to interpret the extent of a study at either end of this continuum. This book has not been concerned with local scale investigations but rather with studies at smaller cartographical scales; the regional to global scales. While it is easy to conceive the coverage of a global scale study, since the term global implies the whole of the Earth's surface or an entire statistical population (e.g., complete land surface), regional scale studies may refer to areas of very different extent; the hierarchy of scales

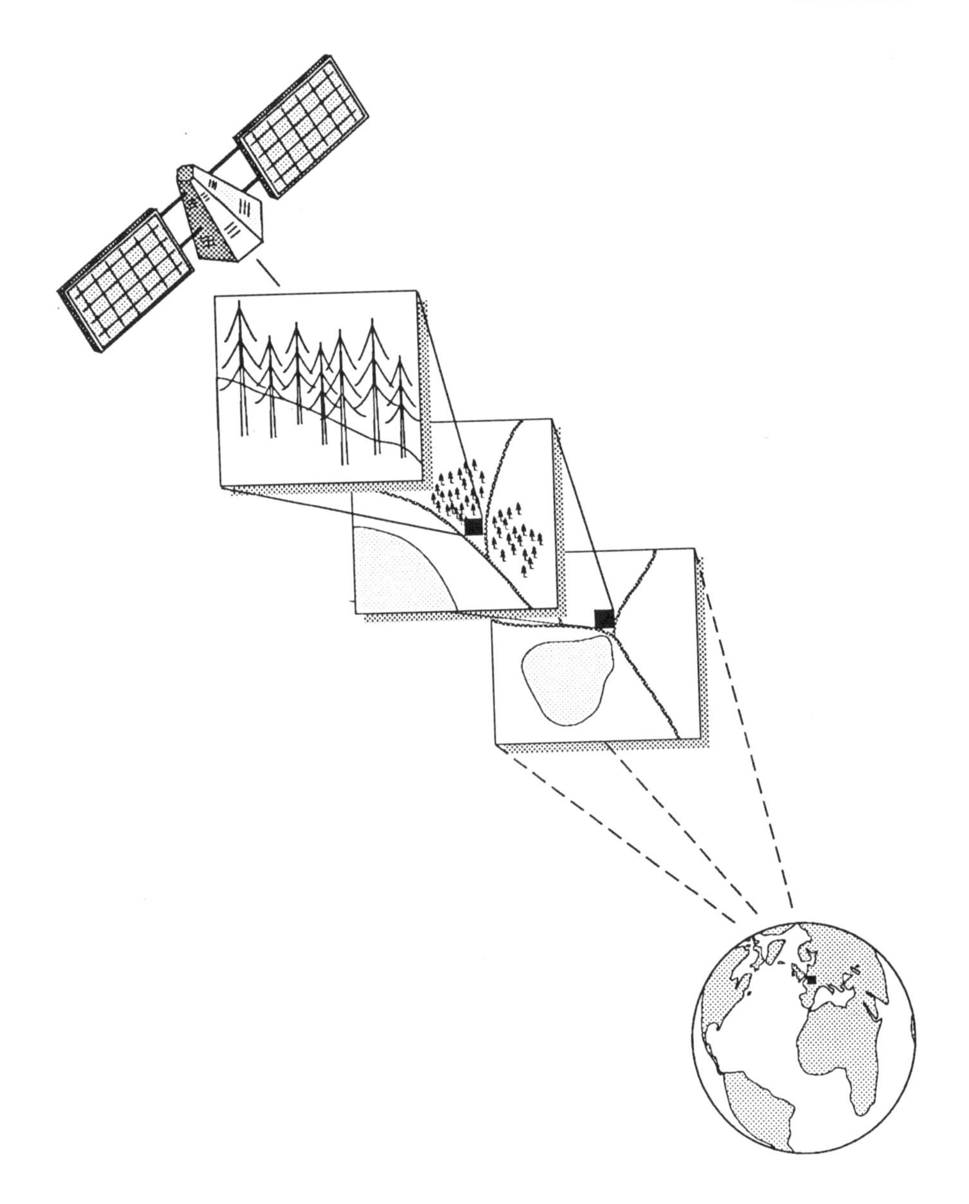

Figure 14.2 A nested hierarchy of remotely sensed data at different scales for environmental research (modified from NRC, 1986).

considered, however, is only an arbitrary division of the scale continuum (Graetz, 1990; Mannion and Bowlby, 1992). In this book the term regional has been used loosely and included relatively small areas such as urban areas (e.g., Weber, 1993) but more typically areas the size of nations (e.g., Lucas and Harrison, 1993; Millington *et al.*, 1993; Smith and Vaughan, 1993). Some of the chapters have illustrated the rôle of remote sensing in environmental studies at a range of scales. These chapters have focused on scientific aspects of remote sensing that aim to increase our understanding of the environment at regional to global scales as well as technical aspects of remote sensing illustrated by reference to practical applications. They also raise three issues in particular which must be addressed if the full value of remote sensing is to be realised in the environmental sciences.

(i) *Need for remotely sensed data.* Problems of existing data sets/data sources have been highlighted. Many spatial models are limited simply by a lack of appropriate data (Baker, 1989), much of which could be derived by remote sensing (Ustin *et al.*, 1991). The synoptic view, coverage and relative consistency of remotely sensed data make them attractive sources of environmental information. Preceding chapters have illustrated the rôle remote sensing may play in the derivation of basic data on land cover (Curran and Foody, 1993; DeFries and Townshend, 1993), snow distribution (Lucas and Harrison, 1993) and fires (Smith and Vaughan, 1993), and derived products such as urban population (Weber, 1993) and forest biophysical and biochemical properties (Curran, 1993). The chapters have also shown that remotely sensed data are a valuable source of environmental information at a range of scales.

(ii) *Use of remotely sensed data within co-ordinated environmental science programmes.* Environmental programmes which utilise remotely sensed data as an integral component are required and a number have been reviewed (e.g., Briggs, 1993; Curran and Foody, 1993). To make full use of remotely sensed data there is also a requirement for efficient data networks and analytical procedures (Mather, 1993), a change in attitude of non-remote sensing scientists to accept more readily remotely sensed data (Allan, 1991; Wickland, 1991b), and a development of tools appropriate for data analysis and integration (Roughgarden *et al.*, 1991; Allan, 1993; Mather, 1993). In addition, further research (preferably within environmental science programmes) on the scaling-up of detailed environmental information to large regions via remotely sensed data is required to help realise the full potential of remote sensing.

(iii) *Development of improved sensors and information extraction and presentation techniques.* Remote sensing systems need to be designed to observe environmentally useful variables. In addition to the selection of appropriate spatial and temporal resolutions, this includes the selection of relevant wavebands and, where appropriate, polarisations (Wickland, 1991a; Curran, 1993). There is also a need to develop information extraction techniques that make fuller use of data acquired by existing remote sensors. This includes the use of information on the angular reflectance dependency of Earth surface features, for example, through the analysis of multiple view angle imagery (Barnsley, 1993; Curran and

Foody, 1993). Approaches for the representation of the environment from remotely sensed data also need to be strengthened (Tong, 1992). A range of approaches may be adopted depending on the nature of the phenomena to be represented. Conventional image classification approaches, for example, may be inappropriate for the representation of basic data such as land cover where the classes are continuous (Foody *et al.*, 1992) or temporally variable (Millington *et al.*, 1993). There is also a general need to analyse the remotely sensed data and derived products in the context of other data and this may be best achieved by the integration of remotely sensed data and other geographical data within geographical information systems (Jackson, 1992; Allan, 1993). To assist this and to help realise the potential of remote sensing, there is a need to develop environmental information systems that are friendly to users, especially those unfamiliar with technical aspects of remote sensing. Although some progress in this direction has been made in recent years, with a general move away from a rather limited access to unprocessed remotely sensed data, there is still much more to be done. Systems such as GENIE (Mather, 1993) or the Commission of the European Communities Centre for Earth Observation (CEO) should help resolve some of the problems by making information on remotely sensed data and associated metadata more accessible.

.... CONCLUSIONS

Remote sensing can be used to provide information on the environment at a wide range of scales. At local scales it can be used to study a small area in considerable spatial detail but, as this book has demonstrated, coarse spatial resolution data sets may also provide valuable data on the environment, especially in the context of global environmental change (Waring *et al.*, 1986). There is a need for increased awareness of scale as a variable in the environmental sciences, and of the potential and limits of remotely sensed data (Greegor, 1986; Roller and Colwell, 1986; Allan, 1991).

It is hoped that this book will help illustrate some of the potential rôles that remote sensing may have in the study of the Earth's environment at regional to global scales.

.... REFERENCES

Allan, J.A., 1984, 'The role and future of remote sensing', *Satellite remote sensing – review and preview*, Remote Sensing Society, Reading: 23–30

Allan, J.A., 1991, 'Global biogeography: who is responsible?' *Journal of Biogeography*, **18**: 121–2

Allan, J.A., 1993, 'Spatial data: data types, data applications and reasons for partial adoption and non-integration', in Foody, G.M., Curran, P.J. (editors), *Environmental remote sensing from regional to global scales*, Belhaven, London (this volume)

Allan, T.D., 1992, 'The marine environment', *International Journal of Remote Sensing*, **13**: 1261–76

Atkinson, P.M., Curran, P.J., 1993, 'Optimal size of support for remote sensing investigations I: choosing an appropriate spatial resolution', *IEEE Transactions on Geoscience and Remote Sensing* (submitted)

Baker, W.L., 1989, 'A review of models of landscape change', *Landscape Ecology*, **2**: 111–33

Barnsley, M.J., 1993, 'Environmental monitoring using multiple-view-angle (MVA) remotely sensed data', in Foody, G.M., Curran, P.J. (editors), *Environmental remote sensing from regional to global scales*, Belhaven, London (this volume)

Bian, L., Walsh, S.J., 1993, 'Scale dependence of vegetation and topography in a mountainous environment', *Professional Geographer*, **45**: 1–11

Briggs, S.A., 1993, 'Remote sensing and terrestrial global environment research – the TIGER programme', in Foody, G.M., Curran, P.J. (editors), *Environmental remote sensing from regional to global scales*, Belhaven, London (this volume)

Cracknell, A.P., 1993, 'Remote sensing and the determination of geophysical parameters for input to global models', in Foody, G.M., Curran, P.J. (editors), *Environmental remote sensing from regional to global scales*, Belhaven, London (this volume)

Curran, P.J., 1993, 'Attempts to drive ecosystem simulation models at local to regional scales', in Foody, G.M., Curran, P.J. (editors), *Environmental remote sensing from regional to global scales*, Belhaven, London (this volume)

Curran, P.J., Foody, G.M., 1993, 'The use of remote sensing to characterise the regenerative states of tropical forests', in Foody, G.M., Curran, P.J. (editors), *Environmental remote sensing from regional to global scales*, Belhaven, London (this volume)

DeFries, R.S., Townshend, J.R.G., 1993, 'Global land cover: comparison of ground-based data sets to classifications with AVHRR data', in Foody, G.M., Curran, P.J. (editors), *Environmental remote sensing from regional to global scales*, Belhaven, London (this volume)

Field, C.B., Ehleringer, J.R., 1993, 'Introduction: Questions of scale', in Ehleringer, J.R., Field, C.B. (editors), *Scaling physiological processes: leaf to globe*, Academic Press, San Diego: 1–4

Foody, G.M., Campbell, N.A., Trodd, N.M., Wood, T.F., 1992, 'Derivation and applications of probabilistic measures of class membership from the maximum likelihood classification', *Photogrammetric Engineering and Remote Sensing*, **58**: 1335–41

Gleick, J., 1988, *Chaos*, Penguin, New York

Golley, F.B., 1989, 'A proper scale', *Landscape Ecology*, **2**: 71–2

Graetz, R.D., 1990, 'Remote sensing of terrestrial ecosystem structure: an ecologist's pragmatic view', in Hobbs, R.J., Mooney, H.A. (editors) *Remote sensing of biosphere functioning*, Springer-Verlag, New York: 5–30

Greegor, D.H., 1986, 'Ecology from space', *BioScience*, **36**: 429–32

Haines-Young, R., 1991, 'Biogeography', *Progress in Physical Geography*, **15**: 101–13

IGBP, 1992, *Global change: reducing the uncertainties*, International Geosphere-Biosphere Programme, Royal Swedish Academy of Sciences, Stockholm

Jackson, M.J., 1992, 'Integrated geographical information systems', *International Journal of Remote Sensing*, **13**: 1343–51

Johnston, R.J., 1981, 'Ecological fallacy', in Johnston, R.J., Gregory, D.,

Haggett, P., Smith, D., Stoddart, D.R. (editors), *The Dictionary of Human Geography*, Blackwell, Oxford: 89

Johnston, R.J., 1984, 'Quantitative ecological analysis in human geography: an evaluation of four problem areas', in Bahrenberg, G., Fischer, M.M., Nijkamp, P. (editors), *Recent developments in spatial data analysis*, Gower, Aldershot: 131–41

Justice, C.O., Townshend, J.R.G., Kalb, V.L., 1991, 'Representation of vegetation by continental data sets derived from NOAA AVHRR data', *International Journal of Remote Sensing*, **12**: 999–1021

Lam, N.S-N., Quattrochi, D.A., 1992, 'On the issues of scale, resolution, and fractal analysis in the mapping sciences', *Professional Geographer*, **44**: 88–98

Lucas, R.M., Harrison, A.R., 1993, 'Snow monitoring in the United Kingdom using NOAA AVHRR imagery', in Foody, G.M., Curran, P.J. (editors), *Environmental remote sensing from regional to global scales*, Belhaven, London (this volume)

Maling, D.H., 1989, *Measurements from maps*, Pergamon, Oxford

Mannion, A.M., Bowlby, S.R., 1992, 'Introduction', in Mannion, A.M., Bowlby, S.R (editors), *Environmental issues in the 1990s*, Wiley, Chichester: 3–20

Mather, P.M., 1993, 'Earth observation data – or information?', in Foody, G.M., Curran, P.J. (editors), *Environmental remote sensing from regional to global scales*, Belhaven, London (this volume)

Millington, A.C., Wellens, J., Settle, J.J., Saull, R.J., 1993, 'Explaining and monitoring land cover dynamics in drylands using multi-temporal analysis of NOAA AVHRR imagery', in Foody, G.M., Curran, P.J. (editors), *Environmental remote sensing from regional to global scales*, Belhaven, London (this volume)

Milne, B.T., Johnston, K.M., Forman, R.T.T., 1989, 'Scale-dependent proximity of wildlife habitat in a spatially-neutral Bayesian model', *Landscape Ecology*, **2**: 101–10

NASA, 1988, *Earth system science: a closer view*, Report of the Earth System Sciences Committee, NASA Advisory Council, NASA, Washington D.C.

NRC, 1985, *A strategy for Earth science from space in the 1980s and 1990s*, National Research Council, National Academy Press, Washington D.C.

NRC, 1986, *Global change in the geosphere-biosphere – initial priorities for an IGBP*, National Research Council, National Academy Press, Washington D.C.

Openshaw, S., 1983, *The modifiable areal unit problem*, Catmog 38, Geo Books, Norwich

Robinson, A.H., Sale, R.D., Morrison, J.L., Muehrcke, P.C., 1984, *Elements of cartography*, fifth edition, Wiley, New York

Roller, N.E.G., Colwell, J.E., 1986, 'Coarse-resolution satellite data for ecological surveys', *BioScience*, **36**: 468–75

Roughgarden, J., Running, S.W., Matson, P.A., 1991, 'What does remote sensing do for ecology?', *Ecology*, **72**: 1918–22

Saunders, R.W., Seguin, B., 1992, 'Meteorology and climatology', *International Journal of Remote Sensing*, **13**: 1231–59

Smith, G.M., Vaughan, R.A., 1993, 'A near-real-time heat source monitoring system using NOAA polar orbiting meteorological satellites', in Foody, G.M., Curran, P.J. (editors), *Environmental remote sensing from regional to global scales*, Belhaven, London (this volume)

Tong, S.T.Y., 1992, 'The use of non-metric multi-dimensional scaling as an ordination technique in resource survey and evaluation', *Applied Geography*, **12**: 243–60

Townshend, J.R.G., 1981, 'The spatial resolving power of Earth resources satellites', *Progress in Physical Geography*, 5: 32–55

Townshend, J.R.G., 1992, 'Land cover', *International Journal of Remote Sensing*, **13**: 1319–28

Ustin, S.L., Wessman, C.A., Curtiss, B., Kasischke, E., Way, J., Vanderbilt, V.C., 1991, 'Opportunities for using the EOS imaging spectrometers and synthetic aperture radar in ecological models', *Ecology*, **72**: 1934–45

Waring, R.H., Aber, J.D., Melillo, J.M., Moore III, B., 1986, 'Precursors of change in terrestrial ecosystems', *BioScience*, **36**: 433–8

Weber, C., 1993, 'Per-zone classification of urban land cover for urban population estimation', in Foody, G.M., Curran, P.J. (editors), *Environmental remote sensing from regional to global scales*, Belhaven, London (this volume)

Wickland, D.E., 1991a, 'Global ecology: the role of remote sensing', in Esser, G., Overdieck, D. (editors) *Modern ecology: basic and applied*, Elsevier, Amsterdam: 725–49

Wickland, D.E., 1991b, 'Mission to planet Earth: the ecological perspective', *Ecology*, **72**: 1923–33

Woodcock, C., Harwood, V.J., 1992, 'Nested-hierarchical scene models and image segmentation', *International Journal of Remote Sensing*, **13**: 3167–87

Woodcock, C.E., Strahler, A.H., 1987, 'The factor of scale in remote sensing', *Remote Sensing of Environment*, **21**: 311–32

INDEX

Note: Page numbers in *italic* refer to figures; those in **bold** refer to tables

5S 183

AATSR (Advanced Along-Track Scanning Radiometer) 203
ADEOS (Advanced Earth Observation Satellite) 186
Advanced Solid-state Array Spectroradiometer (ASAS) 184, 195
Advanced Very High Resolution Radiometer *see* AVHRR
Agropyron smithii 25
airborne thematic mapper 158
AIRSAR system 10, 158
Almaz-1 SARs *55*, 60
Along-Track Scanning Radiometer (ATSR) 186, 195
  ATSR-2 202
Artemisia herba-alba 20
ASAS 184, 195
atmosphere parameters, determination of 168–9, *176*
ATSR 186, 195
ATSR-2 202
AVHRR 5, 10, 184
  Global Area Coverage (GAC) data 94
  heat source monitoring 131–40
  multiple-view-angle data collection 184
  sea surface temperature monitoring 174
  spatial characteristics 224
  tropical forest data 66
AVHRR HRPT (High Resolution Picture Transmission) 112
AVHRR-NDVI
  in analysis of dryland land cover dynamics 16–41
  classifying forest cover using 71–4
  cloud cover and 55–9
  estimating ground data 70
  model of transpiration 23–8
  in monitoring of deforestation 45–8
  snow monitoring using 111–28

backscatter models 10, *11*
BAHC (Biospheric Aspects of the Hydrological Cycle) 9
BATS (Biosphere Atmosphere Transfer Scheme) 151, 204
Bidirectional Reflectance Distribution Function (BRDF) 12, 186–7, 189
Bio-Geo-Chemical cycling 151–3, *152*, *155*
  in Wales 156–8, *157*
Biosphere Atmosphere Transfer Scheme (BATS) 151, 204
Biospheric Aspects of the Hydrological Cycle (BAHC) 9
Boutelouda gracilis 25
BRDF (Bidirectional Reflectance Distribution Function) 12, 186–7, 189, 192–3
  physical models 193–5
BROOK90 model 151

CAESAR 186, 195
Camillo model 12
carbon cycle 9–10, 44–53
carbon dioxide
  annual global balance 45
  tropical forests as sinks and sources 44
Carex eleocharis 25
Carex filifolia 25
CASI (compact airborne spectrographic imager) 158
CCD Airborne Experimental Scanner for Applications in Remote Sensing (CAESAR) 186, 195
Centre for Earth Observation (CEO) 229
Cistus monspeliensis 17
climate modelling 84
cloud cover 115, 134–5, 137, 140
cluster class maps 34
Coastal Zone Colour Scanner (CZCS) 173
Commission of the European Communities (CEC) 5

compact airborne spectrographic imager (CASI) 158
core zone concept 30, *31*
 development and implementation in Pakistan 30–40, *38*, *39*
CZCS (Coastal Zone Colour Scanner) 173

deforestation
 rate of **49**
 remote sensing of 44–8
 socioeconomic causes 84
dryland ecosystems 16–41

Earth observation data 202–11
 obstacles to scientific use of 204–8
  atmospheric correction 206
  geo-referencing 205, 207–8
  illumination and viewing geometry effects 206–7
  radiometric calibration 205–6
 possible solutions 208–11
  expert systems 210–11
  metadata networks 208–10
Earth Observing Mission
 with a morning orbit (EOS AM-1) 4
 with an afternoon orbit (EOS PM-1) 4
Earth Observing System (EOS) 11, 155, 189, 203
 EOS-MISR 187
Earth Observing System Data and Information System (EOSDIS) 203
Economic and Social Research Council (ESRC) (UK) 209
ecosystem simulation models 149–61
 FOREST-BGC model (Bio-Geo-Chemical cycling) 151–3, *152*, *155*
  in Wales 156–8, *157*
 labile carbon model 151, 153
  in Florida 158–61
 preparing to drive with remotely sensed data 156–61
 remotely sensed inputs to 153–6
Enhanced Thematic Mapper (TM) 202
environmental change 1–2
 remote sensing and 2–5
environmental constraint splitting 35
Envisat 4, 203
European Polar Platform (EPOP-1) 203
European Remote Sensing satellite (ERS-1) *5*, 10
 ERS-1 SAR 5, 47, 55, 60, 124, 202
 ERS-2 202
 ERS-ARST 186
European Space Agency (ESA) 48, 202

Food and Agriculture Organisation (FAO) *5*, 24
forest ecosystem process model 204
Forest Resource Assessment 1990 48
FOREST-BGC model (Bio-Geo-Chemical cycling) 151–3, *152*, 155
 in Wales 156–8, *157*
Fuyo-1 (JERS-1) *5*, 10, 47, 55, 60, 202

GENIE (Global Environmental network for Information Exchange) project 209–10, 211, 229
geo-referencing 205, 207–8
geophysical parameters, determination of 167–78
 atmosphere 168–9, *176*
 global models 175–7
 land 174–5
 oceanic global circulation model 177
 sea 169–74, *171*, **172**
  chlorophyll and suspended sediment concentrations 172–3
  near surface windspeeds 173
  sea surface currents 174
  sea surface temperature charts 174
 weather forecast models 175–7
GEWEX 8, *9*
GEWEX Continental-scale International Programme (GCIP) 13
Global Analysis, Interpretation and Modelling (GAIM) IGBP core projects *9*
Global Change and Terrestrial Ecosystems (GCTE) 8, *9*
Global Circulation Models (GCMs) 10
Global Climate Observing System (GCOS) *5*, 8
Global Energy and Water Cycle Experiment (GEWEX) 8, *9*
global land cover 84–110
 comparison of existing data sets 85–7
 verification of land cover information with AVHRR data 87, *88–93*, 94–5, *96–101*
global models 175–7
Global Ocean Observing System (GOOS) *5*

Global Ozone Monitoring Experiment (GOME) on ERS-2 11
Global Resource Information Data Base (GRID) of the United Nations 48
Global Vegetation Index 94
GOES 111
ground control points (GCP) 140

HAPEX-Sahel experiment, Niger *12*
heat source monitoring 131–40
  colour composites 139–40
  hardware system 132–3
  heat sources 132
  performance of system 135–9
    cloud information 137
    extraction of the SAVHRR coverage of UK 135–7
    heat source monitoring 137–9
  software system 134–5
    calculating cloud cover information 134–5, 140
    detecting heat sources 135
    extracting AVHRR coverage of the UK 134
  strawburning 137–9, 140
  use of NOAA AVHRR 132
hemispherical reflectance 193, *194*
High Resolution Imaging Spectrometer (HIRIS) 155, 184, 186
High Resolution Picture Transmission (HRPT) data 17
High Resolution Visible (HRV) 202
High Resolution Visible and Infrared sensor, plus the Vegetation sensor (HRVIV) 202
HIRIS (High Resolution Imaging Spectrometer) 155, 184, 186
hot spot 182
HRVIV (High Resolution Visible and Infrared sensor, plus the Vegetation sensor) 202

Inter-Agency Committee for Global Environmental Change (IACGEC) 209
Intergovernmental Oceanographic Commission (IOC) 5
International Council of Scientific Unions (ICSU) 4
International Forest Inventory Team (IFIT) programme (EC) 10
International Geosphere-Biosphere Programme (IGBP) 5, 8, 47, 149
  IGBP/GCTE 13
International Global Atmospheric Chemistry project (IGAC) 9

Jet Propulsion Laboratory 10

labile carbon model 151, 153, *154*
  in Florida 158–61
land cover classes 85–6, *87*, 105–10, *229*
land cover dynamics, AVHRR-NDVI
  and moisture availability relationships 17–28, 40–1
    modelling vegetation growth and AVHRR-NDVI responses 22–8
    relationships between rainfall, irrigation records and AVHRR-NDVI 17–22
  spatial analysis of multitemporal data 29–40, 41
    core zone concept in Pakistan 30–40, *38*, *39*
    discussion 37–40
    results 35–7
    verification 37
land cover mapping using MVA 189–91, *190*
land parameters, determination of 174–5
Landsat satellites 45, 58, 142
  Landsat MSS *5*, *55*, *59*
    tropical forest data 66–9
  Landsat TM *5*, *55*, *59*
    snow monitoring using 118, 121, 123
    tropical forest data 69–70
  Landsat-6 202
  Landsat-7 202
leaf area index (LAI) 150, 160–1, *160*
Local Area Coverage data 14, 17
LOWTRAN 206
LOWTRAN 7 183

Makkink's radiation method 24
MANE model 151
Marine Observational Satellite (MOS) 202
Measurement of Pollution In The Troposphere (MOPITT) 11
MERIS (Medium Resolution Imaging Spectrometer for oceanographic applications) 203
METEOSAT 111
methane 9

Minnaert model 12
MISR 186, 189, 190, 195
Mississippi Basin continental scale modelling programme 9
MODIS (Moderate Resolution Imaging Spectrometer) 102, 155–6
MODIS-T (Moderate Resolution Imaging Spectrometer – Tilt) 184, 186
Modular Global Change Modelling System (MGCMS) 151
moisture availability relationships 17–28, 40–1
Monte Carlo ray-tracing system 195
MOS 1 MESSR *55, 59*
MSS 202
Multi-angle Imaging SpectroRadiometer (MISR) 186, 189, 195
multiple-view-angle (MVA) data 181–96, 228
  angular reflectance properties of Earth surface materials 182–3
  estimation of Earth surface albedo and 3-D surface structure 192–5
    functional models 192–3
    physical models of the BRDF 193–5
  land cover mapping using 189–91, *190*
  sensor design for 183–9, *185*
    angular sampling capabilities 186–9
multispectral data 181
Multispectral Scanning Systems (MSS) 45
multi-temporal data 181

National Aeronautics and Space Administration (NASA) 158, 170
  Earth Observing System (EOS) programme 11, 155, 189, 203
  Master Directory of the Earthnet system 208–9
National Farmers Union (NFU) (UK) 138
National Oceanic and Atmospheric Administration (NOAA) 168
  'I'-'M' 202
  NOAA-9 187
  NOAA-10 AVHRR 187, *188*
  NOAA-11 187
  see also AVHRR
National Rivers Authority (UK) 111
National Snow Survey (UK) 111, 124
Natural Environmental Research Council (NERC) 8, 158
net primary productivity (NPP) 149, 150, 153
Nimbus-7 172
nitrous oxide 9
Normalised Difference Vegetation Index (NDVI) 16, 51, 94, 103
  application to snow monitoring 117–18, *119*
Numerical Algorithms Group (NAHG) 209

oceanic global circulation model 177
opposition surge 182
Overseas Development Agency (ODA) 55

Picea sitchensis (Sitka spruce) 156–8
Pinus elliottii (slash pine) 158–61, *159–60*
Polar Orbit Earth Observation Mission (POEM) 202–3
Polar Orbiting Environmental Mission (Envisat) 4, 203
POLDER (Polarisation and Directionality of the Earth's Reflectance) 186, 195
population estimation, urban 142–7

Quercus coccifera 17

Radasat 202

SAREX 10
Satellite Pour l'Observation de la Terre *see* SPOT
Satellite Receiving Station (SRS) 133
satellite sensors **46**
satellites 4, **5**
  advantages and disadvantages of **169**
scale 223–9
  nested hierarchy 226, *227*
  spatial and temporal *225*
Scanning Multichannel Microwave Imaging (SMMI) 124
sea parameters, determination of 169–74, *171*, 172
Seasat 170, 202
Simple Biosphere (SiB) model 151, 204
SIR-A/B 55, 60
SMMI (Scanning Multichannel Microwave Imaging) 124

snow monitoring, UK 111–28
classification accuracy 121–3
snow boundary location 121–3, **122**
snow cloud discrimination 121, **122**
description of the snowpack 117–21
application of the NDVI 117–18, *119*
melt/accumulation estimation using AVHRR information 118–21, *120*
detection of snow area using AVHRR data 114–17
generation characteristics of snow, cloud and land surfaces 115
impact of cloud cover on snow observation from satellite sensors 115
procedure of the discrimination of snow covered surfaces 115–17
geometric preprocessing of AVHRR imagery 112–13
between-date registration 113
correction of panoramic distortion 112
integration of ancillary data 113
navigation of imagery 112–13
multispectral snow mapping algorithm 124, *125–8*
operational procedure 123
radiometric preprocessing of AVHRR imagery 113–14
atmospheric correction of thermal infrared data 114
data calibration 114
snow distribution and altitude relationships 123–4
use of elevation data 123–4
Soil-Vegetation-Atmosphere-Transfer (SVAT) modelling 10, 12, 13, 151
Space Shuttle Metric Camera 55, 59
spatial data 214–22
deep and imposed structures 215–16, 218
future needs and likely developments 221–2
spatial segmentations 218
and temporal resolution of maps 216–18
concerns of local communities 217–18
evolution of map-using community and its institutions 216–17
security and related improvements to use of 217
user applications and structures 218–19
vector and raster mapping 219–21, **220**, *221*
Spatial Re-classification Kernel (SPARK) procedures 143
SPOT 142, 202
SPOT-3 202
SPOT-4 202
SPOT HRV 5, 45, 58, 59, 184
multiple-view-angle data collection 184
multispectral 55, 144, 147
panchromatic data 14, 55, 142–3, 144, 147
spatial characteristics 224
Stipa comata 25
Stipa tenacissima (alfa grass) 20
strawburning 137–9, 140
synthetic aperture radar (SAR) 10, 156, 202
estimating leaf biomass, wood biomass and canopy roughness with 51–2
global land cover and 102
use in tropical forests 47

Terrestrial Initiative in Global Environmental Research programme *see* TIGER programme
Thematic Mappers (TM) 45
TIGER programme 8–14
carbon cycle on land 9–10
objective 9
TIGER II: trace greenhouse gases 11
TIGER III 10
water and energy balance 11–13
TIGER IV: impacts on ecosystems 13–14
TIROS Operational Vertical Sounder (TOVS) 168–9
transpiration, AVHRR-NDVI relationships, model of 23–8
sensitivity analysis of model 25–7, **26**, *26*
Tropical Ecosystem Environment Observations by Satellites (TREES) programme 48
Tropical Forest Monitoring Programme 48

tropical forests 9, 44–74
burning 131
carbon balance 10
classifying forest cover with coarse spatial resolution imagery 71–4
correlation between remotely sensed data and ground data 64–70
estimating ground data with remotely sensed data 70
Landsat MSS data 66–9
Landsat TM data 69–70
NOAA AVHRR data 66
deforestation, rate of **49**
estimating canopy roughness with multi-angle optical sensors 52–3
estimating leaf biomass with red and near infrared radiation 50–1
estimating leaf biomass, wood biomass and canopy roughness with SAR 51–2
forest age, correlation with backscatter 52
and the global carbon cycle 44–53
major programmes involved with the remote sensing of deforestation 47–8
remote sensing of deforestation in the tropics 44–7
ground data 60–4
biophysical data 61–4
forest/non-forest map 60
regeneration, remote sensing of 48–50
remotely sensed data 55–60
in microwave wavelengths 60
in optical wavelengths 55–60
study area 53–5
Tropospheric Emission Spectrometer (TES) 11

UARS 5
Universities Global Atmosphere Modelling Programme (UGAMP) (UK) 11
UNESCO scheme for classifying vegetation 85
United Nations Environment Programme (UNEP) 5, 48
urban land cover, per-zone classification 142–7
methodology 144–7
US global change programme 2, *3*

vegetation classification schemes 85–6
Vegetation Sensor 202
VEGIE model 151
VIRSR 203

weather forecast models 175–7
World Climate Research Programme (WCRP) 5, 8
Global Energy and Water Cycle Experiment (GEWEX) 8, 9, 13
World Meterological Organisation (WMO) 4
World Ocean Circulation Experiment (WOCE) 8

Compiled by Annette Musker